The Handbook of Environmental Chemistry

Volume 97

Founding Editor: Otto Hutzinger

Series Editors: Damià Barceló · Andrey G. Kostianoy

In over three decades, *The Handbook of Environmental Chemistry* has established itself as the premier reference source, providing sound and solid knowledge about environmental topics from a chemical perspective. Written by leading experts with practical experience in the field, the series continues to be essential reading for environmental scientists as well as for environmental managers and decision-makers in industry, government, agencies and public-interest groups.

Two distinguished Series Editors, internationally renowned volume editors as well as a prestigious Editorial Board safeguard publication of volumes according to high scientific standards.

Presenting a wide spectrum of viewpoints and approaches in topical volumes, the scope of the series covers topics such as

- local and global changes of natural environment and climate
- anthropogenic impact on the environment
- water, air and soil pollution
- remediation and waste characterization
- environmental contaminants
- biogeochemistry and geoecology
- chemical reactions and processes
- chemical and biological transformations as well as physical transport of chemicals in the environment
- environmental modeling

A particular focus of the series lies on methodological advances in environmental analytical chemistry.

The Handbook of Envir onmental Chemistry is available both in print and online via http://link.springer.com/bookseries/698. Articles are published online as soon as they have been reviewed and approved for publication.

Meeting the needs of the scientific community, publication of volumes in subseries has been discontinued to achieve a broader scope for the series as a whole.

Water Resources in Algeria - Part I

Assessment of Surface and Groundwater Resources

Volume Editors: Abdelazim M. Negm ·
Abdelkader Bouderbala ·
Haroun Chenchouni · Damià Barceló

With contributions by

M. Adjim · R. Alkama · A. A. Assani · M. Attia · D. Barcelo ·
A. S. Belouchrani · F. Bensaoula · S. Bouabdelli · A. Bouderbala ·
S. Bouhsina · S. Bouhssina · H. Chaffai · H. Chenchouni ·
B. Collignon · L. Djabri · L. Djemili · Y. Elmeddahi · L. Ghrieb ·
N. Hadj Mohamed · A. Hani · S. Heddam · A. Henni · M. Hocine ·
L. Houichi · A. Karar · O. Kisi · H. Meddi · M. Meddi · M. Nasr ·
M. A. Neama · A. M. Negm · E.-S. E. Omran · R. Ragab · M. Renima ·
F. Rouaibia · A. Sadeuk Ben Abbes · A. Sebbar · F. Trabelsi ·
F. Z. Trabelsi · A. Zeroual · S. Zeroual

 Springer

Editors
Abdelazim M. Negm
Department of Water and Water
Structures Engineering
Faculty of Engineering, Zagazig University
Zagazig, Egypt

Abdelkader Bouderbala
Department of Earth Sciences
University of Khemis Miliana
Algiers, Algeria

Haroun Chenchouni
Department of Natural and Life Sciences
Faculty of Exact Sciences and Natural
and Life Sciences, University of Tebessa
Tebessa, Algeria

Damià Barceló
Catalan Institute for Water Research
ICRA
Girona, Spain

ISSN 1867-979X   ISSN 1616-864X   (electronic)
The Handbook of Environmental Chemistry
ISBN 978-3-030-57894-7   ISBN 978-3-030-57895-4   (eBook)
https://doi.org/10.1007/978-3-030-57895-4

This Springer imprint is published by the registered company Springer Nature Switzerland AG.
The registered company address is: Gewerbestrasse 11, 6330 Cham, Switzerland

Series Preface

With remarkable vision, Prof. Otto Hutzinger initiated *The Handbook of Environmental Chemistry* in 1980 and became the founding Editor-in-Chief. At that time, environmental chemistry was an emerging field, aiming at a complete description of the Earth's environment, encompassing the physical, chemical, biological, and geological transformations of chemical substances occurring on a local as well as a global scale. Environmental chemistry was intended to provide an account of the impact of man's activities on the natural environment by describing observed changes.

While a considerable amount of knowledge has been accumulated over the last four decades, as reflected in the more than 150 volumes of *The Handbook of Environmental Chemistry*, there are still many scientific and policy challenges ahead due to the complexity and interdisciplinary nature of the field. The series will therefore continue to provide compilations of current knowledge. Contributions are written by leading experts with practical experience in their fields. *The Handbook of Environmental Chemistry* grows with the increases in our scientific understanding, and provides a valuable source not only for scientists but also for environmental managers and decision-makers. Today, the series covers a broad range of environmental topics from a chemical perspective, including methodological advances in environmental analytical chemistry.

In recent years, there has been a growing tendency to include subject matter of societal relevance in the broad view of environmental chemistry. Topics include life cycle analysis, environmental management, sustainable development, and socio-economic, legal and even political problems, among others. While these topics are of great importance for the development and acceptance of *The Handbook of Environmental Chemistry*, the publisher and Editors-in-Chief have decided to keep the handbook essentially a source of information on "hard sciences" with a particular emphasis on chemistry, but also covering biology, geology, hydrology and engineering as applied to environmental sciences.

The volumes of the series are written at an advanced level, addressing the needs of both researchers and graduate students, as well as of people outside the field of "pure" chemistry, including those in industry, business, government, research establishments, and public interest groups. It would be very satisfying to see these volumes used as a basis for graduate courses in environmental chemistry. With its high standards of scientific quality and clarity, *The Handbook of Environmental Chemistry* provides a solid basis from which scientists can share their knowledge on the different aspects of environmental problems, presenting a wide spectrum of viewpoints and approaches.

The Handbook of Environmental Chemistry is available both in print and online via www.springerlink.com/content/110354/. Articles are published online as soon as they have been approved for publication. Authors, Volume Editors and Editors-in-Chief are rewarded by the broad acceptance of *The Handbook of Environmental Chemistry* by the scientific community, from whom suggestions for new topics to the Editors-in-Chief are always very welcome.

Damià Barceló
Andrey G. Kostianoy
Series Editors

Preface

Water resources are the basic elements of life, and they are considered as the best way for durable management in any country worldwide. The increasing demand of water for drinking, irrigation of agricultural crops, and industry needs is the first inquiry of the Algerian authority. The major challenges of constant water supply are related mainly to spatiotemporal water availability and water quality. In water resources management, water quality plays a key role as important as that of water quantity. The Algerian authority faced serious problems in managing its water resources; however, academics and specialized engineers enlightened the road with integrated solutions for policymakers to tackle these issues and make the right decisions to achieve the sustainable management of these life resources.

This book of water resources in Algeria discusses the situation of water resources in this country by assembling an interesting number of works from different specialized Algerian researchers. With the aim to support the sustainable development in Algeria as an arid country in the MENA region, the chapters detail information about the current status of water resources in Algeria including hydrological investigations of surface water and groundwater, the impact of climate change on water resources, characterization of groundwater aquifers, and the assessment and management of water resources. Some chapters focused on surface and groundwater quality assessment using tools and techniques that can sustain the management of water/soil resources in the arid zones of Algeria.

This book serves as a tool, long-awaited and mostly needed, for scientists and readers of miscellaneous disciplines related to water resources. The book is more helpful to the readership that seeks to obtain a clear grasp of key principles and concepts as well as applied and case studies on hydrogeology and water resources management. The authors of the chapters explained the principal methods used for the optimum utilization of water resources with regard to their quality and spatiotemporal availability, the theoretical predictions of some hydrological phenomena, the available tools for hydrological modeling, and water resources protection against deterioration through the mitigation of harmful agents/effects of water and thus ensuring the well-being of human communities and their environment.

The authors share their vision about sustainable water management activities, as well as recent advances, directions for future research, which should be at the same time, the challenges of ensuring adequate water supply, and optimal allocation of different uses that are compounded by changes in climate, land use, and demography. Accordingly, this book represents a unique piece and a reference of interest to students, scientists, engineers, policy- and decision-makers, and water resource managers. The book includes several tools that can be generated to understand similar problems threatening water resources in North Africa, the Middle East, the Mediterranean basin, or in other regions having similar climatic and/or geographical features as Algeria. Based on the subjects treated and discussed in the different chapters, this book will be of interest to a large scientific community in Algeria and in other countries.

Therefore, the book titled "Water Resources in Algeria: Assessment of Surface and Groundwater" is written by experts from Algeria who know well the facts of water resources in Algeria and the factors affecting its demand and supply to update and enhance the existing knowledge on water resources in Algeria concerning two important sources, namely surface water and groundwater. The book consists of 12 chapters in addition to the introduction part, which consists of two chapters, and the conclusion part, which consists of one chapter. The main themes of the book include (a) Current Status of Water Resources in Algeria (3 chapters), (b) Climate Change Impact and Hydrogeological Investigations (5 chapters), (c) Aquifer Characterization and Assessment of Groundwater Resources (2 chapters), and (d) Toward a Sustainable Development (2 chapters).

The editors want to thank all the authors for the contributions to make this book a source of knowledge on the assessment of surface and groundwater resources. Thanks are also extended to the Springer's team who worked hard for a long period to produce this book and make our dream a reality. Special thanks to the editorial board of the *Handbook of Environmental Chemistry* series for their critical evaluation of the book proposal that improved the book contents.

The editors welcome any comments, feedback, and/or new chapters to be included in the next editions. Please send your feedback and your constructive comments and/or your new chapter to the editors via email.

Zagazig, Egypt   Abdelazim M. Negm
Khemis Miliana, Algeria   Abdelkader Bouderbala
Tebessa, Algeria   Haroun Chenchouni
Girona, Spain   Damià Barceló
29 March 2020

Contents

Part I
Introducing the Book

Introduction to "Water Resources in Algeria: Assessment of Surface and Groundwater Resources"

Abdelazim M. Negm, El-Sayed Ewis Omran, Abdelkader Bouderbala, Haroun Chenchouni ⓘD, and Damia Barcelo

Contents

A. M. Negm (✉)
Water and Water Structures Engineering Department, Faculty of Engineering, Zagazig University, Zagazig, Egypt
e-mail: amnegm85@yahoo.com; amnegm@zu.edu.eg

E.-S. E. Omran
Soil and Water Department, Faculty of Agriculture, Suez Canal University, Ismailia, Egypt

Institute of African Research and Studies and Nile Basin Countries, Aswan University, Aswan, Egypt
e-mail: ee.omran@gmail.com

A. Bouderbala
Department of Earth Sciences, University of Khemis Miliana, Khemis Miliana, Algeria
e-mail: bouderbala.aek@gmail.com

H. Chenchouni
Department of Natural and Life Sciences, Faculty of Exact Sciences and Natural and Life Sciences, University of Tebessa, Tebessa, Algeria

Laboratory of Natural Resources and Management of Sensitive Environments 'RNAMS', University of Oum-El-Bouaghi, Oum-El-Bouaghi, Algeria
e-mail: chenchouni@gmail.com

D. Barcelo
ICRA, Catalan Institute for Water Research, Quart, Girona, Spain
e-mail: dbcqam@cid.csic.es

Abdelazim M. Negm, Abdelkader Bouderbala, Haroun Chenchouni, and Damià Barceló (eds.), *Water Resources in Algeria - Part I: Assessment of Surface and Groundwater Resources*, Hdb Env Chem (2020) 97: 3–12, DOI 10.1007/698_2020_565, © Springer Nature Switzerland AG 2020, Published online: 18 June 2020

Abstract This chapter highlights the basic technical elements of the chapters presented in the book according to its four themes. Therefore, the chapter contains information on the current status of water resources in Algeria including surface water and groundwater, the hydrological investigations in Algeria and the impact of climate change on water resources, characterization of groundwater aquifers, and assessment of water resources to support the sustainable development in Algeria as an arid country in the MENA region.

Keywords Aquifer, Climate change, Groundwater, Hydrogeological assessment, Hydrology, Resources, Surface water, Sustainable development

1 Background

Algeria, with its natural resource factors, is a significant country in Africa and the world and is seeking a strong development in the demographic and economic scale. The Algerian authority was facing serious problems in managing its soil and water resources. Therefore, this book should be of special importance to the Algerian authorities and stakeholders who are dealing with water resource issues from assessment and management points of view. While volume II of the book focuses on water quality assessment, wastewater treatment, protection and development [1].

The topics presented in the book cover the assessment of surface water and groundwater resources in four themes, in addition to the introductory section which includes this chapter and the next chapter which focuses on an "Overview of Water Resources, Quality, and Management in Algeria." It provides a general summary of the condition of water resources (e.g., rivers, lakes, wetlands, reservoirs, and groundwater) in Algeria. The chapter also lists the number and type of published documents that have been recently issued to handle the sustainability of water bodies in Algeria. Moreover, the authors present sufficient information about the authors' nationalities, institutional affiliations, and funding agencies and sponsors, considering the water management and research in Algeria [2]. The overview chapter focuses on several learning objectives which include the following:

1. Algeria holds important freshwater resources kept within soil layers as groundwater and in surface water bodies, viz., streams, rivers, lakes, wetlands, and reservoirs.
2. The chapter illustrates that the government, private and public sectors, and policymakers have handled several approaches concerning the country's national water policy, water resources development and planning, and water purification and reclamation.

3. The findings of this chapter would support organizations, stakeholders, government leaders, industrialists, and environmentalists, dealing with all water subjects in Algeria.

On the other hand, the four main themes of the book include:

– Current Status of Water Resources
– Climate Change Impact and Hydrogeological Investigations
– Aquifer Characterization and Assessment of Groundwater Resources
– Toward a Sustainable Development

The next sections present briefly the main technical elements of each chapter under its related theme.

2 Chapters' Summary

The chapters presented in the book are distributed under four themes. Summaries of the chapters are presented under its related theme (section).

2.1 Current Status of Water Resources

Three chapters are presented in this section. The chapter titled "Origin and Quality of Groundwaters of the Taoura Syncline Aquifer (North-Eastern Algeria)" aims to study the current groundwater quality using a hydrochemical statistical approach. Also, it investigates natural and human factors that determine the variations of such quality within the aquifer. The mio-plio-quaternary is composed of heterogeneous material (sand, marl clays, and gravels), and it covers the karst limestone formation which shows important discontinuities related to the tectonics of the region. Indeed, faulting, in the subdivision of the study area, resulted in a variation of the depth of the top of karst basement. The top of karst formation is sometimes close to the ground surface, and sometimes is very far from it. The presence of faults favors the communication of the deep karst aquifer with the superficial aquifer, and water can be drained into the karst aquifer [3].

The observation of stratigraphic logs shows a lithological heterogeneity of the aquifer, which is going to explain the variation of its hydrochemical composition. The piezometric maps carried out indicate that the recharge of the superficial aquifer is by the borders. This lateral contribution influences the chemical composition of groundwater of both aquifers. It revealed that the water quality of the study area remains influenced of natural factors, particularly the geological formations present in this zone. A decrease in piezometric levels in wells is also noticed. This is explained by the overexploitation of the aquifer and/or the climatic conditions. Samples and analysis carried out showed that groundwater of this aquifer is rich in calcium and bicarbonate; however some wells are characterized by the domination of chlorides and sulfates, and these anions are accompanied by sodium or by calcium. So, the impact of geological rocks crossed by water is important in the characterization of groundwater quality. For further detail, audiences are advised to read the full chapter.

The chapter titled "Impact of Toxic Metals on Water Quality Around an Abandoned Iron Mine, Bekkaria, Algeria" focuses on the determination of the impact of abandoned mines on the quality of surface water and groundwater, by analyzing the data of several points in wells and in the wadi. The analysis campaigns concerned major elements and metallic trace elements. In the case of surface water, the effect of leaching on the mineral and the trapping of the chemical elements by the soil has been attempted. For groundwater it is the control of the purifying power of the soil. Samples were collected from points next to the mine and further away from the mine, for both surface and underground water. The results revealed that the contamination by the major chemical elements or the metallic trace elements decreases gradually as it moves away from the mine, which confirms the direct impact of the iron mine in water contamination. The study of the Sr^{2+}/Ca^{2+} ratio gives an outline of the influence of sorted gypso-saliferous on the water salinity. Concerning surface water, the values of Sr^{2+}/Ca^{2+} ratio highlight the influence of sorted gypsiferous on the quality of water [4]. The recourse to modeling constitutes another tool for the description of the impact of the iron mine on the quality of water. The authors examined 697 networks to find the optimal model of the networks of artificial neurons.

On the other hand, the chapter titled "Impacts of Pesticides on Water Resources and Soil in Algeria" focuses on summarizing the main features of the effects of the pesticide on natural resource scarcity and environmental pollution in Algeria, which is exacerbated by features including:

(a) Groundwater and surface water pollution originated from domestic, industrial, and agricultural effluents that exceed the ability of sewage systems and thus significantly reduce the quantity of treated water that can be reused.
(b) Risk of sustainable development in relation to soil and water pollution, where severe issues arose when groundwater samples were evaluated and that showed that impact of water withdrawals on water storage exceeds natural resource renewal boundaries where the need to tap into nonrenewable reserves.
(c) Green chemistry became responsible for finding suitable solutions to all old manufacturing problems by finding alternatives to all previous negatives [5].

2.2 Climate Change Impact and Hydrogeological Investigations

This section is covered into five chapters. The chapter titled "Analysis of flood Characteristics in the Context of Climate Variability in Northern Algeria: Case of

Cheliff Watershed" studies some characteristics of water surface flows in some hydrometric stations to characterize the phenomenon of floods. The authors are interested in the annual floods that marked the watershed and observed in the different stations in the period from the year 1960 to the year 2006, to determine their temporal variability and to find out the peak flows at both annual and monthly levels. The daily average flows data are collected in the gauge stations for the available periods mentioned above. The authors conducted an inventory of all annual flows (daily and instantaneous peak flow values for each year).

To detect exceptional floods and their relationship with average flows, the authors used the observed annual and monthly values of flows and the recurrent values of the flows. The statistical study of the recorded flows of the different stations localized in the watershed of Cheliff shows the highest values of flows which were recorded in the station of Ponteba ($1,300$ m^3/s) and in the station of Ain Hamara (878 m^3/s). Also, the most remarkable events were observed during the 1970s for the stations of Arib Cheliff, Rahouia, and Djnane Ben Ouadha, during the 1980s for the station of Ponteba, and finally during the 1990s in the stations of Ain Amara, Sidi AEK Djilali, and El Ababsa.

The authors reported that the autumn floods characterize the 1990s and the winter floods characterize the 1980s. In general, the different hydrometric stations show that there are two distinct periods in terms of duration and flood power. The first is the most important in terms of duration as it characterizes the spring season, with few floods. The other period is very short. It corresponds to the autumn season, and it is characterized by flash floods. The flow is very abundant in March and October in second place. The analysis of rainfall and the monthly flows showed a relatively temporal concordance, where the rainiest months are usually the most abundant inflow [6].

Furthermore, the chapter titled "Assessing the Climate Change Impact on Water Resources and Adaptation Strategies in Algerian Cheliff Basin" assesses the direct and indirect impacts of climate change on water resources by identifying major trends in precipitation over time, annual flows, and the significant contribution of Cheliff dam in the Cheliff basin in Algeria. The chapter describes its current and future effects on water resources in the Cheliff basin in Algeria, as well as the measures implemented to mitigate the adverse effects through adaptation strategies. Decision-makers thus need objective assessments of the vulnerability of different socioeconomic sectors through the integration of climate information at the national and local levels. The influence of climate change on the water resources of the Cheliff basin in Algeria was assessed with particular emphasis on the major issues related to water reserves. The Cheliff basin, which is one of the largest basins in the north of Algeria, is affected by water scarcity due to the expansion of industrial and agricultural activities with the population growth and to a reduction in water resources caused by extreme droughts. The results indicated that there was a decreasing rainfall in the whole basin, ranging from 14% to 54%, and a reduction in stream flow that exceeds 40% with a break observed at the beginning of the 1980s. This may cause additional stress on public services responsible for water resources management and on the population due to constraints on drinking water supplies.

Furthermore, according to different emission scenarios, several general circulation models (GCMs) predict an increase in temperature of +0.9°C to +5°C on average at the end of the twenty-first century, with a decrease in average rainfall of 10–30%. A conceptual model predicted a flow deficit ranging from 10% to 48% at different periods and in different scenarios [7].

Additionally, the authors of the chapter titled "Assessment of Projected Precipitations and Temperatures Change Signals over Algeria Based on Regional Climate Model: RCA4 Simulations" examine climate in Algeria in terms of monthly mean precipitation and temperature using future projections of regional climate model (RCM) – RCA4 at 0.44° resolution used as part of the CORDEX-Africa program (Coordinated Regional climate Downscaling Experiment) under the RCP4.5 and RCP8.5 forcing scenarios. The authors proceeded to the spatiotemporal variability analysis of precipitations and temperatures according to the two stages:

(a) In the first stage, they produced mean climate maps and estimated linear trends over the historical period (1951–2005) of monthly precipitation and temperature obtained from observation and simulation models. Then the authors computed the mean surface area of each climate zone based on a precipitation and temperature dataset.

(b) As a second step, they calculated the mean surface area of each climate zone and linear trends for the two future periods (2006–2060 and 2045–2100) based on modeled monthly precipitation and temperature under the two RCP scenarios. Here, it must be recalled that the outputs of the RCA4 simulation in the future have been corrected using the quantile mapping (QM) bias correction algorithm.

On the other hand, the chapter titled "Comparison of Evolving Connectionist Systems (ECoS) and Neural Networks for Modelling Daily Pan Evaporation from Algerian Dams Reservoirs" (1) presents the proposition of a new data-driven model for predicting daily pan evaporation from dam reservoirs in Algeria, using two versions of the dynamic evolving neural-fuzzy inference systems named (DENFIS), and (2) compares the accuracy of the DENFIS models with those of the ANN models.

Accurate calculation of the component of the water budget for the dam reservoirs is critical and helps for rational water planning and management. Estimation of evaporation from surface water is governed by several factors, among them are the weather variables. Evaporation called as pan evaporation (EP) is measured using (1) empirical equations in which measurement of several weather variables are needed and (2) direct measurement using evaporimeter pan. One of the accurate and reliable alternative approaches used for the estimation of the pan evaporation is the application of the data-driven models that has gained much popularity during the last few years. The aims of the present chapter are the application of a new data-driven model called dynamic evolving neural-fuzzy inference systems named "DENFIS" for predicting daily pan evaporation using measured weather variables at two dam reservoirs in Algeria. Results obtained proved that the proposed DENFIS models were compared to those of the artificial neural networks (ANN) and multiple linear regression (MLR) models. By comparing the accuracy of the models using several statistical indices, it was concluded that the performances of the models varied depending on the inputs variables, and DENFIS model is more suitable for predicting pan evaporation in the arid region compared to the ANN and MLR models, while the ANN was more suitable for the humid region.

The next chapter is titled "New Formulation for Predicting Daily Reference Evapotranspiration (ET_0) in the Mediterranean Region of Algeria Country: Optimally Pruned Extreme Learning Machine (OPELM) Versus Online Sequential Extreme Learning Machine (OSELM)." The authors present (1) the estimation of the reference evapotranspiration (ET_0) using data-driven techniques and (2) the propositions of a new kind of model based on the extreme learning machines and (3) compare the accuracy of the OPELM and OSELM models. One of the most important components of the hydrological cycle is certainly the reference evapotranspiration (ET_0) that has received great importance by researchers worldwide. ET_0 is calculated using the standards FAO56 Penman-Monteith model. However, several alternative methods have been proposed and applied in many places of the world, among them are radiation- and temperature-based methods. With the development of the machine learning models, there have been several studies since several decades that have tried to estimate the ET_0 at different time scales, in particular to see whether fewer climatic variables can help to accurately quantify the ET_0. In the present chapter, two relatively data-driven models called optimally pruned extreme learning machine (OPELM) and online sequential extreme learning machine (OSELM) were proposed as new alternatives to the FAO56 Penman-Monteith method for estimating ET_0 in Algeria using fewer climatic variables. The two models were compared, and the obtained results demonstrated that the OPELM model was more accurate compared to the OSELM model. It was also demonstrated that both minimal and maximal temperatures were the most important climatic variables and must be included as input to the models [8].

2.3   Aquifer Characterization and Assessment of Groundwater Resources

This section is covered into two chapters. The chapter titled "Water resources in Coastal Aquifers of Algeria Face Climate Variability: Case of Alluvial Aquifer of Mitidja in Algeria" presents the evaluation the relationship between climate change and water resources in the coastal aquifer of Mitidja, particularly with regard to the changes affecting the sustainability and availability of groundwater resources.

Algeria is considered as a vulnerable country regarding its water resource availability, especially in front of climate change conditions [9]. The water supply is the main challenging task for public institutions.

In the last decades, the trend of rainfall was decreased at about 20%, with an important annual irregularity in time, which had a negative impact on groundwater resources. The analysis of the piezometric map showed a drawdown level of groundwater for more than 10 m on average, and in the coastal sector, the wells have high salinities due to seawater intrusion after overexploitation of groundwater in the catchment field. Also, the analysis of physicochemical parameters of groundwater showed high concentrations of nitrate for the major part of the plain. NO_3 levels are moderately higher than the standard value (50 mg/L), which is due to the anthropogenic activities in Mitidja Plain such as the intensification agriculture, primarily linked to overuses of chemical fertilizers. Also, the authors reported that the discharge of urban sewage without treatment played a key role in the pollution of the the plain, which is a very common practice observed and connected to water pollution of many hydrosystems in Algeria [10–12].

Moreover, the chapter titled "Assessment of Groundwater Resources in the Jurassic Horst (Western Algeria)" is presented to identify, characterize, evaluate, and propose protection measures for the water resources of Jurassic Horst (Western Algeria). It describes the geological structure in western Algeria and the karst hydrogeology, water reserves, vulnerability, and constraints for groundwater resource management and sustainable water management in the Mounts of Tlemcen.

The authors state that the main groundwater resource of western Algeria is provided by the karstic aquifers of the Tlemcen Mountains. This water mobilized by more than 270 drillings is ensuring a total production capacity of 40 million of cubic meters of water per year. Consequently, the increasing water demand due to the development of the region and the population rapid growth coupled with the decrease of rainfalls over the last decades resulted in a significant drop in the water table level. This overexploitation was solved by using a desalinated seawater as an alternative way in order to avoid the overuse of the very slowly renewable groundwater. Additionally, an integrated water resource management was developed by modeling the water withdrawals based on the annual recharge of the aquifers. The study suggested that the artificial recharge of these aquifers during heavy rainfall periods would be a wiser alternative [13].

2.4   Toward a Sustainable Development

Also, this section is covered into two chapters. In the chapter titled "Participatory Approaches to Sustainable Development and Management of Soil Resources in Arid Zones of Algeria", the authors evaluated the chemical fertility status of the soils in the irrigable perimeter of Tadjmout (Laghouat) and the realization of the chemical fertility maps of these soils. The management of soil resources through a participatory approach for the sustainable development in arid zones of Algeria aims to preserve soil resources and water resources; these two resources are natural resources which are non- or the very slowly renewable [5].

In order to elaborate chemical fertility maps of arid soils, and with the aim of improving the yield of soil cultivation in arid zones located in the irrigated area of Tadjmout (Laghouat), a detailed description of pedological profiles and characterization of the chemical parameters (pH, cation exchange capacity (CEC), electrical conductivity (EC), total limestone ($CaCO_3$), exchangeable phosphorus (P_2O_5), and exchangeable potassium (K_2O)) were required.

The authors showed an interest in the chemical parameters of the soil because they have a very important influence on plant mineral nutrition; in other words it is called the crop yield. The crops grown in this region are vegetable crops (especially potato, tomato, lettuce, carrot, turnip); fruit trees (apricot, pomegranate, apple, vine); and date palm and cereals (barley, bread, and durum wheat). All these crops are irrigated except the cereal crops which are rainfed. These crops are practiced for the self-consumption of local populations as well as for the nutritional supplementation of animals. The surplus of the production is conveyed to the market. The aridity of climate and poor soil fertility in arid and semiarid regions of Algeria limits considerably the production of the crops mentioned above and consequently has negative effects on food security [14]. The prevention and development of these types of soils (dryland soils) require integrated management. Our main goal is to increase agricultural production through fertilization while protecting the environment.

Furthermore, since water plays a central role in human activities, socioeconomic development, and the ecological balance, the chapter titled "Scale Inhibition in Hard Water System" is about scaling in natural hard water as it is a major concern in different facets of industrial processes and domestic installations. Various studies and methods were developed to prevent the scale formation in water such as the inhibitors. There are many inhibitors for $CaCO_3$ precipitation reported in the literature, and the ideal inhibitor would be a compound in a solid form whose solubility would be very low but largely sufficient to ensure a total scaling inhibition. Thus, environmental requirements impose many challenges in the field of water treatment. For this, the use of green inhibitors has become a necessity. The inhibitory effect of these substrates is mainly due to adsorption and subsequent blocking of the active growth site. The water distribution of some Algerian town resulting from the drilling water is supersaturated with respect to calcium carbonate. The authors aim to give an overview of the different antiscalant property in hard water. They also reviewed the inhibitors used and the researches done on Algerian water.

The book ends with the conclusions and recommendations chapter.

Acknowledgments The writers of this chapter would like to acknowledge the efforts of authors of the chapters during the different phases of the book including their inputs to this chapter.

References

1. Negm A, Bouderbala A, Chenchouni H, Barcelo D (2020) Water resources in Algeria: II-: water quality, treatment, protection and development, handbook of environmental chemistry. Springer, Berlin
2. EUWI (2007) Mediterranean groundwater report (technical report on groundwater management in the Mediterranean and the water framework directive), 120 p

3. CEDARE (2014) Algeria water sector M&E rapid assessment report", monitoring & evaluation for water in North Africa (MEWINA) project, Water Resources Management Program, CEDARE
4. Water Quality Management, Algeria (2005) http://siteresources.worldbank.org/EXTMETAP/Resources/WQM-AlgeriaP.pdf
5. Laoubi K, Yamao M (2012) The challenge of agriculture in Algeria: are policies effective. Bull Agric Fish Econ 12(1):65–73
6. WWF (2005) Climate change impacts in the Mediterranean resulting from a 2°C global temperature rise. A report for WWF, by C. Giannakopoulos, M. Bindi, M. Moriondo, T. Tin
7. Drouiche N, Ghaffour N, Naceur MW, Lounici H, Drouiche M (2012) Towards sustainable water management in Algeria. Desalin Water Treat 50(1–3):272–284
8. Laoubi K, Yamao M (2009) Brebbia CA, Popov V (eds) Irrigation schemes management in Algeria: an assessment of water policy impact and perspectives on development. Water Resources Management V. WIT Press, Southampton, pp 503–514
9. Benabderrahmane MC, Chenchouni H (2010) Assessing environmental sensitivity areas to desertification in eastern Algeria using Mediterranean desertification and land use "MEDALUS" model. Int J Sustain Water Environ Syst 1(1):5–10. https://doi.org/10.5383/swes.01.01.002.5
10. Bouaroudj S, Menad A, Bounamous A, Ali-Khodja H, Gherib A, Weigel DE, Chenchouni H (2019) Assessment of water quality at the largest dam in Algeria (Beni Haroun Dam) and effects of irrigation on soil characteristics of agricultural lands. Chemosphere 219:76–88. https://doi.org/10.1016/j.chemosphere.2018.11.193
11. Belabed BE, Meddour A, Samraoui B, Chenchouni H (2017) Modeling seasonal and spatial contamination of surface waters and upper sediments with trace metal elements across industrialized urban areas of the Seybouse watershed in North Africa. Environ Monit Assess 189 (6):265. https://doi.org/10.1007/s10661-017-5968-5
12. Guemmaz F, Neffar S, Chenchouni H (2020) Physicochemical and bacteriological quality of surface water re-sources receiving common wastewater effluents in drylands of Algeria. In: Negm A, Bouderbala A, Chenchouni H, Barcelo D (eds) Water resources in Algeria: part II: water quality, treatment, protection and development. The handbook of environmental chemistry series. Springer Nature, Cham. https://doi.org/10.1007/698_2019_400
13. Algeria MWR, CEDARE, and Demmak A (2005) Algeria state of the water reporting, monitoring and evaluation operational framework and guidelines", monitoring & evaluation for water in North Africa (MEWINA) project, Ministry of Water Resources, Algeria – MWR, Water Resources Management Program – CEDARE
14. Oustani M, Halilat MT, Chenchouni H (2015) Effect of poultry manure on the yield and nutriments uptake of potato under saline conditions of arid regions. Emirates J Food Agric 27 (1):106–120. https://doi.org/10.9755/ejfa.v27i1.17971

Overview of Water Resources, Quality, and Management in Algeria

Mennat Allah Neama, Michael Attia, Abdelazim Negm, and Mahmoud Nasr

Contents

M. A. Neama and M. Nasr (✉)
Sanitary Engineering Department, Faculty of Engineering, Alexandria University, Alexandria, Egypt
e-mail: mennaneama@gmail.com; mahmmoudsaid@gmail.com; mahmoud-nasr@alexu.edu.eg

M. Attia
Irrigation Engineering and Hydraulics Department, Faculty of Engineering, Alexandria University, Alexandria, Egypt
e-mail: eng.michael_george@yahoo.com

A. Negm
Water and Water Structures Engineering Department, Faculty of Engineering, Zagazig University, Zagazig, Egypt
e-mail: amnegm@zu.edu.eg; amnegm85@yahoo.com

Abdelazim M. Negm, Abdelkader Bouderbala, Haroun Chenchouni, and
Damià Barceló (eds.), *Water Resources in Algeria - Part I: Assessment
of Surface and Groundwater Resources*, Hdb Env Chem (2020) 97: 13–26,
DOI 10.1007/698_2020_522, © Springer Nature Switzerland AG 2020,
Published online: 9 May 2020

Abstract Recently, Algeria has encountered various concerns in the sector of "water resource management" because of the rapid growth in the domestic, agricultural, and industrial activities. This chapter offers a summary of the essential surface water (e.g., rivers, lakes, and reservoirs) and groundwater resources in Algeria. The information is covered regarding the recently published articles, chapters, books, and conference proceedings. Moreover, the chapter lists the peer-reviewed journals, academic and institutional affiliations, and research funding agencies presented in the Scopus library, covering the water features in Algeria. Feasible and practical strategies needed for sustainable management of the water shortage concerns in Algeria are suggested. The chapter illustrates that the government, private and public sectors, and policymakers have handled several approaches concerning the country's national water policy, water resources development and planning, and water purification and reclamation. The findings of this chapter would support organizations, stakeholders, government leaders, industrialists, and environmentalists, dealing with all water subjects in Algeria.

Keywords Algeria, Scopus library, Sustainable water strategies, Water development and research, Water resources

1   Introduction

Algeria is situated in the northwest part of Africa, belonging to the southern shores of the Mediterranean region. Algeria is bordered by Morocco from west, Tunisia and Libya from east, Niger and Mali from south, and Mauritania and Western Sahara from southwest [1]. In Algeria, the peak rainfall event occurs during November–December, whereas the rainy period ranges from October to March, with an average rainfall of about 600 mm per annum. The climate of Algeria is described by an irregular monthly rainfall pattern; e.g., 5–9 mm/month during June–August, 80–90 mm/month during November–January, and 40–60 mm/month during March–May [2]. Due to the variation of water supply and demand throughout the country, a summary of the available water resources in Algeria should be reported.

Recently, the water consumption profile in Algeria has increased due to the rapid population and economic growth, urban land expansion, climate change condition, and industrial transformation and upgrading [3, 4]. Moreover, the rapid increase in agricultural activities and other human and domestic practices has led to the deterioration of Algerian water bodies used for drinking requirements. This situation results in decreasing the country's per capita water availability (i.e., below 300 cubic meters per year). To sort out the problems of water shortage and to meet the growing water needs, most Algerian towns have undertaken a program of water rationing adopted by the Company of Production of Water "Algerian Des Eaux" (ADE) [5]. In addition, the Algerian Water Organization (ADE) carries out the public drinking water facilities and services of the entire country. The Hydraulic Resources Directorate of Wilaya (DHW) is also responsible for various water projects in the country.

Algeria contains several water resources (surface water and groundwater) that have been used for various household, drinking, agricultural, industrial, and environmental purposes [6]. However, the availability of water resources has been negatively influenced by the irregular and uncontrolled withdrawal of water from surface and subsurface water bodies [7]. For instance, most of the total water demands (about 60%) are supplied to the agricultural sector. In this context, additional governmental, environmental, and engineering actions have been developed to handle the current and future challenges belonging to the water resources in Algeria [5, 8].

This chapter gives a general summary of the condition of water resources (e.g., rivers, lakes, wetlands, reservoirs, and groundwater) in Algeria. The chapter also lists the number and type of published documents that have been recently issued to handle the sustainability of water bodies in Algeria. Moreover, the chapter provides sufficient information about the authors' nationalities, institutional affiliations, and funding agencies and sponsors, considering the water management and research in Algeria.

2   Algerian's Water Statistics from Scopus Database

Figure 1 shows the number of documents retrieved from the Scopus database, addressing various water concerns in Algeria (https://www.scopus.com/search/ form.uri?zone=TopNavBar&origin=AuthorProfile&display=basic). The cumulative number of documents found in Scopus was 135 in 2001–2010, when searching the keywords "Water," "Resources," and "Algeria." This number increased by threefold during 2011–2019 (Fig. 1a), suggesting that scientists are conducting more researches to handle the water problems in Algeria. Moreover, the total numbers of manuscripts obtained using the keywords "Water," "Management," and "Algeria" were 94 and 283 documents during 2001–2010 and 2011–2019, respectively (Fig. 1b). Furthermore, by using the search keywords "Water," "Quality," and "Algeria" in the Scopus database, the total number of papers was 104 during 2001–2010. This number reached over a threefold increase (i.e., 360 documents) during 2011–2019 (Fig. 1c). The improving pattern in the number of published

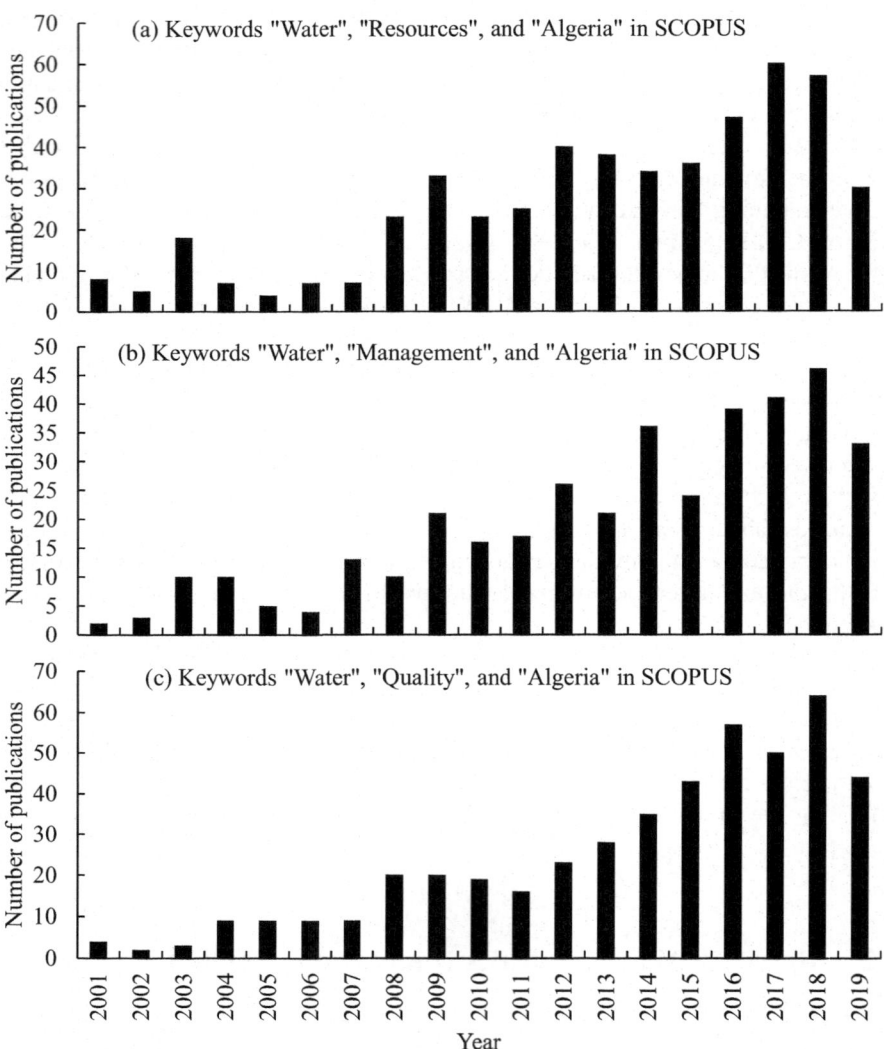

Fig. 1 Increasing pattern of published documents retrieved from Scopus database using research keywords (**a**) "Water," "Resources," and "Algeria"; (**b**) "Water," "Management," and "Algeria"; and (**c**) "Water," "Quality," and "Algeria"

documents during the last 10 years suggests that the topic of "Water Resources, Quality, and Management in Algeria" has become a critical domain of research. The documents were published in several international journals such as Arabian Journal of Geosciences, Energy Procedia, Desalination and Water Treatment, Desalination, Environmental Earth Sciences, and Journal of Water and Land Development. These journals cover the following subject areas: Applications of Desalination to Seawater, Groundwater, and Wastewater; Environmental Interaction between Humans and

Natural Resources; Natural Resources Management; Earth System Science; GIS and Remote Sensing; Environmental Sciences and Sustainable Development; Development of Monitoring Systems; and Water and Soil Contamination. Moreover, various publishers such as Elsevier, Springer, Taylor & Francis Group, and Wiley handled these peer-reviewed and highly reputable international journals. Algeria was the top country followed by France, Morocco, the United States, and Tunisia that contributed to the study of "Water Resources, Quality, and Management in Algeria" in the Scopus database during 2001–2019. Most of the published documents (about 70%) was an article type, followed by conference manuscripts, review papers, and book chapters. The main funding sponsors of these articles were the Ministry of Higher Education and Scientific Research; Agence Nationale de la Recherche; Ecole Nationale d'Ingénieurs de Tunis; European Commission; and Agence Nationale pour le Développement de la Recherche Universitaire. Based on this short survey, the public and private sectors in Algeria are working to address the existing and anticipated water-related issues in the country.

3　Rivers in Algeria

Algeria contains a number of rivers that have been used for drinking purposes, navigation, and agricultural irrigation [9, 10]. Some of these rivers are listed as follows:

3.1　Cheliff River

Cheliff River is located in the northwest part of the Algerian territory, and it is considered an important water resource for irrigation and drinking water supply. However, the river has recently suffered from high anthropogenic inputs [11].

3.2　Tafna River

Tafna River is located in the northwest part of Algeria, originating from the Tlemcen Mountains at about 1.1 km above sea level. After a 170-km course, the river joins the Mediterranean Sea near the Beni Saf town [12]. The Tafna basin covers an area of 7,245 km^2 (32° 40′ and 35° 20′ N; 1° 00′ and 1° 45′ W). The basin is bordered by Traras Mountains from north, Tlemcen Mountains from south, Beni-Snassen Mountains from west, and Sebaa-Chioukh Mountains from east [13].

3.3   Oued Isser River

The Oued Isser is a river of Algeria that reaches the Mediterranean Sea near the coastal town of Djinet in Lower Kabylia. The river forms a watershed with a total area of approximately 4,141 km^2 (36°52′N~35°52′ N and 3°56′E~2°52′ E), having an annual rainfall of about 800 mm [9].

3.4   Seybouse River

Seybouse River originates due to the confluence of Cherf River and Bouhamdane River, and it reaches the Mediterranean Sea near Annaba [10]. The river has a basin that covers an area of 6,471 km^2 in the northeast region of Algeria with annual rainfall values of 735 and 450 mm in its north and south areas, respectively [14].

3.5   Soummam River

Soummam River is located in northern Algeria, reaching the Mediterranean Sea at Béjaïa. The river has a watershed that extends between 35°75′ and 36°75′N of latitude and between 3°60′ and 5°55′E of longitude [8].

3.6   El Harrach River

El Harrach River is situated in the northern part of Algeria, covering a surface area of about 1,236 km^2. Surface runoff, rain (annual rainfall of 805 mm), and infiltration are the input water sources to the river [15].

4   Lakes and Wetlands in Algeria

Algeria contains a number of lakes that have been employed for (a) supplying water to industry and irrigation, (b) preserving the ecological biodiversity of aquatic and wildlife species, (c) minimizing the impact of floods and droughts, and (d) attaining various recreational activities [16, 17]. Moreover, some lakes have historical and traditional values; however, they are currently suffering from soil erosion, water-soluble salt content, rock weathering, and a huge amount of untreated wastewater. In addition, Algeria has about 50 water sites designed and operated as wetlands, covering a total surface area of 30,000 km^2. In Algeria, wetlands tend to provide various ecological and economic benefits, in which water can present above or near the soil surface during a portion of the year. Some of these lakes and wetlands can be listed as follows:

4.1 Oubeira Lake

Oubeira Lake is located in northeastern Algeria at approximately $36°49'-36°51'$ N and $8°22'-8°25'$ E. This large freshwater lake has been used as an essential source of drinking water for the residents of the east region of Algeria. Oubeira Lake can also be considered a natural wetland, hosting a diversity of rare flora and fauna species and receiving wild migratory and resident water birds [17].

4.2 Mellah

Mellah is considered an important lagoon situated in the northeast region of Algeria ($36.54°$ N and $08.20°$ E). The lagoon has a total surface area of 8.65 km^2 and a maximum depth of 5 m [18].

4.3 Tonga

Tonga Lake is a significant wetland situated in northeastern Algeria, having an area of 25 km^2. The lake offers vital ecological sites and breeding areas for aquatic biota habitats and communities. It also provides shelter and food to water birds and species [16].

4.4 Chott Merouane

Chott Merouane is a large salt lake located in the Oued Righ region, providing high economic benefits such as mineral and salt extractions, agricultural practices, and tourism activities. The wetland is also essential for zooplankton, flora, and fauna biodiversity [19].

4.5 Reghaia Lake

Reghaia Lake is located in the northeastern part of the plain of Mitidja (latitude 36° 45′–36° 48′ North and longitude 3° 19′–3° 21′ East). The lake has an area of 0.75 km^2, length = 2.5 km, and a 7 m depth [20].

4.6 Megarine Lake

Megarine Lake is a large water body separated into two small lakes, viz., Zerzaim Lake (latitude 33°12′12″ North and longitude 06°05′50″ East), and Lella Fatma (latitude 33°12′21″ North and longitude 06°05′54″ East) [21].

4.7 Other Wetlands

A previous study by Bellagoune et al. [22] has mentioned a number of wetlands that are located in Batna, Hauts Plateaux, and East of Algeria (Table 1). These wetlands can offer habitat for several wildlife species, reduce erosion, remove pollutants from surface water, attain flood control, provide ecosystem services, and recharge aquifers. However, the recent farming and urban developments have negatively influenced wetland conservation.

Table 1 Main Algerian wetlands located in Batna, Hauts Plateaux, East of Algeria [22]

Name	Location	Area (km^2)
Garaet Gemot	35°38.303′N; 07°00.506′E	0.57
Chott Zehar	35°36.135′N; 07°03.314′E	0.76
Chott Melah	35°36.446′N; 07°05.136′E	0.85
Garaet El Maghssel	35°49.581′N; 06°43.529′E	1.1
Ougla Touila	35°47.829′N; 07°04.494′E	1.75
Lac Boulhilet	35°44.542′N; 06°47.222′E	1.8
Etang de Timerganine	35°39.241′N; 06°57.468′E	2.5
Garaet Ouled M'barek	35°20.261′N; 07°15.429′E	3.4
Garaet Ouled Amara	35°23.378′N; 07°20.315′E	9.5
Chott Tinsilt	35°53.975′N; 06°29.581′E	36
Garaet Djendli	35°41.466′N; 06°31.193′E	37.5
Garaet Guellif	35°45.225′N; 06°54.442′E	55.25
Garaet Ank Djemel	35°45.225′N; 06°54.442′E	85.5
Garaet Tarf	35°38.42′N; 07°01.281′E	255

5 Dams and Reservoirs in Algeria

Algeria contains several dams that have found essential applications in water storage, flood control, electrical generation, and irrigation [23, 24]. The dams in Algeria are operated under the umbrella of National Agency for Dams and Transfers, Algeria.

5.1 Cheffia Dam

Cheffia Dam is located in the El Taref Wilaya, the northeastern part of Algeria ($36°07'$ N and $8°03'$ E). The dam covers an area of about 10 km^2 with a maximum depth of 30 m [25].

5.2 Beni Haroun Dam

Beni Haroun Dam is located in Mila Province, northeastern Algeria. The dam serves a catchment area of 8,815 km^2, and its reservoir has a storage capacity of 795×10^6 m^3. The dam is used for (a) irrigation with 333×10^6 m^3/year and (b) supplying drinking water with 255×10^6 m^3/year [26].

5.3 Zit Amba Dam

Zit Amba Dam is located in Kebir West catchment, northeastern Algeria, occupying an area of 1,900 km^2. The reservoir has a storage capacity of 120×10^6 m^3 with an annual rainfall of approximately 700 mm, mainly used for irrigation and drinking purposes. The upstream and central zones of the reservoir basin are situated in the Wilaya of Guelma, whereas the downstream part is located in the Wilaya of Skikda [27].

5.4 Mexa Dam

Mexa Dam is located in the extreme northeast of Algeria, near El Kala town and Wadi Kebir [28]. Initially, the dam basin covered an area of 560 km^2, which have recently reduced to 393 km^2 due to the installation of a dam on Wadi Barbara, Tunisia. Lastly, the final basin area is estimated as 158 km^2 after the construction of the Bougous Dam [23].

5.5   *Foum El Gherza Dam*

Foum El Gherza Dam is a reservoir with an initial volume of about 47×10^6 m^3 situated at 18 km east of Biskra Province, Algeria. The reservoir lost approximately 65% of its initial capacity mainly because of silting, leading to a current capacity of 16.5×10^6 m^3 [7].

5.6   *Babar Reservoir*

Babar Reservoir was constructed in 1989 with an initial capacity of 41×10^6 m^3, which reduced to 38×10^6 m^3 in 2004. It is located in the southeast of Khenchela Province for meeting domestic and agricultural water requirements [24].

6   Groundwater in Algeria

Groundwater is considered an essential source of drinking water in addition to its use in irrigation. In Algeria, the best estimate of groundwater storage reaches up to 91,900 km^3, ranging between 56,000 and 243,000 km^3 [29]. Generally, groundwater is protected from pathogenic contamination, and hence the treatment of groundwater requires standard systems. The treatment of groundwater depends on the removal of fluoride, phenol, nitrate, arsenic, and chlorinated organic compounds [30]. For instance, Kut et al. [31] reported that groundwater in the southern region of Algeria contains 0.4–2.3 ppm of fluoride concentration. The quantity of groundwater relies mainly on rainfall patterns, weathering conditions, and geology and geochemistry of soils. The National Agency of Hydraulic Resources (ANRH) in Algeria authorizes the quantitative and qualitative assessments of groundwater resources. However, increasing aquifer recharge is required to avoid over-exploitation of groundwater resources.

7   Conclusions

This work aims at giving a general overview of the essential water resources in Algeria. It can be concluded that:

- Algeria holds important freshwater resources kept within soil layers as groundwater and in surface water bodies, viz., streams, rivers, lakes, wetlands, and reservoirs.

- Recently, Algeria has coped with various issues in the sector of "water resource management" due to the rapid rise in population growth, a large increase in urbanization, and improvement in agricultural and industrial practices.
- The cumulative number of published documents covering the status and condition of water resources in Algeria during 2011–2019 was about threefold during 2001–2010.
- The river system (e.g., Cheliff, Tafna, Oued Isser, Seybouse, Soummam, and El Harrach) is recognized as an essential input of drinking, domestic, and irrigation activities in Algeria.
- Algeria contains vital lakes and wetlands such as Oubeira, Mellah, Tonga, Chott Merouane, Reghaia, and Megarine.
- Algeria contains several dams and reservoirs (e.g., Cheffia, Beni Haroun, Zit Amba, Mexa, Foum El Gherza, and Babar) employed for different objectives including irrigation, water storage and balance, hydroelectric energy supply, and protection of downstream areas from flooding and drought.
- The quantity of groundwater storage in Algeria ranges between 56,000 and 243,000 km^3, supporting major drinking and agricultural purposes.
- The findings obtained from this chapter would support environmental organizations responsible for all problems and concerns of water bodies in Algeria.

8 Recommendations

Based on the chapter findings, a number of suggestions should be followed to protect the valuable water resources in Algeria:

- Regular spatial and temporal assessment of different physicochemical factors (e.g., temperature, pH, total and dissolved solids, dissolved oxygen, biological oxygen demand, nutrients, and ions), bacteriological parameters (e.g., *E. coli*), and heavy metals in water bodies should be conducted.
- Unconventional mathematical models and computational tools should be employed to handle large and interconnected data about water quality.
- Further procedures and actions should be performed to reduce the release of wastewater from the industrial and agricultural sectors into water bodies. In addition, appropriate and suitable waste disposal methods and wastewater treatment techniques should be applied to maintain human health during water consumption.
- Locals and farmers should be provided with proper awareness, strategies, and scientific communications that cover most aspects of environmental and health concerns.
- The government should offer special strategies to the concepts of "Wastewater Recycling and Reuse" and "Rainwater Collection and Harvesting" to meet the water supply targets, especially in rural parts of Algeria.

Acknowledgments The authors would like to acknowledge Nasr Academy for Sustainable Environment (NASE).

References

1. Benmecheta A, Belkhir L (2016) Oil pollution in the waters of Algeria. In: Carpenter A, Kostianoy A (eds) Oil pollution in the Mediterranean Sea: part II. The handbook of environmental chemistry, vol 84. Springer, Cham, pp 247–262
2. Abdellah M, Mohamed H, Farouk D (2018) The implication of climate change and precipitation variability on sedimentation deposits in Algerian dams. Arab J Geosci 11(23):733
3. Ramdani H, Laifa A (2017) Physicochemical quality of Wadi Bounamoussa surface waters (Northeast of Algeria). J Water Land Dev 35(1):185–191
4. Daifallah T, Hani A (2018) Water demand management is solution of water stress?: a case study of the Kebir-west river basin in northern Algeria. Water Energy Int 60(11):62–66
5. Habi M, Harrouz O (2015) Domestic water conservation practices in Tlemcen City (Algeria). Appl Water Sci 5(2):161–169
6. Amira AB, Bougdah M (2018) Influence of Mafragh and Seybouse inputs (sediment and salts) on the productivity of Annaba Bay. AACL Bioflux 11(3):653–665
7. Benfetta H, Ouadja A, Hocini N (2017) Enhancement of the study of water leaks in the Algerian dam of Foum El Gherza. Arab J Geosci 10(22):482
8. Sahli Y, Mokhtari E, Merzouk B, Laignel B, Vial C, Madani K (2019) Mapping surface water erosion potential in the Soummam watershed in Northeast Algeria with RUSLE model. J Mt Sci 16(7):1606–1615
9. Tachi SE, Ouerdachi L, Remaoun M, Derdous O, Boutaghane H (2016) Forecasting suspended sediment load using regularized neural network: case study of the Isser River (Algeria). J Water Land Dev 29(1):75–81
10. Reggam A, Bouchelaghem E-H, Hanane S, Houhamdi M (2017) Effects of anthropogenic activities on the quality of surface water of Seybouse River (northeast of the Algeria). Arab J Geosci 10(10):219
11. Benkaddour B, Abdelmalek F, Addou A, Noguer T, Aubert D, Vouvé F (2019) Assessment of anthropogenic and natural factors on Cheliff River waters (north-west of Algeria) at two contrasted climatic seasons. Int J Environ Res 13(6):925–941
12. Haddou K, Bendaoud A, Belaidi N, Taleb A (2018) A large-scale study of hyporheic nitrate dynamics in a semi-arid catchment, the Tafna River, in Northwest Algeria. Environ Earth Sci 77 (13):520
13. Benabdelkader A, Taleb A, Probst JL, Belaidi N, Probst A (2019) Origin, distribution, and behaviour of rare earth elements in river bed sediments from a carbonate semi-arid basin (Tafna River, Algeria). Appl Geochem 106:96–111
14. Talbi H, Kachi S (2019) Evaluation of heavy metal contamination in sediments of the Seybouse River, Guelma – Annaba, Algeria. J Water Land Dev 40(1):81–86
15. Bouragba S, Komai K, Nakayama K (2019) Assessment of distributed hydrological model performance for simulation of multi-heavy-metal transport in Harrach River, Algeria. Water Sci Technol 80(1):11–24
16. Draidi K, Bakhouche B, Lahlah N, Djemadi I, Bensouilah M (2019) Diurnal feeding strategies of the ferruginous duck (Aythya nyroca) in Lake Tonga (northeastern Algeria). Ornis Hungarica 27(1):85–98
17. Messerer Y, Retima A, Amira AB, Djebar AB (2019) Climatic changes, hydrology and trophic status of lake oubeira (extreme northeast of Algeria). AACL Bioflux 12(4):1442–1457
18. Draredja MA, Frihi H, Boualleg C, Gofart A, Abadie E, Laabir M (2019) Seasonal variations of phytoplankton community in relation to environmental factors in a protected meso-oligotrophic

southern Mediterranean marine ecosystem (Mellah lagoon, Algeria) with an emphasis of HAB species. Environ Monit Assess 191(10):603

19. Benhaddya ML, Halis Y, Lahcini A (2019) Concentration, distribution, and potential aquatic risk assessment of metals in water from Chott Merouane (Ramsar site), Algeria. Arch Environ Contam Toxicol 77(1):127–143

20. Metna F, Lardjane-Hamiti A, Merabet S, Boukhemza-Zemmouri N, Rakem K, Boukhemza M (2016) Ecology of the Coot's Fulica atra reproduction (Linnaeus, 1758) in the nature reserve of Lake Réghaïa (Algiers, Algeria). Zool Ecol 26(3):166–172

21. Khellou M, Laifa A, Loudiki M, Douma M (2018) Assessment of phytoplankton diversity in two lakes from the northeastern Algerian sahara. Appl Ecol Environ Res 16(3):3407–3419

22. Bellagoune S, Maazi M-C, Saheb M, Bara M, Bouslama Z, Houhamdi M (2015) Ecology of wintering of common Shelduck (Tadorna tadorna) in Sebkhet Djendli (Batna, Hauts Plateaux, east of Algeria). Adv Environ Biol 9(3):395–402

23. Bahroun S, Chaib W (2017) The quality of surface waters of the dam reservoir Mexa, northeast of Algeria. J Water Land Dev 34(1):11–19

24. Tebbi FZ, Dridi H, Kalla M (2018) Performance analysis of a reservoir in arid region case study: Babar reservoir, Aurès region, Algeria. J Water Land Dev 39(1):141–146

25. El Herry S, Nasri H, Bouaïcha N (2009) Morphological characteristics and phylogenetic analyses of unusual morphospecies of Microcystis novacekii forming bloom in the Cheffia Dam (Algeria). J Limnol 68(2):242–250

26. Bouaroudj S, Menad A, Bounamous A, Ali-Khodja H, Gherib A, Weigel DE, Chenchouni H (2019) Assessment of water quality at the largest dam in Algeria (Beni Haroun Dam) and effects of irrigation on soil characteristics of agricultural lands. Chemosphere 219:76–88

27. Bougdah M, Amira AB (2017) Water and sediment retention in a reservoir (Zit Amba, Algeria). AACL Bioflux 10(3):534–542

28. Boukhrissa ZA, Khanchoul K, Le Bissonnais Y, Tourki M (2013) Prediction of sediment load by sediment rating curve and neural network (ANN) in El Kebir catchment, Algeria. J Earth Syst Sci 122(5):1303–1312

29. MacDonald AM, Bonsor HC, Dochartaigh BÉÓ, Taylor RG (2012) Quantitative maps of groundwater resources in Africa. Environ Res Lett 7(2):024009

30. Fu F, Dionysiou DD, Liu H (2014) The use of zero-valent iron for groundwater remediation and wastewater treatment: a review. J Hazard Mater 267:194–205

31. Kut KMK, Sarswat A, Srivastava A, Pittman CU, Mohan D (2016) A review of fluoride in african groundwater and local remediation methods. Groundw Sustain Dev 2–3:190–212

Part II
Current Status of Water Resources

Origin and Quality of Groundwater of the Taoura Syncline Aquifer in North-Eastern of Algeria

L. Djabri, S. Bouhsina, A. Hani, H. Chaffai, F. Z. Trabelsi, F. Rouaibia, and A. Bouderbala

Contents

L. Djabri (✉), A. Hani, and H. Chaffai
Water Resource & Sustainable Development Laboratory, University of Badji Mokhtar Annaba, Annaba, Algeria
e-mail: djabri_larbi@yayoo.fr

S. Bouhsina
Unit of Environmental Chemistry and Interactions on the Living (UCEIV), University of the Littoral Opal Coast, Dunkerque, France
e-mail: saad.bouhsina@univ-littoral.fr

F. Z. Trabelsi
Higher School of Engineers of Medjez El-Bab, University of Jendoub, Jendouba, Tunisia
e-mail: trabelsifatma@gmail.com

F. Rouaibia
University of Constantine, Constantine, Algeria
e-mail: rf_0310@hotmail.fr

A. Bouderbala
Department of Earth Sciences, University of Khemis Miliana, Khemis Miliana, Algeria
e-mail: bouderbala.aek@gmail.com

Abdelazim M. Negm, Abdelkader Bouderbala, Haroun Chenchouni, and
Damià Barceló (eds.), *Water Resources in Algeria - Part I: Assessment
of Surface and Groundwater Resources*, Hdb Env Chem (2020) 97: 29–52,
DOI 10.1007/698_2020_523, © Springer Nature Switzerland AG 2020,
Published online: 20 July 2020

Abstract The water quality of aquifer is conditioned by geological formations crossed by water during its flow. So water crossing calcareous rocks becomes rich in bicarbonates; also water crossing saliferous rocks becomes rich of sulphates or chlorides. This affirmation can be applied to water of the karst aquifer of Taoura. The recharge of this aquifer comes from rainfall or by water of different draining zones of the watershed cross-border of Medjerda-Mellegue in south part of the city of Souk Ahras. The geological studies showed that the outcrop formations in the syncline contain groundwater in low level, which make the aquifer captive. This last is exploited by several wells (T1, T3, T7, D3, D5).

The observation of stratigraphic logs shows a lithological heterogeneity of the aquifer, which is going to explain the variation of its hydrochemical composition. Samples and analysis carried out showed that groundwater of this aquifer is rich in calcium and bicarbonate; however some wells are characterized by the domination of chlorides and sulphates; these anions are accompanied by sodium or by calcium.

So, the impact of geological rocks crossed by water is important on the characterization of groundwater quality. Also, the risks of pollution in unconfined parts of the aquifer are quasi unavoidable, and they are confirmed by concentration of nitrates >50 mg/l.

Keywords Algeria, Hydrochemical characterization, Taoura syncline, Water quality

1   Introduction

The syncline of Taoura occupies a subbasin of 826 km^2. It is located on the Algerian-Tunisian border, in the southeast of Souk Ahras. The Taoura region includes several aquifers as well as many cold springs and few hot springs.

The research concerns the Taoura system, which is characterized by the interference of two aquifers located in the syncline. The first is formed by sedimentary formations: it overlays a karstic/fractured aquifer. The covers filling the syncline hide the karstic formation.

The groundwater of the Taoura syncline aquifer is stored in the system aquifers. Also the first level is porous, and it is contained in the Mio-Plio-Quaternary formations, while the second aquifer is the fractured type, which refers to the deep karstic aquifer.

The karst aquifers are little developed in the Maghreb. In Algeria the karst has a limited extension. They are present in some regions in the North of Algeria, such as

Tlemcen [1], Kabylie [2], Constantine [3], Tebessa [4], Chérea [5] and Souk Ahras [6]. In southern Algeria, they are present in the regions of Biskra [7] and Ouargla [8]. In general, karst reservoirs are characterized by karst forms and discontinuities as in the karst scheme developed by C. Drogue [9].

The genesis, the structure and functioning of karst systems are known through various research works [10–13].

Groundwater from Taoura karst aquifer is used to supply drinking water for several localities and to irrigate numerous perimeters. So, karst groundwater of this aquifer is subject to a double risk, the first is related to its over-exploitation and the second is due to the exposure of the groundwater to pollution by return flows from irrigation.

The tectonic history of the area, the structure of the system and the human activities in this region propose some questions about the quality of karst groundwater. This research aims to study the current groundwater quality by using a hydrochemical-statistical approach and to understand the natural and human factors which have an impact on the variation of the water quality within the aquifer.

2   Geological and Hydrogeological of the Study Area

The Taoura area is located in the eastern of Algeria and in the borders of Tunisia, at approximately 30 km in south of the city of Souk Ahras (Fig. 1).

The study area is bounded in the north by the city of Souk Ahras, in the south by the city of Drea, in the east by Haddada and the Algerian-Tunisian border and in the west by the cities of Zarrouria and M'Daourouch.

2.1   Geological and Structural Features

Geological and lithostratigraphic data from wells indicate that the study area is characterized by the superposition of numerous formations, with different permeabilities, which promote the storage of groundwater. The paleogeology of the region shows that these formations were formed during the orogeny of the Tello-Rifean or Maghrebian chains, at the period of the Alpine cycle. As a result, the formations of Mesozoic and Cenozoic ages have been affected by the successive tectonic phases of the Souk Ahras region. The study area includes in its northern part, allochthonous lands composed by the Flysch domain. In the south and the east, the Algerian-Tunisian border till is an intensely wrinkled and fractured and formed the Atlas foreland outcrops [14].

The Taoura region is characterized by SW-NE anticlinal structures. These structures are of age ranging from Cretaceous to Eocene [15]. The NW-SE and NE-SW faults (Fig. 2) show rejections. Vertical faults structured all the lands in collapses

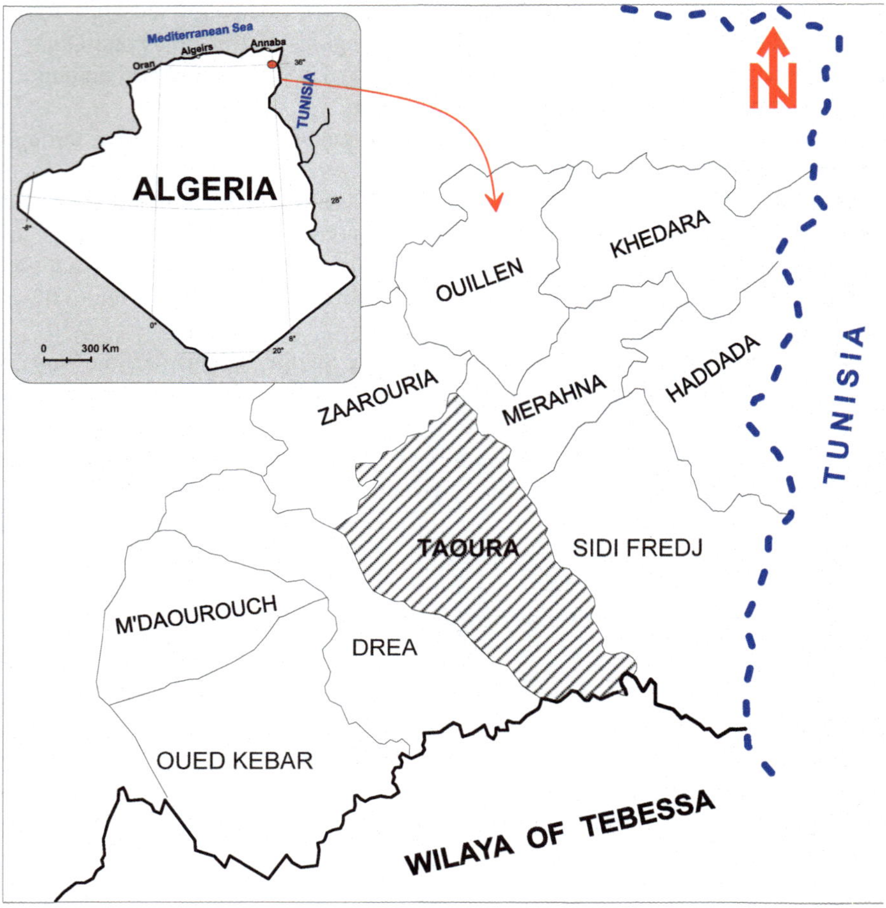

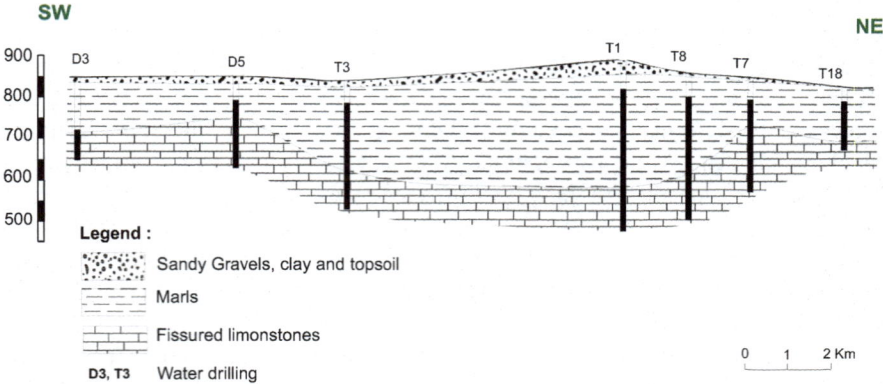

Fig. 1 Geographical location of the Taoura city

Fig. 2 Hydrogeological cross-section in the Taoura region

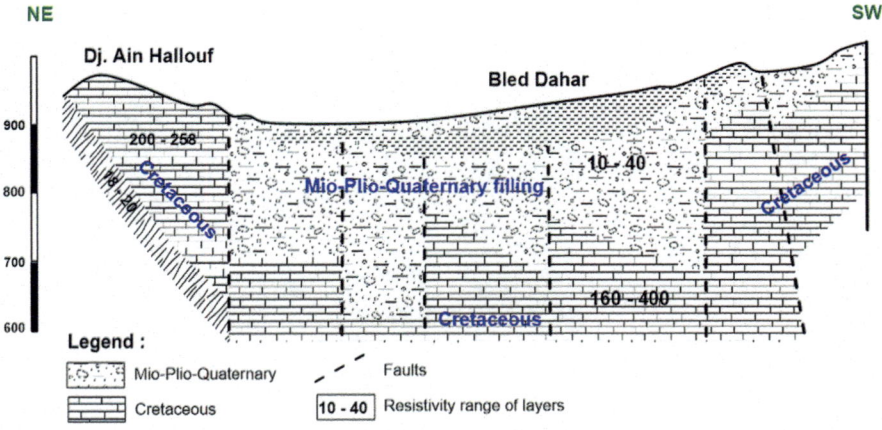

Fig. 3 Tectonics causing the discontinuity of layers in the study area

during the Plio-Quaternary. This explains the discontinuity of the aquifers and the disappearance of the karstic aquifer at the Drea-M'Daourouch plain.

The geo-electric section (Fig. 3) shows a subdivision of the aquifer system by the many faults that affect the syncline [16]. This discontinuity in the reservoirs explains the disappearance of the aquifers in the extreme zones.

Figures 2 and 3 show the presence of two superimposed aquifers. The first has groundwater levels near to 40 m in the Mio-Plio-Quaternary aquifer, which is rich in marl, conglomerates, sands and gravels constituting the filling of the Taoura syncline. This aquifer is mainly exploited by domestic wells in the north-eastern part towards Merahna [17]. The resistivities of this aquifer are ranged between 10 and 40 Ω m.

The second aquifer is considered as confined, and it is composed of marls and fractured limestones of low Maastrichtian and the middle Campanian. The aquifer occupies the centre of Taoura and disappears towards the southwest part due to the formation of the anticline chain towards Madour. We note that the resistivities oscillate between 50 and 100 Ω m and the hydrodynamic exchange between the main aquifers is possible due to the very important faults [18].

2.2 Hydrogeological Features

In the piezometric measurements carried out in July 2011 and December 2012 on wells of the first aquifer, they allowed us to see the general flow of groundwater in this aquifer for the two periods, as well as the recharge area and the over-exploitation zones. Figures 4 and 5 show groundwater flows, which they directed from the borders towards the centre of the plain, indicating a lateral contribution of recharge. The convergence of the flow to the west of Haddada and Merahna should reflect the

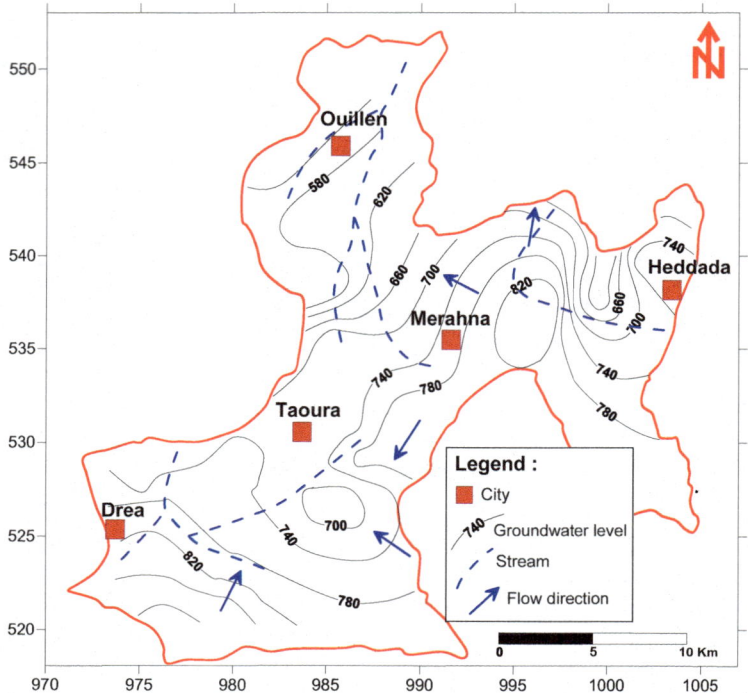

Fig. 4 Piezometric map of the superficial aquifer (July 2011)

effect of the aquifer exploitation in these two regions. In the south of the area, the presence of the piezometric dome is explained by the existence of a groundwater-divide, which could be at the origin of the two different flows present in the region. The recharge comes from the area between the reliefs and the north of the plain in the syncline, which is bordered by the mountain.

– In first, a notable decrease of groundwater levels is generate by the over-exploitation in the plain.
– The low recharge of about 12 mm/year is probably due to the ascending drain of the aquifer.

The interpretation of pumping data tests (six pumping of long periods for more than 48 h) are referred to wells. The transmissivity values are varied from 1.2×10^{-3} to 0.7×10^{-2} m²/s [19]. The highest values are observed in the drillings T1, T3 and T15, which drain the Turonian, Maastrichtian and Cenomanian formations. These wells are located along the synclinal axis of the Taoura basin. In the downstream part of Taoura, at Ms1, Ms2 wells and in Bir Louhichi drillings, the calculated transmissivity values are between 3.4×10^{-4} and 0.2×10^{-5} m²/s [20]. These values are lower than previous ones, presumably indicating a variation of the lithology. The values of the storage coefficient range between 1.9×10^{-3} in the area of Drea

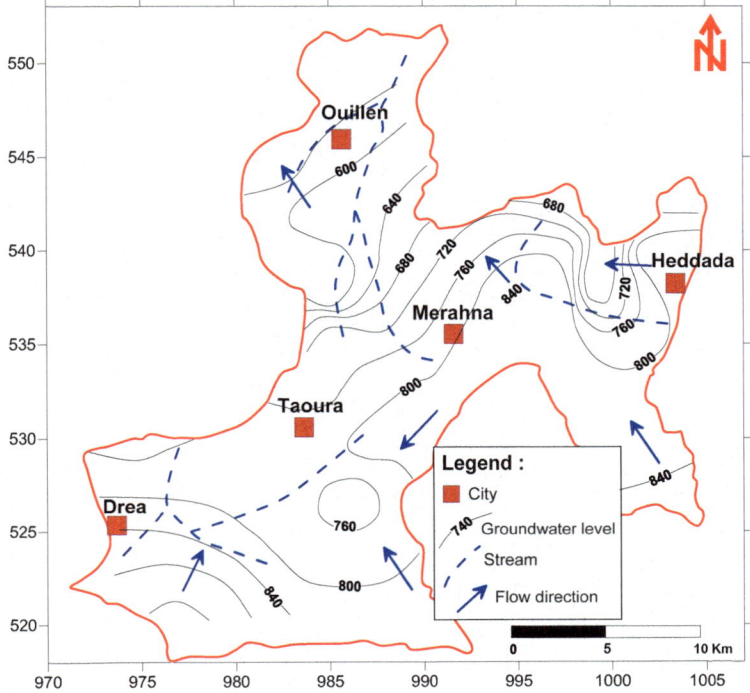

Fig. 5 Piezometric map of the superficial aquifer (December 2012)

(D3) and 2.6×10^{-1} (T7 and T8) in the part of the Taoura syncline [21]. The storage coefficient of 2.6×10^{-1} characterizes the karst aquifer. The value of the permeability resulting from the analysis of the pumping tests varies between the feeble value 1.2×10^{-5} m/s and the high value 12.9×10^{-4} m/s.

3 Materials and Methods

108 groundwater samples were taken during 6 months (18 samples per month) for the years 2011 and 2012.

The physicochemical parameters (temperature, electrical conductivity (EC) and pH) were measured in situ using a WTW multivariable instrument (P3 pH/LF-SET) [22]. Anions (Cl^-, HCO^-_3, SO_4^{-2}, NO^-_3) have been analysed by colorimetry method. The cations (Ca^{2+}, Mg^{2+}, Na^+, K^+) were performed by atomic absorption spectrophotometry.

For each sample, we took two bottles, the first one acidified for the determination of the cations and the second non-acidified for the anions. The bottles were stored in

Table 1 Statistics of the measured parameters in water (mg/l)

Variable	Min	Max	Mean	Standard deviation
EC	705.0	3500.0	1871.2	624.4
pH	6.8	8.3	7.8	0.3
T°	9.0	46.0	16.2	6.4
Ca^{2+}	8.3	376.0	115.8	70.1
Na+	5.3	186.0	60.5	40.2
K^+	0.2	94.2	8.8	9.9
Mg^{2+}	0.5	106.0	28.7	17.7
HCO_3^-	8.9	1167.0	285.1	217.3
SO_4^{2-}	15.2	341.0	135.1	72.9
Cl^-	6.1	367.0	95.8	59.8
NO_3^-	0.001	128.8	31.8	26.3
NO_2^-	0.0	6.8	0.685	1.2

Electrical conductivity is µS/cm at 20°C

a fridge at 4°C before being transported to the Algerian Water Company (ADE) laboratory.

Table 1 shows the minimum, maximum and average values and the standard deviation of all samples in the study area.

4   Results and Discussions

Table 1 shows that the electrical conductivity (EC) varies over a wide range, suggesting a variation in the geological nature of the aquifer. It is admitted that the geological formations directly influence the water chemical composition.

The groundwater has a near than neutral pH, which may denote a basicity of water; also the data of cations and anions show a dominance of the bicarbonate and calcium concentrations.

4.1 Hydrochemical Types

The piper diagram of Fig. 6 refers to the whole groundwater samples of the six surveys (2011–2012) showing five main hydrochemical types:

- Calcium-magnesium chloride-sulphate water type (45 samples)
- Sodium-potassium or sodium-sulphate water type (9 samples)
- Calcium and magnesium bicarbonate water type (46 samples)
- Sodium and potassium carbonate water type (2 samples)
- Hyper-chlorinated and hyper-sulphated calcium water type (6 samples)

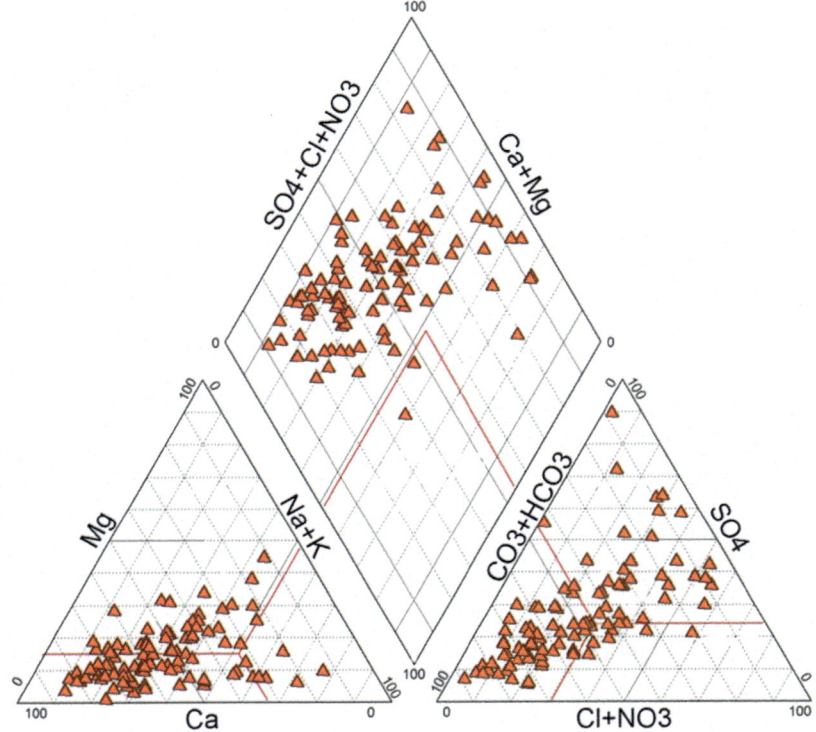

Fig. 6 Piper diagram referred to all groundwater samples in the six surveys of 2011 and 2012

Figure 6 shows a competition between bicarbonate water and water rich in chlorides or sulphates. This competition is favoured by the local geology and the human activities.

The two piper diagrams of Figs. 7 and 8 refer, respectively, to the samples of the months of July 2011 and December 2012, which are considered as two extreme periods of the climate conditions.

Figure 7 shows a distribution of hydrochemical types as follows:

– Chloride and sulphate calcium and magnesium water type (8 samples)
– Calcium and magnesium bicarbonate water type (9 samples)
– Hyper-chlorinated and hyper-sulphated calcium water type (1 sample)

Figure 8 shows a different distribution of hydrochemical types:

– Calcium and magnesium bicarbonate water type (7 samples)
– Chlorinated, sulphated calcium and magnesium water type (7 samples)
– Sodium chlorinated water type (01 sample)
– Strongly mineralized water type (3 samples)

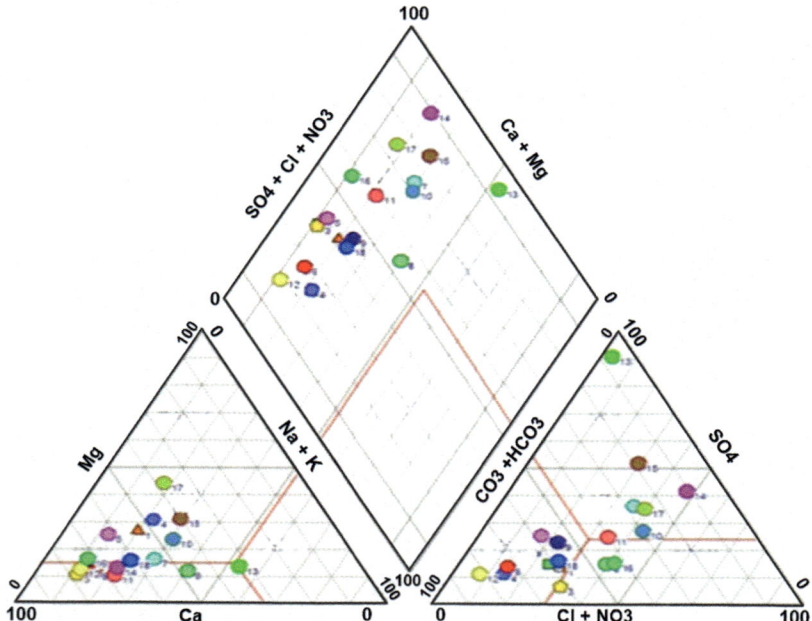

Fig. 7 Piper diagram, July 2011

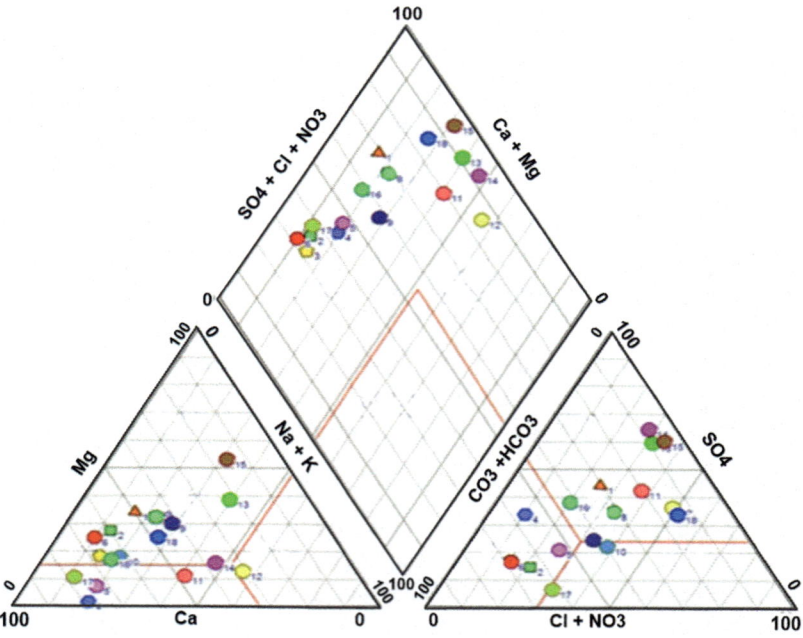

Fig. 8 Piper diagram, December 2012

From July 2011 (dry season) to December 2012 (wet season), there is a clear diminution of bicarbonate concentrations, which is compensated by the increase of chloride or sulphate concentrations (Table 2). This situation can be explained by a decrease in the dilution caused by the lack of recharge during July 2011. This occurrence, coupled with the increase of exploitation during the dry season, suggests that wells draw water of higher time residence compared to that drawn during a wet season.

4.2 Principal Component Analysis

The principal component analysis (PCA) has been applied for all the chemical data (2011 and 2012). The acceptable correlation coefficient (Table 3) is of the order of $r = 0.55$, which allows recognizing the associations shown in Table 4.

Data of Tables 3 and 4 show a good correlation between the EC and bicarbonates and calcium. There is also a good correlation between bicarbonates and calcium, probably reflecting the same origin of these two elements. Good correlations are between chlorides, sulphates and sodium, which suggests the same origin for these elements.

Table 2 Number of samples belonging to the different hydrochemical types in the wet and dry seasons

Chemical type\period	July 2011	December 2012	Variation from the dry to wet seasons
Calcium-magnesium bicarbonate	7	9	Decrease of bicarbonate concentrations
Calcium-magnesium chloride-sulphate	7	8	Slight depletion of calcium and magnesium concentrations
Sodium chloride	1	0	Increase of sodium concentrations
Highly mineralized	3	1	Increase of mineralization

Table 3 Taoura correlation matrix

	EC	Ca^{2+}	Na^+	K^+	Mg^{2+}	HCO_3^-	SO_4^{2-}	Cl^-	NO_3^-
EC	**1.00**								
Ca^{2+}	**0.80**	**1.00**							
Na^+	0.13	0.14	**1.00**						
K^+	0.05	0.30	**0.60**	**1.00**					
Mg^{2+}	0.44	0.32	**0.62**	0.46	**1.00**				
HCO_3^-	**0.86**	**0.96**	0.28	0.27	0.34	**1.00**			
SO_4^{2-}	0.28	0.27	**0.65**	0.38	**0.85**	0.27	**1.00**		
Cl^-	0.11	0.05	**0.91**	0.45	**0.62**	0.18	**0.59**	**1.00**	
NO_3^-	0.11	0.17	**0.61**	**0.65**	**0.68**	0.16	**0.72**	**0.52**	**1.00**

Bold values indicate significant correlations

Table 4 Associations between chemical elements according to the value of the correlation coefficient (Table 3)

Couple	r value	Couple	r value	Couple	r value
EC-Ca	0.80	Na-SO$_4$	0.65	Mg-Cl	0.62
EC-HCO$_3$	0.86	Na-Cl	0.91	Mg-NO$_3$	0.68
Ca-HCO$_3$	0.96	Na-NO$_3$	0.65	SO4-Cl	0.59
Na-K	0.60	K-NO$_3$	0.65	SO4-NO$_3$	0.72
Na-Mg	0.62	Mg-SO$_4$	0.85	Cl-NO$_3$	0.55

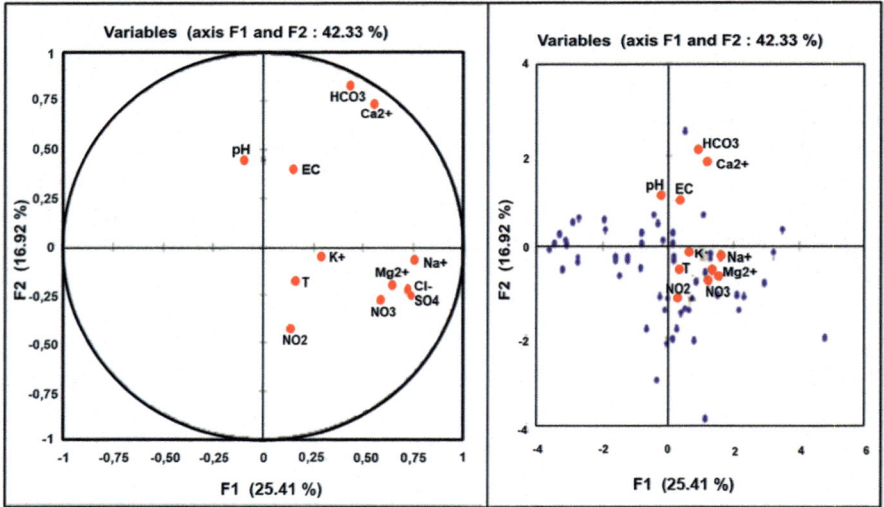

Fig. 9 Results of PCA of all data

The observation of the PCA (referred to the whole of chemical data) circle (Fig. 9) formed by the axes F1–F2 (42.33%) shows along the axis F1 (25.41%) an opposition between the highly mineralized water characterized by the presence of all the main ions and the weakly mineralized water occupying the negative part of the axis.

According to the F2 axis (16%, 92%), there is an opposition between calcium-bicarbonate water (positive part of the axis) and water rich of sodium, chlorides, sulphates and magnesium. These samples are polluted by nitrites, which suggest the influence of wastewater contamination. In general, we find the following distribution:

- Water from evaporitic and/or salt formations (anhydrite and gypsum), it is rich on sulphates and calcium.
- Water from the dissolution of carbonates (Ypresian limestones and Campanian-Maastrichtian. It is present in the south-west and north-east of the study area. Water in these sectors is rich on bicarbonate.

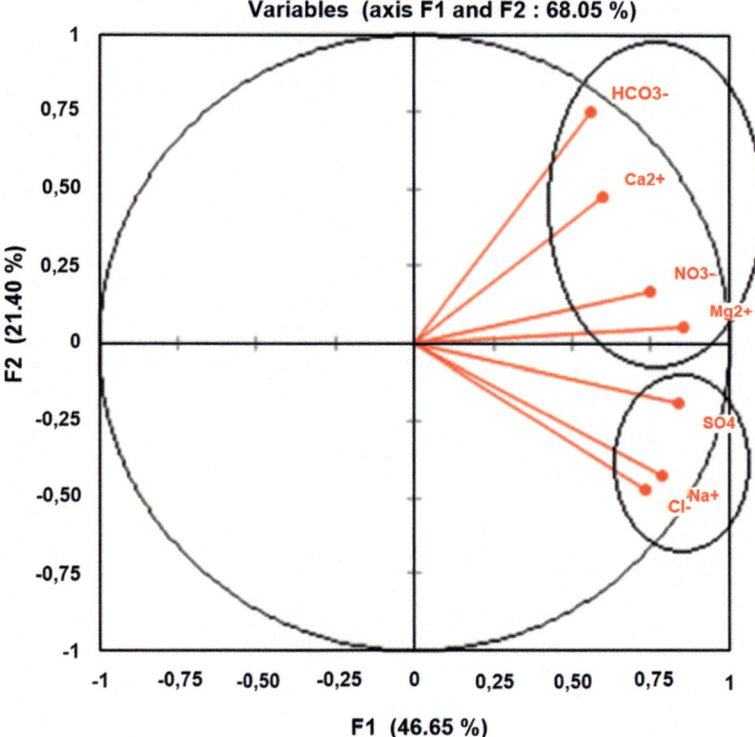

Fig. 10 PCA circle (analyses of December 2012)

The PCA circle referred to December 2012 analyses (Fig. 10) is formed by two axes F1 and F2 (68.05%). It shows along the F1 axis (46.65%) an opposition between the highly mineralized water (positive part of the axis) and the weakly mineralized water (negative part of the axis). According to the F2 axis (26.40%), there is an opposition between the natural elements of endogenous origin, resulting from the water-rock interaction with the geological formations, and the exogenous elements of human origin (NO_3^-, NO_2^-, SO_4^{2-}), which could be linked to the agricultural practices. In addition, we note that sulphates can come from either rock or inputs used in agriculture.

The PCA circle referred to analyses of March 2011 (Fig. 11) is formed by the F1 and F2 axes and provides 65.52% of the information. According to the F1 axis (47.63%), there is opposition between the highly mineralized water and the weakly mineralized ones.

By observing the F2 axis (16.89%), there is an opposition between the bicarbonate water and the sodium-sulphate or sodium-chloride water type. This distribution results from the dilution that occurs during this particularly wet period.

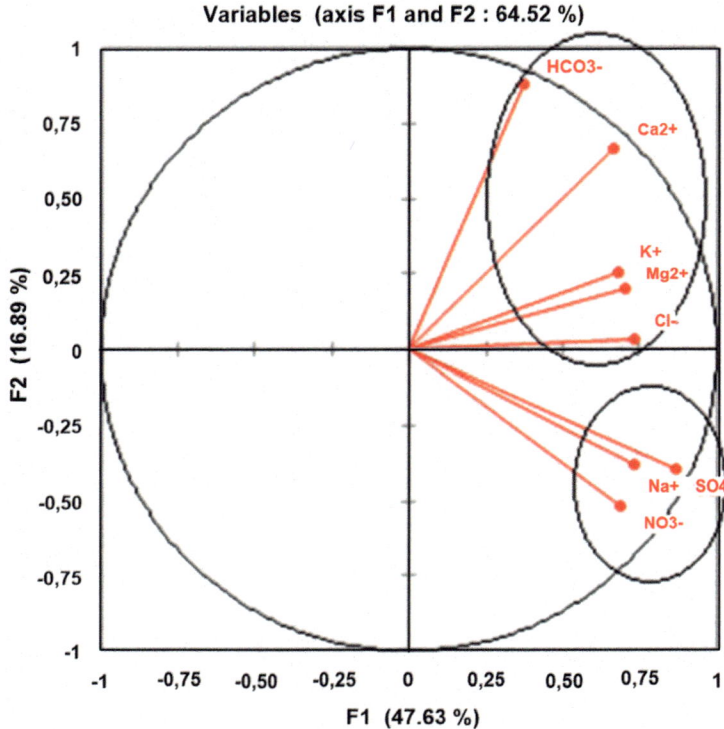

Fig. 11 PCA circle (analyses of March 2011)

4.3   Cross-Plot Analysis

The diagram Ca^{2+} versus HCO_3^- (Fig. 12a) shows that the majority of samples are under the line 1:1 (that indicates pure $CaCO_3$ dissolution), and only a few points suggest an excess of calcium compared to bicarbonates. The excesses of bicarbonate suggest another source for these elements.

When comparing the sum of HCO_3^- and SO_4^{2-} with Ca^{2+} concentrations (Fig. 12b), the relationship shows that in most cases, the calcium concentrations exceed this sum, while only in the range from 1 to 8 meq/L, calcium concentrations are less than the sum. The presence of sulphates suggests gypsum dissolution, which should overlap to carbonate dissolution. As first result, it can be defined that calcium concentrations are defined by the competition between at least two water-rock interaction processes [16].

The relationship between sodium and chloride concentrations (Fig. 12c) shows dispersion around the line 1:1, indicating dissolution of NaCl. Some points are belonging to the first family, indicating a common origin of the two elements. A large number of samples (65%) are located above the 1:1 line (second family)

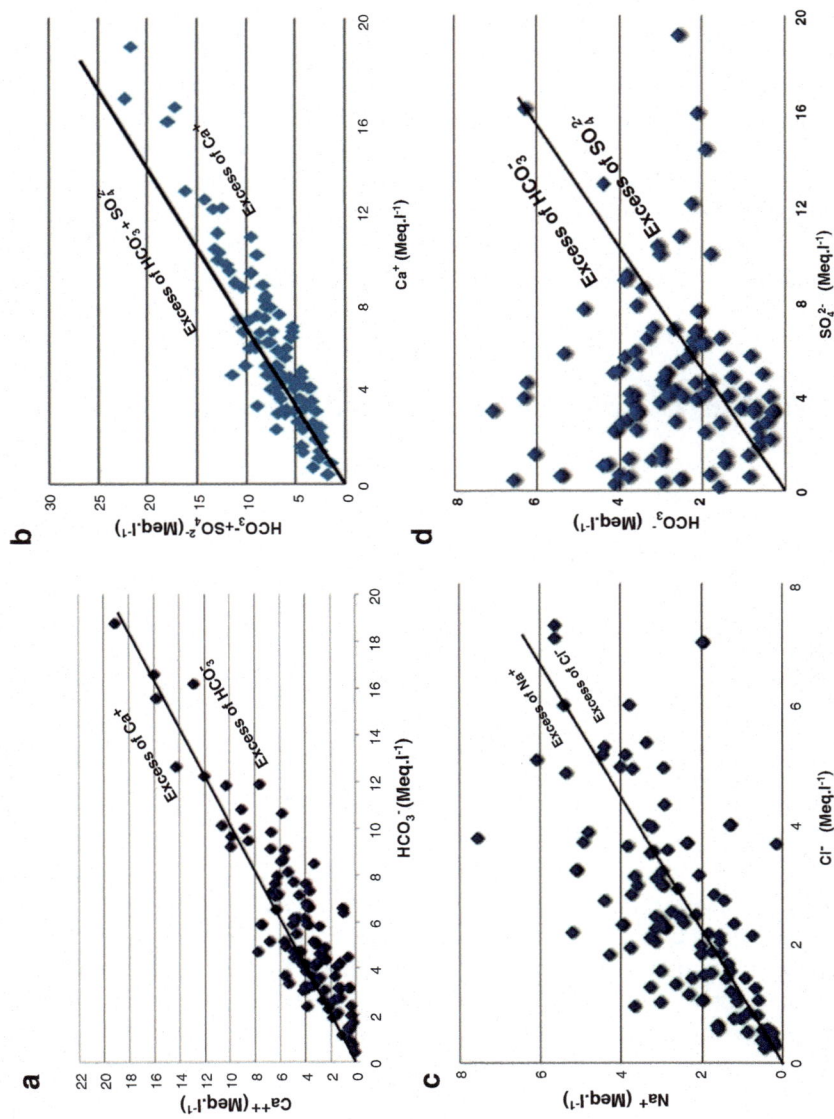

Fig. 12 (**a**) Relationships between Ca^{2+} and HCO_3^-; (**b**) $HCO_3^- + SO_4^{2-}$ and Ca^{2+}; (**c**) Na_2^+ and Cl^-; (**d**) HCO_3^- and SO_4^{2-}

showing an excess of sodium compared to NaCl solution, while the points under the 1:1 line (third family) show sodium and chloride concentrations, respectively, lower and higher than those defined by pure NaCl solution. Thus, Na^+ and Cl^- may have a common origin (first family), as the dissolution of Triassic formations rich in salt [6] or they can derive from other sources or processes such as base exchange, infiltration of wastewater or recycling of irrigation water.

The relationship between bicarbonates and sulphates (Fig. 12d) shows a large dispersion compared to the 1:1 line, suggesting that the two major constituents and origin of groundwater are not the same. This situation could derive from the dissolution of evaporitic rocks related to the outcropping formations [23].

The relationships show that the elements intervening in the chemical composition have various origins. Some elements came from the dissolution of the formations, but their presence could also be related to the human activities. The base exchange indices and the saturation indices can help in removing ambiguities.

4.4 The Base Exchange Index (BEI) and Saturation Indices (SI)

The base exchange index is unitless. It defines the direction of the ionic exchange between the water and rocks. It is given by the following expression:

$BEI = Cl^- - (Na^+ + K^+)/Cl^-$, with concentrations are expressed in meq/L.

- If $BEI = 0$, then there are no exchanges.
- If $BEI < 0$, then Ca^{2+} and Mg^{2+}, exchanged by Na^+ and K^+.
- If $BEI > 0$, then Na^+ and K^+, exchanged with Ca^{2+} and Mg^{2+}.

The BEI has been calculated for all the samples. Table 5 shows the results for two periods of each survey year.

Table 5 shows the dominance of BEIs < 0, which is the case of 62.5% of analysed samples. These values of $BEI < 0$ indicate a direct exchange of the alkaline earths (Ca and Mg) by the alkalis (Na and K). Therefore, sodium and potassium are displaced from the exchange sites and enrich in solution: this explains the presence and excesses of sodium in the water. To confirm this hypothesis, the map of BEIs distribution in Fig. 13 shows the presence of two zones: the first, located around the city of Taoura, is characterized by negative values of BEI; and the second occupies the area of Merahna where the BEIs are positive and the exchange is inverse, i.e. the alkaline ions displace the earthy alkaline ions from the exchange sites. The

Table 5 Values of the exchange index

BEI \period	Low period 2011	High period 2011	Low period 2012	High period 2012	Observations
BEI > 0	11	04	06	07	28/72
BEI < 0	09	14	12	11	45/72

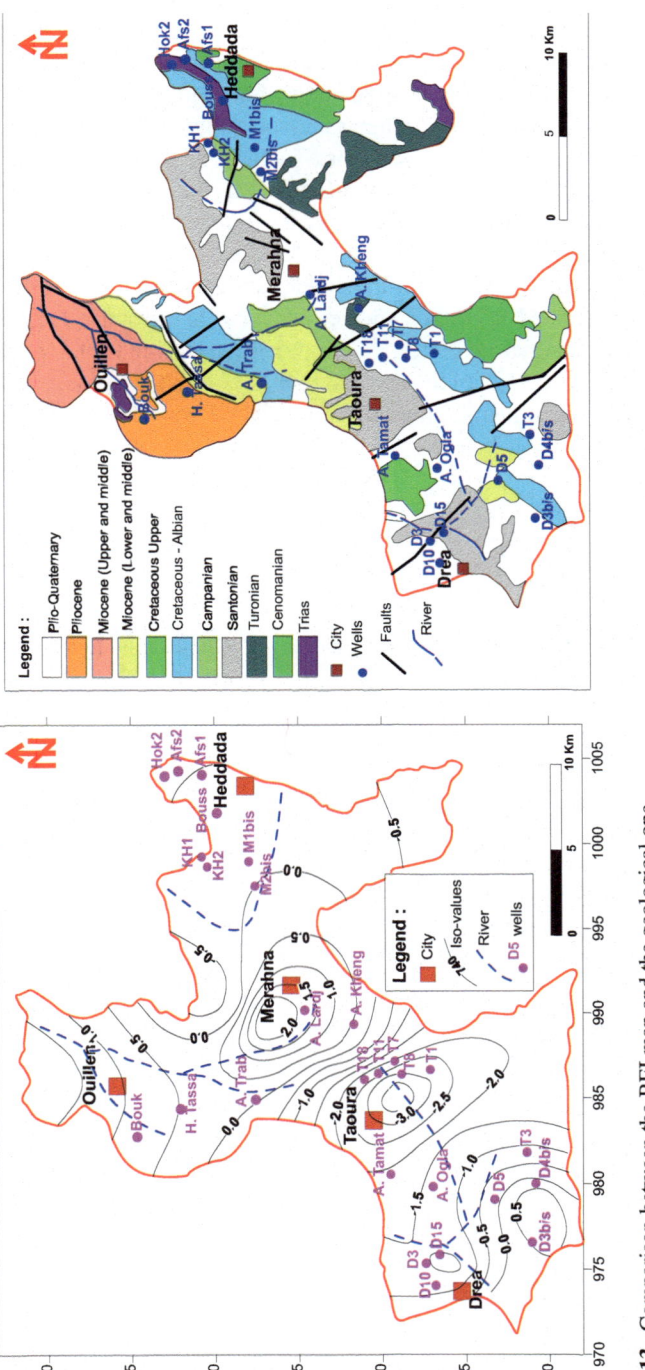

Fig. 13 Comparison between the BEI map and the geological one

comparison of the BEI map with the geological map containing the works carried out at the level of the study area leads to the following considerations:

(1) The Merahna zone, located on the top of the syncline, includes many drillings that capture the deep and heterogeneous formations, such as the limestone and the evaporitic rocks of Trias. The water-rock interaction processes with evaporitic formations enrich groundwater with sodium, which can be then exchanged with calcium occupying the exchange complex. (2) The Taoura zone, located in the centre of the syncline is composed by quaternary formations (sands and sandy clays). Wells intercepting groundwater of this zone have a variable chemical composition, which derive from the heavy role of the direct base exchange.

The calculated saturation indices indicate that groundwater are undersaturated compared to anhydrite ($-1.3 <$ SI < -2.4), aragonite ($-0.3 <$ SI < -0.9), calcite ($-0.2 <$ SI < -0.7), dolomite ($-0.8 <$ SI < -1.8) and gypsum ($-1.3 <$ SI < -2.3). Thus, all such minerals have the potential to contribute to the mineralization of groundwater.

5 Neuronal Model Contributions

The MLP_{BFGS} model has a very good performance in both phases (learning and testing) with a standard deviation of 630.16 and 609.35, respectively (Tables 6 and 7). The respective values of the determination coefficient (R^2) and the precision factor (A_f) for the phases are 0.928 and 1.420 for the learning phase and 0.974 and 1.180 for the test phase.

Figure 14 shows an EC dispersion diagram simulated by the MLP_{BFGS} model with respect to the observed EC. The error graph indicates the contrast of the observed and the simulated CE value. Residual error values for each observation (water point) ranged between -452.54 and 241.53 μS/cm. Both figures show that the overall agreement between the observed and the simulated EC values was satisfactory (Fig. 15).

Table 6 Performance criteria in various MLP neural networks

Training data sets		Training data sets					
ANN	Architecture	RMSE	R^2	A_f	RMSE	R^2	A_f
MLP (CG 51)	11-12-1	0.098	0.784	1.945	0.049	0.801	1.401
MLP (CG 55)	11-11-1	0.085	0.845	1.471	0.066	0.855	1.332
MLP (BFGS 70)	11-13-1	0.078	0.928	1.420	0.058	0.974	1.180

Table 7 Statistical parameters of the electrical conductivity output (EC)

Data sets	Moy.	Min.	Max.	Ecart	RMSE	Correlation
Training	1889,44	705,00	3500,00	630.16	0.078	0.981
Testing	1796,81	820,00	3040,00	609.35	0.058	0.965

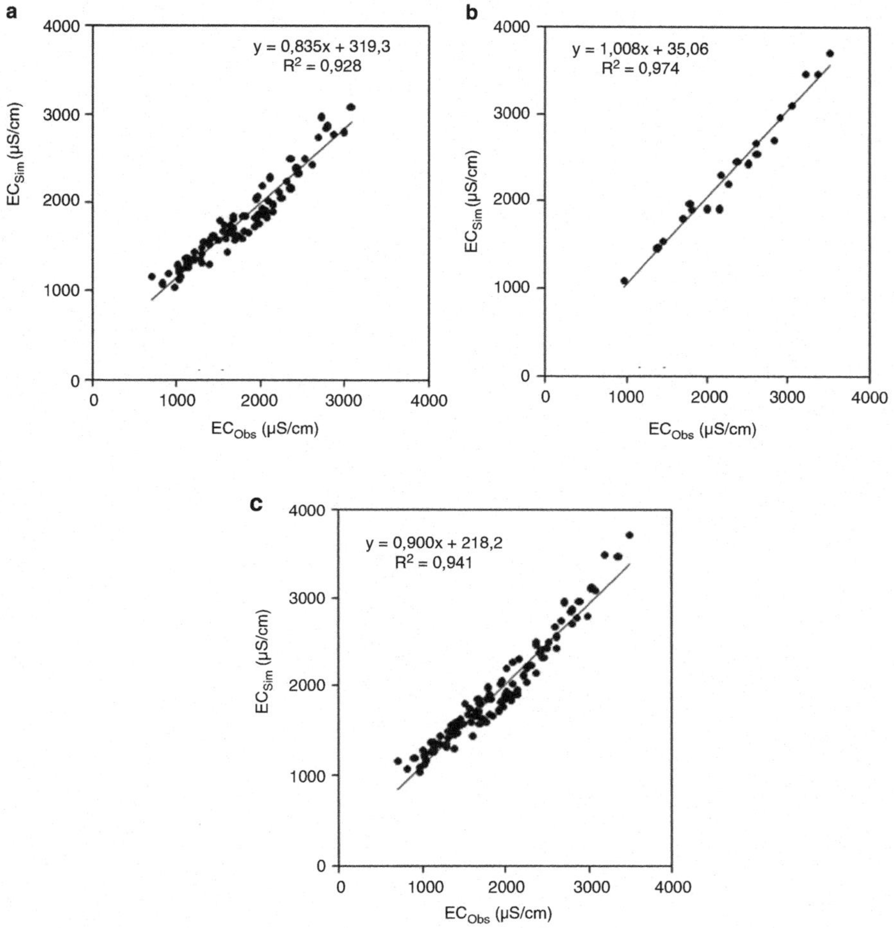

a
b
c
y = 0,835x + 319,3
R^2 = 0,928
y = 1,008x + 35,06
R^2 = 0,974
y = 0,900x + 218,2
R^2 = 0,941
EC_{Sim} (µS/cm)
EC_{Obs} (µS/cm)
0
1000
2000
3000
4000

Fig. 14 Observed EC versus simulated EC: (**a**) training, (**b**) testing and (**c**) all data sets

EC Observed
EC Simulated
Residual
4000
3000
2000
1000
0
500
0
- 500
- 1000
- 1500
- 2000
- 2500
- 3000
- 3500

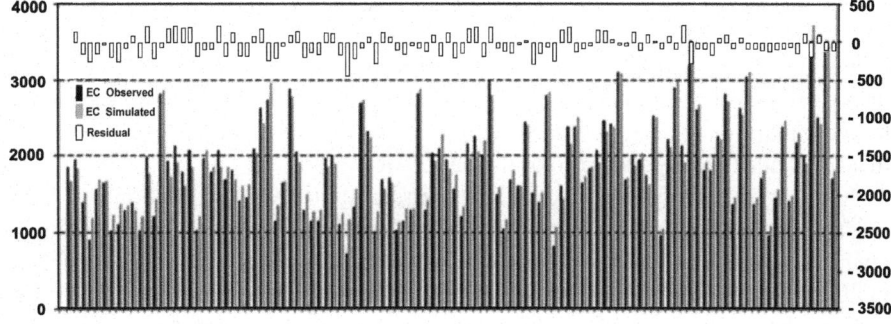

Fig. 15 Observed EC versus simulated EC (in µS/cm)

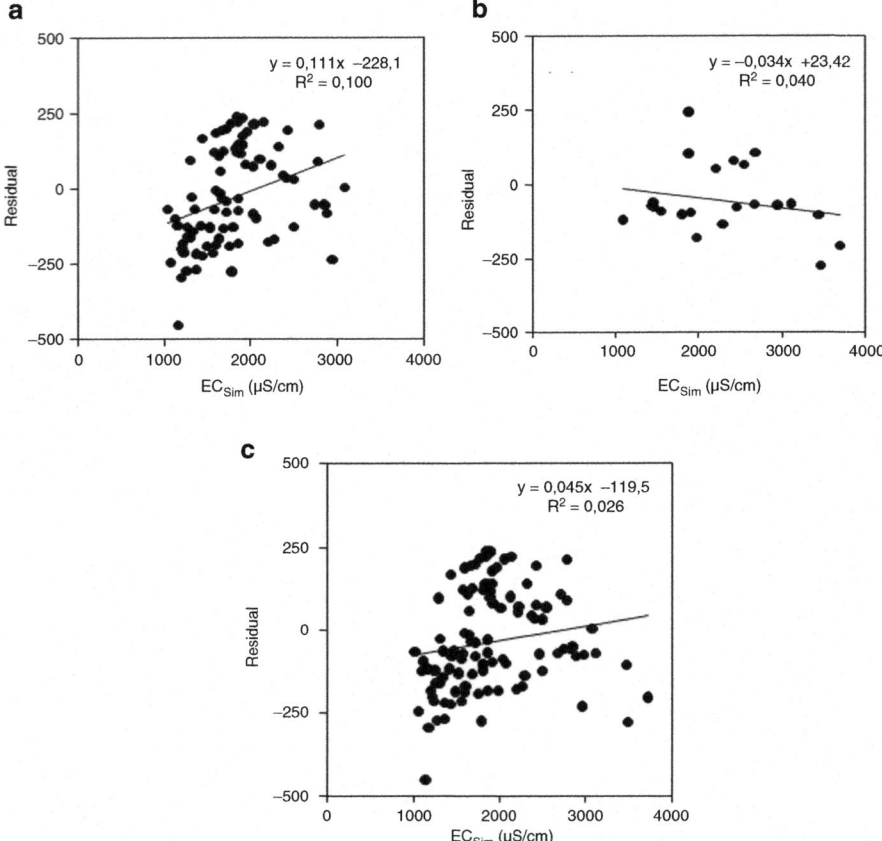

Fig. 16 Plot of the residuals versus MLP$_{BFGS}$ CE values: (**a**) learning, (**b**) testing and (**c**) whole database

The following figures show a dispersion diagram of the simulated EC, by MLP$_{BFGS}$. The observed relationship between the EC values simulated by the MLP$_{BFGS}$ and the residual error for the both phases shows a complete independence and random distribution. In addition, it is supported by the negligible correlations ($R^2 = 0.100$ for learning, $R^2 = 0.040$ for the test and $R^2 = 0.026$ for the entire database). Figure 16 explains that the points are well distributed on both sides of the horizontal line of ordinate zero representing the mean of the residual error suggesting that the model corresponds well to the data.

In order to identify the effect or the contribution of each input variable (water quality parameters) on the output (EC), the sensitivity analysis of the MLP$_{BFGS}$ neural network was calculated in the two phases (learning and testing). Table 8 indicates that the top four most contributing on water quality parameters on the EC are Na$^+$, HCO$_3^-$, Cl$^-$ and Ca^{++}. The remaining water quality parameters according to their classification in the test phase are water temperature, pH, sulphates,

Table 8 Sensitivity analysis of model MLP$_{BFGS}$ to water quality variables

	pH	T	Ca^{++}	Na$^+$	K$^+$	Mg^{++}	HCO$_3^-$	SO$_4^-$	Cl$^-$	NO$_3^-$	NO$_2^-$
Rang	6	3	4	2	8	9	1	5	10	7	11
Ratio	1,068	1,192	1,158	1,275	1,047	1,0311	1,431	1,096	1,019	1,054	1,010
Rang	6	5	4	1	8	9	2	7	3	10	11
Ratio	1,134	1,180	1,148	1,253	0,988	0,971	1,236	1,019	1,165	0,963	0,110

Underline values indicate most contributing parameters on water quality

potassium, magnesium, nitrates and nitrites. In the light of these findings, the management responsible for water quality monitoring could give priority attention to the first four most contributing parameters in the EC.

6 Conclusions

The research concern an area characterized by the presence of two superimposed aquifers, of which the first is porous (Mio-Plio-Quaternary formations) and the second (deeper) is fractured and karstic. The Mio-Plio-Quaternary is composed of heterogeneous material (sand, marl clays and gravels), and it covers the karst limestone formation which shows important discontinuities related to the tectonics of the region. Indeed, faulting resulted in the subdivision of the study area caused a variation of the depth of the top of karst basement. The top of karst formation is sometimes close to the ground surface and sometimes is very far from it. The presence of faults favours the communication of the deep karst aquifer with the superficial aquifer, and water can be drained into the karst aquifer.

The piezometric maps carried out indicate that the recharge of the superficial aquifer is by the borders. This lateral contribution influences the chemical composition of groundwater of both aquifers.

The water quality of groundwater of the karst aquifer is studied by PCA method. It is proved an opposition between the calcium bicarbonate water in the origin of the observed mineralization of water.

The interpretations of cross-plots of main ions show good correlations between $HCO_3^- + SO_4^{-2}$ and $Ca^{2+} + Na^+$ and between Na^+ and Cl^-. The presence of sodium in water is linked to the process of ion exchange. This tendency is confirmed by the BEI indices.

The present study revealed that the water quality of the study area remains influenced of natural factors, particularly the geological formations present in this zone. We also noticed a decrease in piezometric levels in wells. This is explained by the over-exploitation of the aquifer and or the climatic conditions.

7 Recommendations

In point of view of integrated water resources management, the Taoura syncline aquifers play several roles: water supply to different cities located this area and development of agricultural and industrial sectors.

It is important to maintain the good working of wells and to define the protection perimeters for these wells, taking into account vulnerability of the aquifer from the surface pollution.

References

1. Collignon B (1986) Applied hydrogeology of karstic aquifers of the Tlemcen Mountains (Algeria). PhD thesis, University of Avignon, France, p. 282 (In French)
2. Abdesselam M, Mania J, Mudry J, Gélard JP, Chauve P, Lami H, Aigoun C (2000) Hydrogeochemical arguments in favor of non-flush evaporitic Triassic in the Djurdjura massif (Kabylie dorsal, element of the Maghrebids). Rev Sci Eau 13(2):155–166. (In French)
3. Djebbar M (2005) Characterization of the Constantine-Hamma Bouziane-Salah Bey hydrothermal karst system in Central Constantine (North East Algeria). PhD thesis, University of Constantine, Algeria, p 250 (In French)
4. Zerrouki H (2013) Quantitative and qualitative aspects of Bouakkous spring: impact of the capturing field of ain chabro (semi-arid zone of Tebessa). PhD thesis, University of Annaba, Algeria, p 168 (In French)
5. Baali F (2001) Hydrogeological and hydrochemical study of the karst region of Algerian Cherian. Magister's thesis, University of Annaba, Algeria, p 100 (In French)
6. Bouroubi OY, Djebbar M (2014) Characterization of Taoura's multilayers system by the geochemical tracers of carbonates and the evaporites (oriental extreme Algeria). Am J Sci Ind Res 3(5):305–314
7. Kardache R, Lounis R, Abdesselam M, Hannachi N, Djabri L (2013) Karstology in arid zones: karst formations of the Southeast Algeria. Methods and tools. Revue Courrier du Savoir 17:71–76
8. Djidel M (2008) Alteration of the aquifer water in hyperarid climate, by wastewater: cases of groundwater from Ouargla (Northern Sahara, Algeria). Am J Environ Sci 4(6):569–575
9. Mangin A (1975) Contribution to the hydrodynamic study of karstic aquifers. PhD thesis in sciences, University of Dijon, France, p 601 (In French)
10. Bakalowicz M (2002) Karst hydrogeology. DEA course of the University Paris-6. Features and concepts. Exploration, exploitation and active management methods, p 278 (In French)
11. Marsaud B (1996) Structure and operation of the embedded zone of karsts from the experimental results. PhD thesis in sciences, University of Paris XI Orsay, France, p 301 (In French)
12. Peyraube N, Lastennet R, Denis A (2012) Geochemical evolution of groundwater in the unsaturated zone of a karstic massif, using the PCO2— sics relationship. J Hydrol 430:13–24
13. Senani S (2011) Estimation and management of groundwater in the region of Souk-Ahras, Taoura, Magister's thesis in Hydraulics, University Mohamed Cherif Messaadia, Souk Ahras, Algeria, p 182 (In French)
14. Chadi M (2004) Geological and structural context of the Neritic Cretaceous series of Constantinois (East-Algerian). PhD thesis, University of Mentouri, Constantine, p 180 (In French)
15. Dahdouha N (2012) Diagnostic study of drilling, pumping test part. Final report "Drilling on the aquifer of Taoura, Souk Ahras. Algeria", p 30 (In French)
16. Mudarra M, Andreo B (2011) Relative importance of the saturated and the unsaturated zones in the hydrogeological functioning of karst aquifers: the case of Alta Cadena (Southern Spain). J Hydrol 397(3–4):263–280
17. Lannani K, Abdouni L (2008) Geo-electrical study by vertical electrical sounding in the Taoura region. Engineering thesis, University of Annaba, Algeria, p 81 (In French)
18. Bourouga M (2015) Demineralization of drilling water from the Ouled Abbés region and its impact on the environment (Souk Ahras, extreme northeastern Algeria). Magister's thesis in Hydrogeology, Badji Mokhtar University, Annaba, Algeria p 120 (In French)
19. Strojexport Prague & Progress (1977); modefie Progress (2010) Geophysical prospection on the Taoura-Bordj M'raou syncline, Souk-Ahras. Algeria p 87 (In French)
20. Bousnoubra H (2002) Water resources in the regions of Skikda, Annaba, El Tarf, Guelma Souk Ahras (Algeria N-E) (assessment, management and perspective, vulnerability and protection). PhD thesis, p 159 (In French)

21. Athmani AS (2011) Assessment of the quality of surface water in the Oued Medjerda watershed. Souk Ahras. Magister's thesis. University Mohamed Cherif Messaadia Souk-Ahras, p 162, (In French)
22. Rodier J (2009) L'Analyse de l'eau, 9ème édition, DUNOD, p 1527
23. Danquigny C, Emblanch C, Blondel T, Garry B, Roche A, Sudre C (2010) Influence of great flood on the functioning of Karst aquifer: example of the Fontaine de Vaucluse Karst System (SE France). In: Andreo B, Carrasco F, Durán J, LaMoreaux J (eds) Advances in research in Karst media. Environmental earth sciences. Springer, Berlin, pp 115–121

Impact of Toxic Metals on Water Quality Around an Abandoned Iron Mine, Bekkaria, Algeria

Larbi Djabri, Lassaad Ghrieb, Azzedine Hani, Saad Bouhssina, Hicham Chaffai, and Fatma Trabelsi

Contents

L. Djabri (✉), A. Hani, and H. Chaffai
Laboratory of Water Resources and Sustainable Development, Université Badji Mokhtar
Annaba, Annaba, Algeria
e-mail: larbi.djabri@univ-annaba.dz; azzedine.hani@univ-annaba.dz; hicham.chaffai@univ-annaba.dz

L. Ghrieb
University of 8 Mai 1945, Guelma, Algeria
e-mail: ghrieb.lassaad@univ-guelma.dz

S. Bouhssina
Unité de Chimie Environnementale et Interactions sur le Vivant (UCEIV), Maison de la Recherche en Environnement Industriel, Université du Littoral Côte d'Opale, Dunkerque, France
e-mail: saad.bouhssina@univ-littoral.fr

F. Trabelsi
UR-Sustainable Management of Water and Soil Resources (GDRES), Higher School of Engineers of Medjez El Bab (ESIM), University of Jendouba, Jendouba, Tunisia
e-mail: fatma.trabelsi@iresa.agrinet.tn

Abdelazim M. Negm, Abdelkader Bouderbala, Haroun Chenchouni, and
Damià Barceló (eds.), *Water Resources in Algeria - Part I: Assessment
of Surface and Groundwater Resources*, Hdb Env Chem (2020) 97: 53–68,
DOI 10.1007/698_2020_524, © Springer Nature Switzerland AG 2020,
Published online: 8 July 2020

Abstract The mine is situated in the East Algerian near the frontiers of Tunisia. The exploitation of iron is stopped until 1967.The water samples were collected from groundwater and surface water. Physicochemical parameters were measured during fieldwork. Water samples were analyzed for major ions and metallic trace element.

We noted two aquifers, the first one is situated at 3 m under the soil, and it is connected with the Oued, and the second one is situated at 20 m deep. The water levels in the mine waste dump indicated the occurrence of a losing stream during the period of peak streamflow was a result of snowmelt runoff facilitates the displacement of pollutants.

The analysis realized permitting to study the evolution of the MTE in the two aquifers.

The electrical conductivity is very high near the mine. This situation explains the high concentrations of sulfate, chlorides, calcium, and sodium. The concentrations observed are generated by the phenomenon of dissolution of gypsiferous formations. The graphics realized shows hilt concentrations in the first aquifer, but in the deep, the concentrations become low. This repartition explains the retention of elements by the soil.

Keywords Algeria, Contamination, Groundwater, Iron, Mine

1   Introduction

The impacts of base-metal mining activities on groundwater and surface water have been widely studied around the world [1].

Many observations reported that the deposits of slag heaps and mine waste around the mine are also a major source of contamination and are easily mobilizable [1–5].

The ore from the mine will reach the aquifer via the wadi, either through wind or through runoff due to precipitation. Ignorance of the extension of the impact of the mine also increases pollution and therefore the degree of pollution on the soil and on the water [6, 7].

In reality the mines are generally abandoned without rehabilitation after the cessation of the operation, which leads to an extension of the pollution process, so the quality of the water is threatened even after the cessation of exploitation, which leads to a change in the quality of the water in contact with the mine sites.

Some elements that are present in crystalline networks of minerals and which are generally hardly released are put in solution and dispersed in the environment [8, 9].

The main objective of this study was to determine the impact of abandoned mines on surface water and groundwater.

2 Characteristics of the Study Area

Tebessa frontier city with Tunisia is located at the extreme Algerian Northeast (Fig. 1), at the front of the desert, approximately 230 km in the South of Annaba on the Mediterranean coast. The area is limited to the south by the sector of Biskra, to the west by the region of Constantine and in the east by the Tunisian border. It has a semiarid climate [10].

The climatic factors contribute to the propagation of the pollutants, and the study of the climatic factors proves to be essential. For that, we considered 2 extreme years. The first is referred to the year 1972–1973, considered as most wet with a precipitation of about 625.3 mm. During this year, the wet season is spread out over 10 months with a fall of precipitations in November. Conversely, year 1996/1997 with 207.4 mm is supposed to be driest, the wet season is spread out over 2 months (December and January), and it starts again mid-Mars until mid-May [10].

In studying zone the outcrop formations are of the sediment type; it is characterized by the appearance of Triassic formations which will constitute our interest.

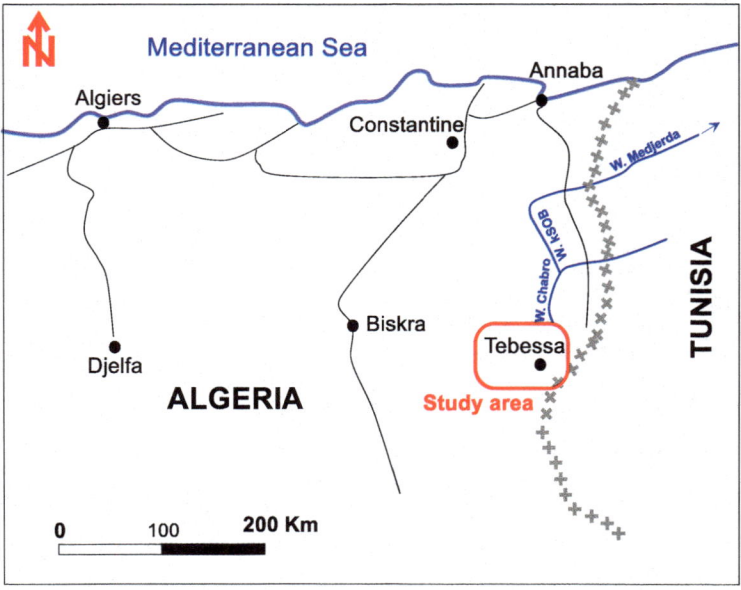

Fig. 1 Situation of the area studied [10]

In Tébessa area the most important Triassic outcrops, are those of Djebissa, Ouenza, Boukhadra, Mesloula, Boujaber, northern Hameimat, southern Hameimat, and other solid masses.

This material, moreover, (saliferous) is also characterized by the evolution of a structure to several mineralogical zonations accompanied in the majority by metalliferous concentrations Pb-Zn Ba-Sr. Djebel Djebissa contains indices of polymetallic ores and iron-bearing of which a layer apart from our study area (Iron mine of Khanguet). The Pb-Cu index located close of the contact Cenomanian-Turonian on the southeast side; carbonated rocks (limestone) contain a mineralization with crystal in dissemination, in particular clusters, and in nests. We also meet the epigenese products: cerusite, limonite, and hypogene minerals of copper represented by grey copper ore and digenite hypergenes represented by malachite and azurite.

The shallow aquifer with low depth (maximum 10 m) remains the most exposed to pollution; this is why we will be interested in his study.

In general, the piezometric surface has the same morphology as topography (Fig. 2). The flow is directed southeastern or northwestern. We note the appearance of a much accentuated depressive zone located at the north of Aïn Chabro. This situation is generated by the exploitation of drillings and wells in this part. More than 30 wells in exploitation are listed in the study area [10].

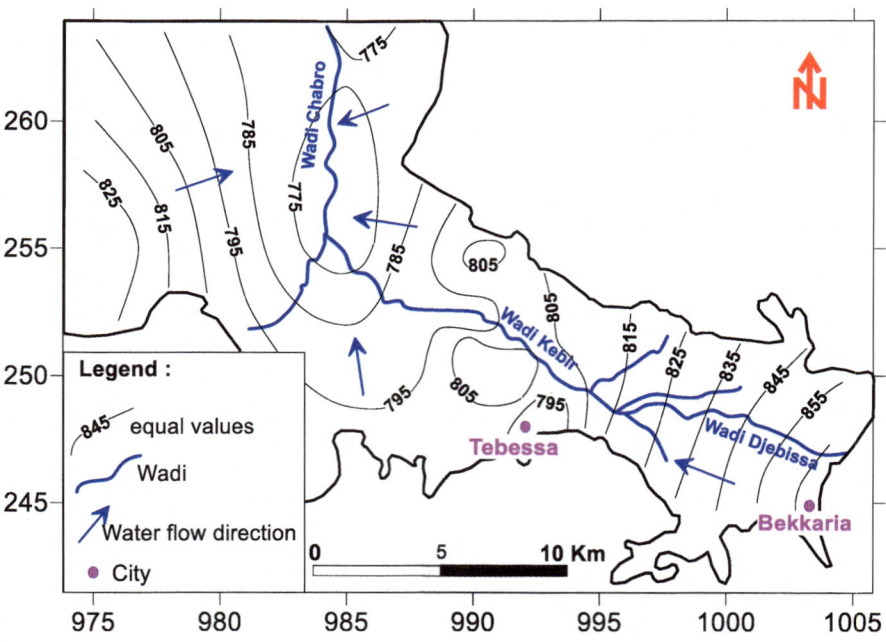

Fig. 2 Piezometric map of the shallow aquifer (July 2006)

3   Impact of the Abandoned Mine on Water Quality

Once mining stopped no initiative of environmental protection was taken. This had reflected negatively on the environment; indeed during long years the spoil heaps remained deposited on the soil surface, upstream of the wadi, and the aquifer, directly exposing to the effects of pollution. To highlight the effects of these spoil heaps on the water quality of this area, we will study successively the quality of water of the wadis and wells [11]. Work will carry mainly on the analyzed metals and on the saliferous formations outcropping upstream the study zone.

3.1   *Impact on Surface Water Quality*

The evolution of surface water chemistry was the aim of this study. The analyses related to eight points are distributed on the two wadis; Wadi Djebissa and Wadi El Kebir, according to the direction of the flow determined by the piezometric map (Fig. 3).

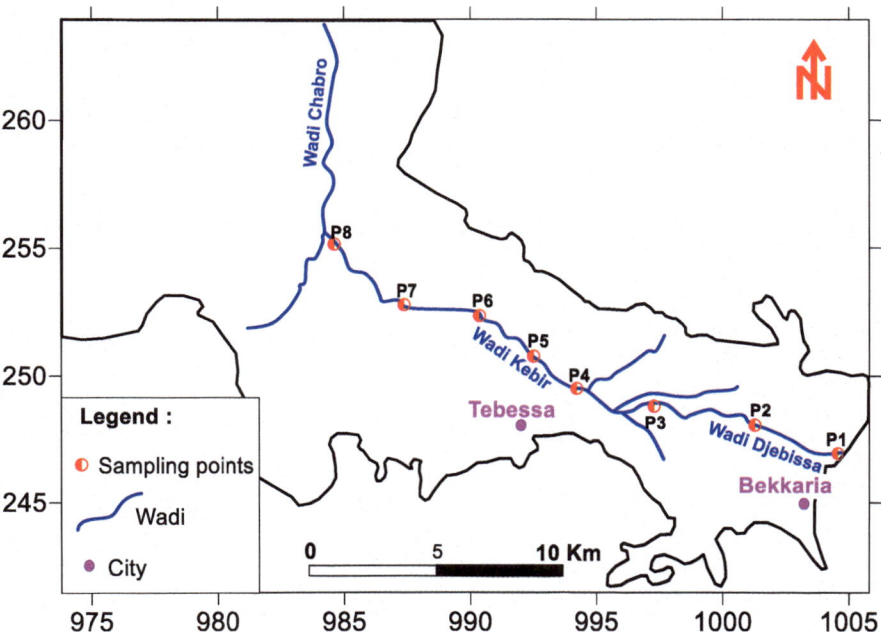

Fig. 3 Distribution of sampling points of surface water [10]

In this fact, analyzed elements are major cations and anions (Ca, Mg, Na+K, Cl, SO_4, HCO_3), trace elements, and Sr^{2+}/Ca^{2+} ratio

3.1.1 Evolution of Major Elements in Surface Water

The graph observation showing the evolution of the major elements in surface water reveals that for the first point close to Djebel Djebissa (P1) important concentrations in sulfates and chlorides are present. These two elements move simultaneously (Figs. 4 and 5). This evolution is accentuated by the climate; indeed during the wet period, the dissolution of gypsiferous formations enriched water by sulfate on the other hand, and during the dry seasons, the evapotranspiration increases the concentrations enriching water by chloride [12]. Sodium evolves in the same way as chlorides and sulfates. The other elements are more or less stable.

Fig. 4 (Left–P1). Evolution of major elements in surface water [10]

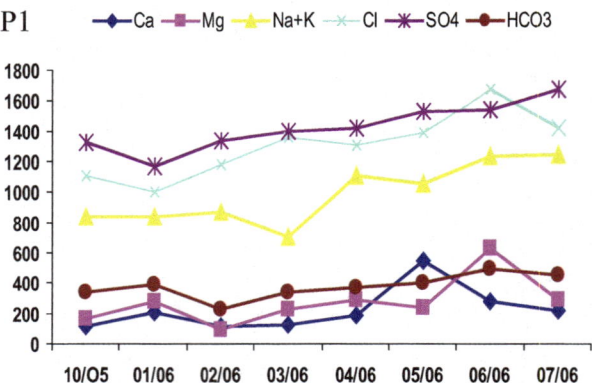

Fig. 5 (Right–P8). Evolution of major elements in surface water [10]

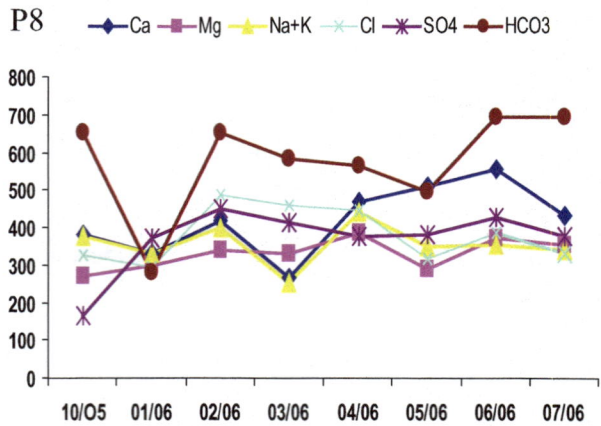

In the center of the plain, we notice a Ca, Mg, and HCO$_3$, increase. Sulfates, and chlorides concentrations decrease but remain important during the dry period. The bicarbonates present an increase for the last points; this is explained by the contribution of carbonate border [13].

3.1.2 Evolution of Trace Elements in Surface Water

The variation of the traces elements in surface water is irregular. Iron and manganese evolve together, especially for the wet period. The contribution in these two elements is probably due to the dissolution of iron from an abandoned mine. Copper especially presents a light increase for the dry period; zinc has a certain stability for the whole graphs. The observation of graphs shows a decrease in the concentrations of the trace elements toward the mine direction (Figs. 6 and 7). This tendency highlights a probable trapping of the ETM by the soil.

Fig. 6 (P1). Evolution of trace elements in surface water [10]

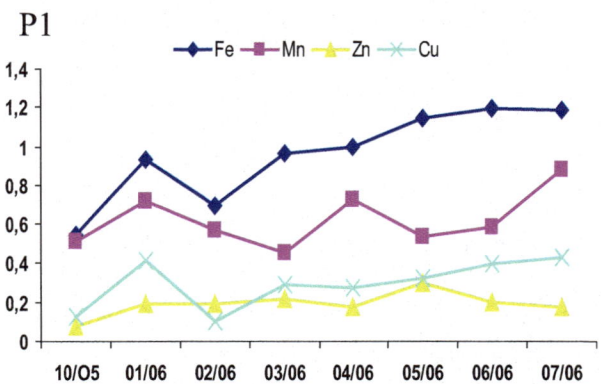

Fig. 7 (P8). Evolution of trace elements in surface water [10]

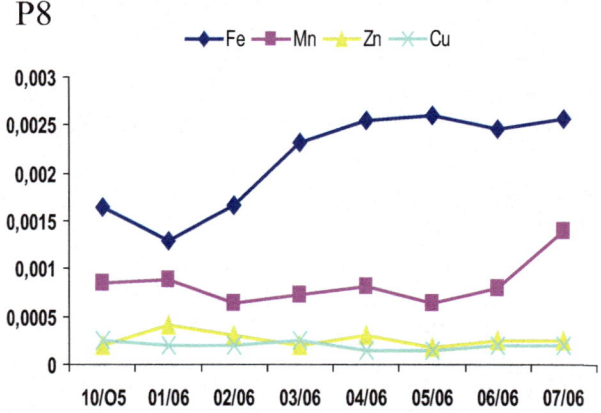

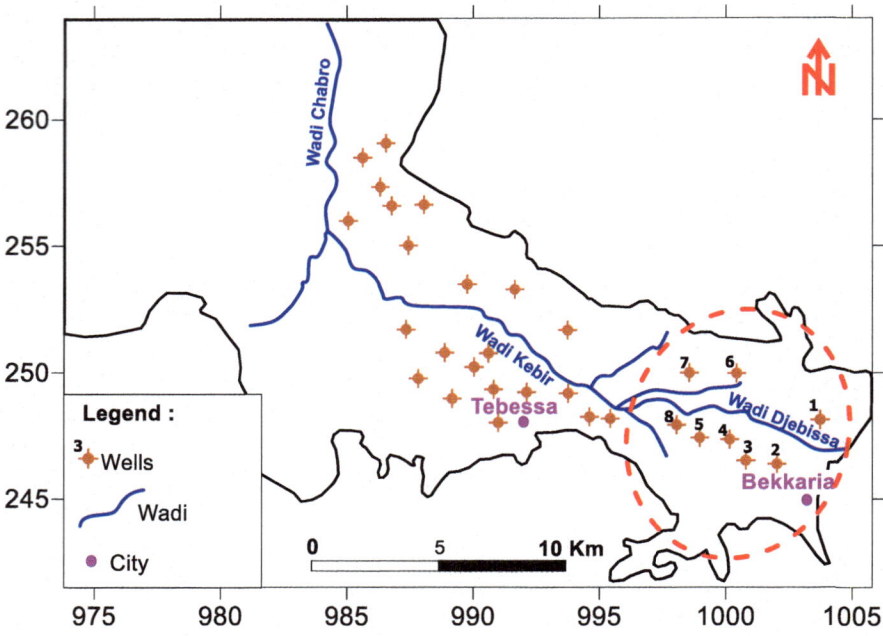

Fig. 8 Distribution of the sampled water wells [10]

3.2 Impact on Groundwater Quality

The waters of eight wells was analyzed. These wells are distributed on both sides of the banks of the wadi (Fig. 8).

3.2.1 Evolution of Major Elements in Groundwater

The examination of the graphs carried out shows that water is characterized by important concentrations, particularly of chlorides, sulfates, and sodium. At the well 1, the concentrations of the three elements oscillate between 700 (Na+K) and 1,600 mg/l for chlorides (Figs. 9 and 10). In well 8, we note a very noticed fall of these concentrations, the maximum reached is about 700 mg/l and the minimum borders 400 mg/l.

3.2.2 Evolution of Trace Elements in Groundwater

In groundwater samples collected from wells, the concentrations remain low and decrease as one moves away from the mine. Indeed, for iron on the level of the wells,

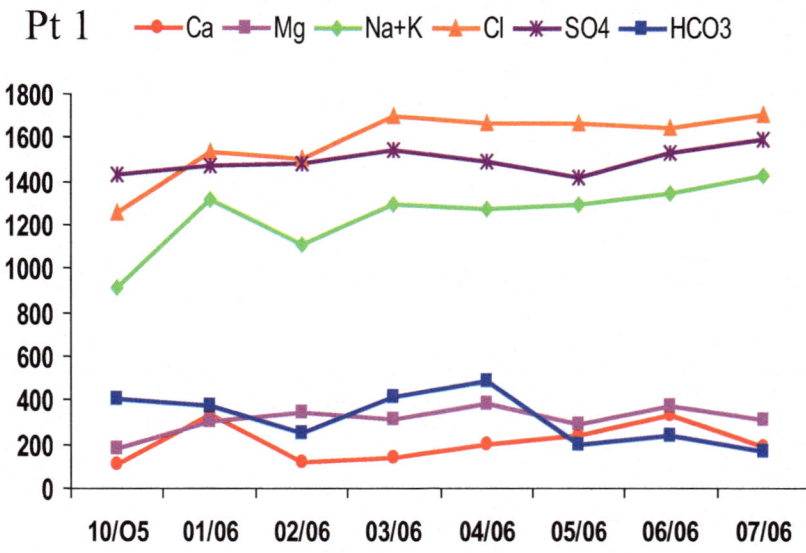

Fig. 9 (Pt 1). Evolution of major elements in groundwater [10]

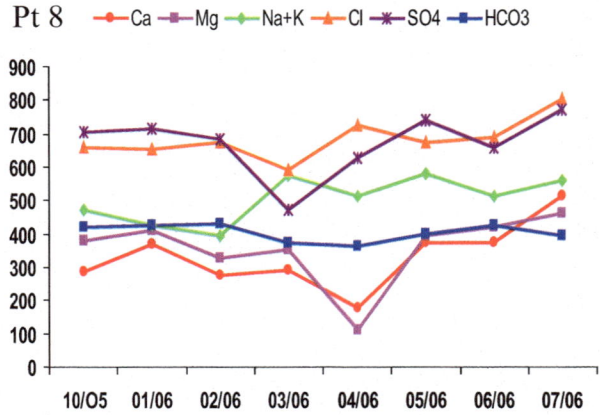

Fig. 10 (Pt 8). Evolution of major elements in groundwater [10]

the concentrations are about 0.2 mg/l indicating water pollution on the other hand in well 8, and the concentrations are low, even unimportant (0.05 mg/l) (Figs. 11 and 12).

This interpretation highlighted a double variation of the concentrations:

- The first being done in the horizontal direction, indicating a decrease of the concentrations toward the source of pollution (the mine) [14]
- The second vertical showing that water of the aquifer is not contaminated, which highlights trapping of metals by the sediments [14, 15]

Fig. 11 (Pts 1). Evolution of traces elements in groundwater [10]

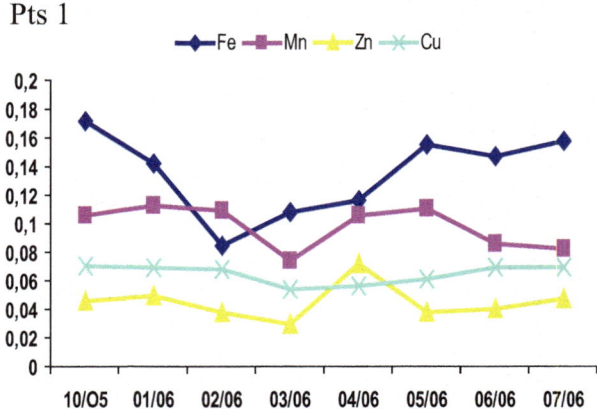

Fig. 12 (Pts 8). Evolution of traces elements in groundwater [10]

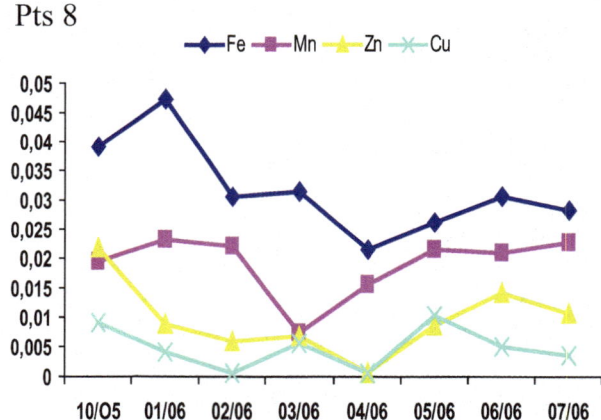

4 Confirmation of Contamination by Sr^{2+}/Ca^{2+} Ratio

The study of the Sr^{2+}/Ca^{2+} ratio gives an outline of the influence of sorted gypso-saliferous on the water salinity. Strontium is related to the evaporites. High concentrations of strontium in water can have several natural or anthropogenic origins. In this region the presence of strontium is of natural origin essentially resulting from the dissolution of celestite (Sr SO_4); mineral associated with the gypsum which is a good indicator of evaporites [16, 17].

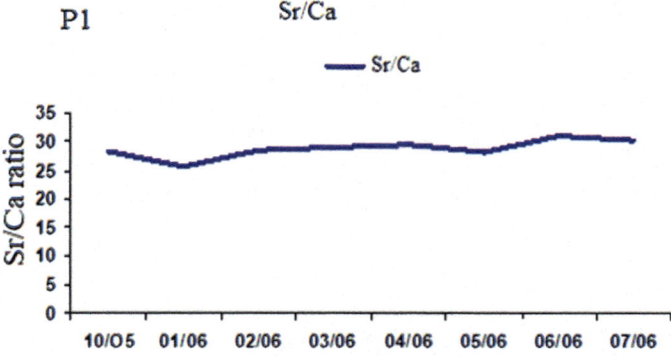

Fig. 13 (P1). Evolution of Sr^{2+}/Ca^{2+} ratio in surface water [10]

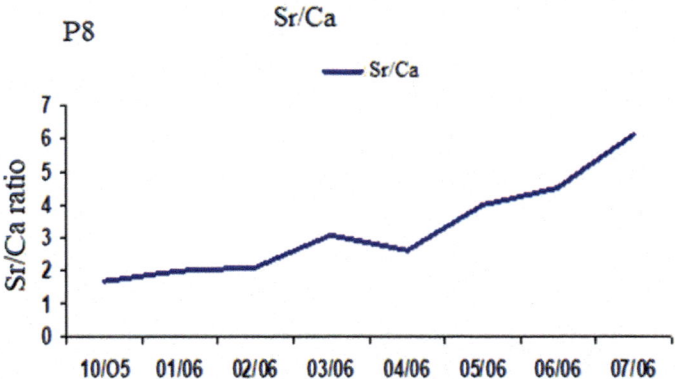

Fig. 14 (P8). Evolution of Sr^{2+}/Ca^{2+} ratio in surface water [10]

4.1 Surface Water

Concerning surface water, the Sr^{2+}/Ca^{2+} ratio reached important values highlighting the influence of sorted gypsiferous on the quality of water. Indeed the dissolution of minerals contained in the formations enriched water in element traces (Figs. 13 and 14) [18, 19].

5 Confirmation of Contamination by Modeling

The recourse to modeling constitutes another tool for the description of the impact of the iron mine on the quality of water. To complete the work, we chose the model based on the networks of artificial neurons [20].

Fig. 15 Artificial neuron [20]

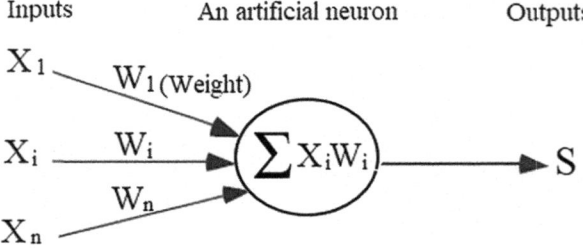

Inputs   An artificial neuron   Outputs

5.1 Presentation of the Method

The networks of artificial neurons RNA or ANN are a nonlinear empirical model. It is composed of interconnected elements of treatment (neurons) working jointly to solve a specific problem. Hecht Nielsen 1990 gives the following definition: a network of neurons is a system of calculation made up of strongly interconnected simple elements of treatment, which process the data by their change of dynamic state in response to an external entry.

5.2 Connections Between the Neurons

The networks of neurons (Fig. 15) are organized in layers; these layers are composed of a certain number of interconnected neurons, which contain a function of activation.

5.3 Creation of the Model

In this work, a multilayer network of Perceptron was selected like a model of the system where the network treats a vector of entry being composed of the variables including/understanding Ca, Mg, Na, K, Cl, SO_4, HCO_3, NO_3, pH, M, and Sr/Ca. This vector of entry produced a vector of output (left) which is electric conductivity (EC). The network of multilayer network of Perceptron can be represented by the following compact form: {EC} = ANN [Ca, Mg, Na, K, Cl, SO_4, HCO_3, NO_3, pH, Mineralisation, Sr/Ca].

5.4 Choice of the Execution Criteria

The data of the parameters of quality of subsoil waters analyzed for the year 2006 were employed to create the model of the RNA by using software STATISTICA neural network version 4.0. The parameters of quality of water include concentration in ion of calcium (Ca^{2+}), magnesium (Mg^{2+}), sodium (Na^+), potassium (k^+), chloride (Cl^-), sulfate (SO_4^{2-}), bicarbonate (HCO_3^-), nitrate (NO_3), hydrogen (pH), mineralization (M), and the strontium report/ratio of calcium (Sr^{2+}/Ca^{2+}). These parameters, which represent the quality of water, are regarded as a variation of entry, while the variable of the output of a target (left) is electric conductivity (EC). The statistical parameters used in this work are the average error of the square RMSE (Root Mean Public garden Error) and the coefficient of R^2 determination.

5.5 Modeling Results

The types of networks considered are MLP (three and four layers), RBF, GRNN, and linear. During the analysis, 697 networks were examined. The best optimal model of the found RNA is the MLP (three layers) with six hidden nodes (Fig. 16). The minimal error of 0.3125517 is compared with the other types of networks RNA (Table 1).

Results presented in Table 2 show that the model has an excellent performance in the checking with a report/ratio of regression of 0.016661 and one coefficient of correlation higher than 99% for the training. The sensitivity analysis of the variable water quality of RNA in phases of the training and of checking indicates that mineralization (M) and the strontium report/ratio on calcium (Sr^{2+}/Ca^{2+}) are the most important factors influencing electric conductivity in groundwater [21, 22].

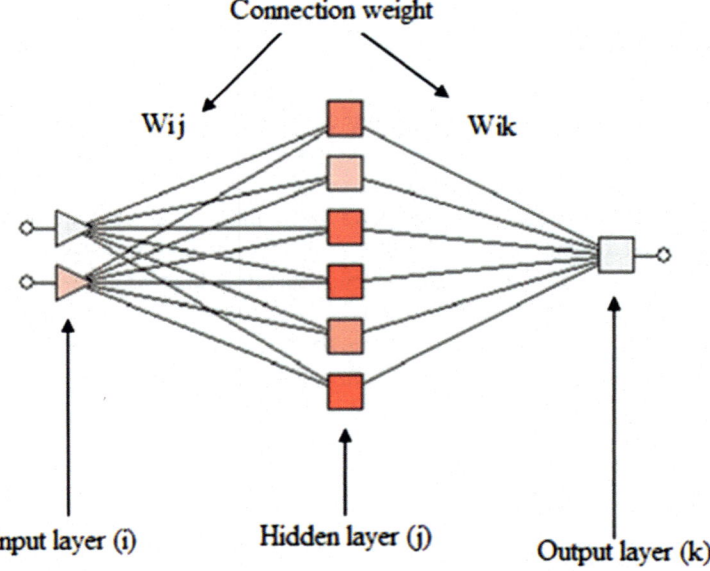

Fig. 16 Illustration of the three-layer network (MLP) [10]

Table 1 Error RMS in various networks of neuron

Type of network	Error (RMS)
GRNN	3.312591
RBF	3.085885
Linear	2.149379
MLP (4 layers)	1.169872
MLP (3 layers)	**0.3125517**

Bold value represents the Root Mean Square Error. In this model we tested several types of network and the multi-layers perceptron with 3 layers turns out to be the network with the least error compared to other types of networks

The database of this model has been divided into three phases: The learning phase, the test phase and the validation phase

Table 2 Statistical parameters of regression

	Tr. CE	Ve.CE	Te.CE
Data mean	3046.842	2511.111	2251.111
Data S.D	1433.147	1712.779	612.7896
Error mean	9.065143	-5.944085	24.99978
Error S.D	183.7065	28.53741	117.1496
Abs E. mean	82.72142	18.07498	65.60018
S.D. ratio	0.128184	**0.016661**	0.1911742
Correlation	0.9920684	**0.9999169**	0.9933773

Bold values represent the ratio of regression (0.016661) and the coefficient of correlation for this testing phase that turns out to be the best

6 Conclusion

The work carried out concerns the effects of the spoil heaps deposited upstream of a wadi and an aquiferous system. The samples showed that the surface water taken from the wadi and the surface aquifer are the most charged with metallic trace elements. The concentrations observed of water in the wadis remain however very high compared to water of the wells; this distribution would be due to the trapping of the ETM which is made at the level of soils separating the two levels from water. To confirm the origin of the ETM, we studied the Sr^{2+}/Ca^{2+} ratio, which shows the influence of the gypsiferous formations in the water quality. The results obtained by the mathematical model carried out confirm this relation well.

7 Recommendation

These mines will be sources of environmental threats unless all necessary measures are taken to reduce their impact.

Abandoned mining sites around the world pose a major problem, both on the environment and on humans.

After cessation of operations, mines are generally not rehabilitated, which may have an impact on water quality.

The abandoned Bekkaria mine in Algeria is considered as a source of pollution for the immediate environment, especially by a former ore deposit on the ground. For that it is strongly recommended to:

- Remove the mineral just around the mine
- Make a good drainage of surface water from the mine
- Cleaning of the wadi nearby that can avoid the trapping of chemical elements in the soil
- Reforestation companions around the mine that can increase the soil's purifying power
- Undertake an abandoned mine management strategy in the region

References

1. Alhamed M, Wohnlich S (2014) Environmental impact of the abandoned coal mines on the surface water and the groundwater quality in the south of Bochum, Germany. Environ Earth Sci 72(9):3251–3267
2. Banks D, Younger PL, Arnesen RT et al (1997) Mine-water chemistry: the good, the bad and the ugly. Environ Geol 32(3):157–174
3. Jahanshahi R, Zare M (2015) Assessment of heavy metals pollution in groundwater of Golgohar iron ore mine area, Iran. Environ Earth Sci (74)1:505–520
4. Atanacković N, Dragišić V, Stojković J et al (2013) Hydrochemical characteristics of mine waters from abandoned mining sites in Serbia and their impact on surface water quality. Environ Sci Pollut Res 20:7615–7626
5. Dold B, Fontbote L (2001) Element cycling and secondary mineralogy in porphyrycopper tailings as a function of climate, primary mineralogy, and mineral processing. J Geochem Explor 74(1–3):3–55
6. Djabri L (1987) Contribution to the hydrogeological study of the alluvial aquifer in the plain of Tebessa. Modeling test. Thesis of Doc. Ing., Université de Franche-Comté, Besançon, p 176
7. Djabri L, Hani A, Mania J, Mudry J (2001) The salinity process: verification by the CPA in the Annaba-Bouchegouf and Guelma sector. Water Tribune 610(54):29–43
8. Moran RE, Wentz DA (1974) Trace element content of a stream affected by metal mine drainage, Bonanza, Colorado. International symposium on water-rock interaction, Prague, 3 Aug 2010, p 22
9. Rapantova N, Grmela A (2002) Environmental impact of mine liquidation on groundwater and surface water. In: Geoenvironmental models for resource exploitation and environmental security, vol 80. Springer, Berlin, pp 365–384
10. Ghreib L (2007) Impact of the Triassic formations on the water and the soil of the plain in semi-arid zone: case of the plain of Bekkaria-Tebessa (East of Algeria). Magister memory, University of Annaba, Algeria, p 126
11. Papp DC, Cociuba I, Baciu C, Cozma A (2017) Composition and origin of mine water at Zlatna gold mining area (Apuseni Mountains, Romania). Procedia Earth Planet Sci 17:37–40
12. Lottermoser BG (2007) Mine wastes: characterization, treatment, environmental impacts.2nd edn. Springer, Berlin

13. Rapantova N, Licbinska M, Babka O et al (2013) Impact of uranium mines closure and abandonment on groundwater quality. Environ Sci Pollut Res 11(20):7590–7602
14. Laga L, Blaga LM, Ciobotaru T (1975) The origin and evolution of some mineral water sources estimated from their deuterium content. Isotopenpraxis 11:297–301
15. Alvarez E, Fernandez Marcos ML, Vaamonde C, Fernandez-Sanjurjo MJ (2003) Heavy metals in the dump of an abandoned mine in Galicia (NW Spain) and in the spontaneously occurring vegetation. Sci Total Environ 313(1-3):185–197
16. Koros I (1998) Problems of mine water after the termination of coal exploitation in Lower Silesian basin (in Czech). 10th conference on mining and geology proceedings, Straz pod Ralskem, Czech Republic, pp 52–56
17. Collon P, Fabriol R, Buès M (2005) Evolution of water quality in the abandoned iron mines of Lorraine: towards a semi-distributed modelling approach. C R Geosci 337:1492–1499
18. Grmela A (1997) Hydrogeochemical alterations of mine water owing to selected waste disposing into abandoned mines-case study. Proceedings of Hydrogeochemia'97, Comenius University, Bratislava, Slovakia, 28 May 1997, pp 20–29
19. Rösner U (1998) Effects of historical mining activities on surface water and groundwater - an example from northwest Arizona. Environ Geol 4(33):224–230
20. Lallehem S, Mania J (2002) A linear and nonlinear rainfall-runoff model using neural network technique: example in fractured porous media. J Math Comput Modell 1:55
21. Bowell RJ, Bruce I (1995) Geochemistry of iron ochres and mine waters from Levant Mine, Cornwall. Appl Geochem 10(2):237–250
22. Milu V, Leroy JL, Peiffert C (2002) Water contamination downstream from a copper mine in the Apuseni Mountains, Romania. Environ Geol 42:773–782

Impacts of Pesticides on Soil and Water Resources in Algeria

El-Sayed Ewis Omran and Abdelazim Negm

Contents

Abstract Algeria, with its natural resource factors, is a significant country in Africa and the world and is seeking a strong development in the demographic and economic scale. Algeria, with an area of 2.4 million km^2, is North Africa's biggest nation. Sahara occupies most of this surface, unfit for farming, but rich in mineral resources. Over 90% of the population lives in the north, including a coastal land along the

E.-S. E. Omran (✉)
Soil and Water Department, Faculty of Agriculture, Suez Canal University, Ismailia, Egypt

Department of Natural Resources, Institute of African Research and Studies, Aswan University, Aswan, Egypt
e-mail: ee.omran@gmail.com

A. Negm
Water and Water Structures Engineering Department, Faculty of Engineering, Zagazig University, Zagazig, Egypt
e-mail: amnegm85@yahoo.com; amnegm@zu.edu.eg

Abdelazim M. Negm, Abdelkader Bouderbala, Haroun Chenchouni, and
Damià Barceló (eds.), *Water Resources in Algeria - Part I: Assessment
of Surface and Groundwater Resources*, Hdb Env Chem (2020) 97: 69–92,
DOI 10.1007/698_2020_468, © Springer Nature Switzerland AG 2020,
Published online: 19 March 2020

Mediterranean Sea, plains, hills, and highlands. In the north, the annual quantity of rain ranges from 300 to 1,000 mm. The annual amount of rain in the Sahara and the Saharan Atlas in the south is less than 100 mm. Algeria has 17 main hydrographic basins and shares with Tunisia the basin of Medjerda and with Morocco the basins of Tafna, Draa, Guir, and Daoura. Agriculture continues to play a dominant role in the economy of the country. Twenty years ago, agriculture accounted for more than 75% of the active population in the south. This has now dropped to about 20%. It is another tale in the country's south. The population was only 0.9 million in 1967, but by 1987 it increased to nearly two million, and by 2010, it is over three million, and in 2019, it is around 43 million. About 40% of the inhabitants now rely for their livelihood on agriculture. The Algerian authority was facing serious problems in managing its soil and water resources. This chapter offers an overview of the present issues in pesticides that harm animal and human health and cause natural resource scarcity and environmental pollution by accumulating in soil and leaching into water bodies. Naturally, the current situation in Algeria is exacerbated by two important constraints:

1. Groundwater and surface water pollution, which domestic, industrial, and agricultural waste far exceeds the ability of sewage systems, significantly reducing the quantity of treated water that can be used.
2. Risk of sustainable development in relation to soil and water pollution, which severe issues arose in groundwater evaluated samples that exceed natural resource renewal boundaries and need to tap into nonrenewable reserves.

This chapter also highlights the urgent need to develop new branches of chemistry that are less dangerous to human health and the environment. Therefore, we must pursue the goals of green chemistry. Green chemistry became responsible for finding suitable solutions to all old manufacturing problems by finding alternative solutions to all previous negatives.

Keywords Algerian government, Climate change, Environmental problems, Green chemistry, Pesticides, Pollution, Soil, Water resources

1 Introduction

More than half of the world's population development is anticipated to occur in Africa between now and 2050, according to the United Nations study [1]. Food security is, therefore, a key issue for African nations. In the face of this challenge, the Algerian government has implemented a Sahara agricultural extension strategy covering more than 90% of the nation [2]. Over the previous three decades, Algeria has produced remarkable but uneven progress. However, the environmental harm that accompanied it threatens this advancement. Lack of water, degraded arable soil, polluted air and water, and insufficient hygiene threaten the ability of the region to sustain financial development and absorb growing populations.

Through disease and early death, they also enforce huge financial and human expenses. The region is poorly endowed with two important natural resources, although blessed with big petroleum and gas reserves: productive land and affordable, renewable water resources. Only 6% of the land in the region is arable, and there is restricted to freshwater supply accessible [3]. Consequently, human settlements were focused in a comparatively tiny portion of the land mass, and food production was strongly dependent on irrigated farming. A paradigm of growth centered on the search for food self-sufficiency and fast industrialization has placed ever more unmanageable pressure on natural resources. Therefore, soil and water information in the country is crucial.

On the one hand, soil is an important compartment that each year receives a substantial quantity of pollutants from various sources. Soil not only acts as a reservoir for chemical contaminants but also acts as a natural buffer by regulating the transport to the atmosphere of chemical elements and substances [4]. Contamination of agricultural soil with toxic components such as heavy metals draws people's interest not only because metals can build up in the soil but also because metals can be accumulated in plants that pose important potential risks to human health [5, 6]. On the other hand, water is an indispensable natural resource not only for the conservation of human life and health but also for the conservation of all ecosystems and economic operations. At the global level, water is threatened by multiple pollution levels, including uncontrolled urban and industrial waste, excessive use of chemical fertilizers and pesticides in agriculture, and exploitation.

Pesticides (herbicides, insecticides, fungicides, etc.) are vital instruments for agriculture; they assist in combating damaging insects and weed and thus contribute to economic food production in large numbers [7]. By comparison, if these herbicides are misused, their residues can be very harmful to soil, water, and the environment and ultimately to human health. This is because most insecticides are persistent and, therefore, toxic owing to their lipophilic characteristics [8]. Glyphosate and 2.4-D, particularly in the irrigable perimeter of Bou Namoussa from 1968, are the most common herbicides in Algeria [9]. The importance of fertilizing components like nitrogen and assimilable phosphorus greatly affects crop output. For these reasons, studying the effect of herbicides on soil and water is essential. It is necessary to have an overview of the present natural resource (soil and water) scarcity and environmental pollution in Algeria. In Algeria, the occurrence of pesticides in the various environmental compartments is not systematically evaluated. The objective of this research is, therefore, to shed light on the impact of pesticides in Algeria on water and soil resources.

2 Man's Relationship with the Environment and Pesticides

The man's relationship with the environment began on earth with fear of its dangers, cruelty, and ignorance in dealing with its secrets. He started working and struggling to adapt to his requirements and secure his life to protect himself from the danger.

As his life developed, he began to delve deeper into the secrets of the universe and the manifestations of the environment, and a relationship of harmony and mutual accommodation has invested in the environment, giving it the sources of life as much as it is done in it. He reached the age of science and technology, guided to many of the secrets of nature and to the interactions of matter and energy in the environment, and became important to his knowledge to employ the environment in his service to achieve the best level for himself.

Then this harmony between man and environment turned into hostility again. Human beings have been deeply distressed by the exploitation of many things that have been disrupted by this harmony. It has exploited natural resources from fuel sources, mineral ores, and others. Even its sources began to wipe out his hands. The agricultural land was spoiled by its beds in the use of fertilizers and pesticides. Until 1962, the *Silent Spring* textbook by author Rachel Carson, spoke extensively about the impact of certain pesticides on the eggs of multiple birds. How pesticides and other insecticides caused fatal effects by their decay into the food chain. How these pesticides are chemically constant and require several years to break down.

The United States listed in 1986 a large number of toxic chemicals released by various industrial sectors. However, this number is a small part of the remaining 75,000 types of chemicals currently in use in the industry, which is growing in number 1 day and a large part of which is certain to carry certain toxicity. In the United States alone, 1994 recorded the launch of more than 2.26 billion pounds for more than 300 dangerous substances to the environment. In order to understand the amount of this, it is stated that when reading a page of this book, one ton of dangerous substances is released to the environment. Chemical industries are the most dangerous chemicals in all than other industrial sectors. The chemical industry is more than four times the waste from the industry that follows it, the metal industry.

3 General Presentation of the Study Area

Algeria is the Mediterranean Sea's biggest nation and the African continent's second-biggest nation. Algeria can be divided in three regions: the Tell (Atlas mountain chains), the High Plains (steppe plains lying between the Tell and Saharan Atlas ranges), and the Sahara desert (Fig. 1). It includes 238 million km^2 of which 8.3 million km^2 are cultivated. It borders the southern Sahara desert. So most of the nation (84%) has a desert environment. This is in disparity to the north of the country (16%) which enjoys a Mediterranean climate (Fig. 1). The northern region is mountainous and hilly, with a fertile plain running parallel to the shoreline between the coast and the 1,500-km-long Tell Atlas mountain chain. The Saharan Atlas range runs south of the Tell Atlas and parallel to it.

With an average annual precipitation of 350 mm in the west and as much as 1,000 mm in the northeast, precipitation is variable across this area. Rainfall is quickly declining south of the Saharan Atlas and in the direction of the Sahara desert.

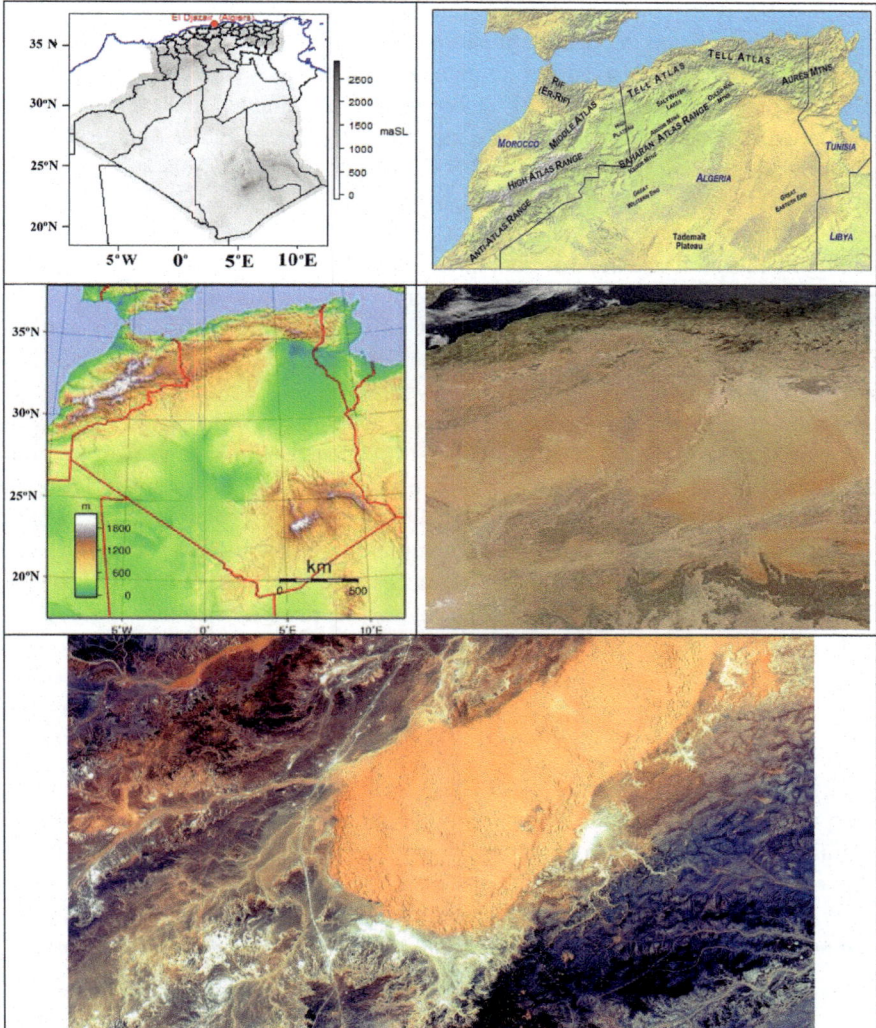

Fig. 1 The geography and topography of Algeria

The average annual rainfall here over this area is less than 100 mm, which is generally focused over a short period. Oases dominate the population's life and culture in this area. In the latest decades, prolonged and persistent droughts have also led in less supply of surface water, and this has encouraged groundwater exploitation.

Northern Algeria lies within the temperate area, and its environment resembles that of other Mediterranean nations, although the variety of relief offers sharp heat contrasts. The coastal region has a pleasant climate, with average winter temperatures between 10 and 12°C and average summer temperatures between

24 and 26°C. There is an abundance of rainfall in this region, which is 38–69 cm per year and up to 100 cm in the eastern part except in the area around Oran (Ouahran), where mountains form a barrier to rain-bearing winds. When heavy rains fall (often more than 3.8 cm/1.5 in 24 h), big regions are flooded and then so rapidly evaporated that they are of little assistance in agriculture. Climate changes farther inland; winters average between 4 and 6°C, with considerable frost and occasional snow on the massifs; summers average between 26 and 28°C. The prevailing winds in this region are west and north in winter and east and northeast in summer, leading in a general rise in precipitation from September to December and a decline from January to August. In the summer months, there is little or no rainfall.

Temperatures range from −10 to 34°C in the Sahara Desert, with extreme 49°C peaks. There are daily variations over 44°C frequent and violent winds. The precipitation is irregular and dispersed unevenly.

Agriculture in Algeria is subject to natural restrictions, which thwart its growth and restrict its productivity. Due to the harsh weather conditions and their irregular distribution across the land, Algeria is defined by the weakness and fragility of its natural assets (soil and water). These limitations significantly restrict farming opportunities and highly distinguish agricultural place (see Fig. 1).

The sandy and rocky terrain of the Sahara desert in central Algeria (Fig. 1, image: ESA/NASA) was captured in 10/07/2015 by the Sentinel-2A satellite.

The geology map (see Fig. 2) shows a simplified overview of the geology at a national scale.

As far as geomorphology is concerned, Algeria is defined by three contrasting areas:

- The Northern "Tell" area (4% of the land), which is the most favored by environment and natural resources but remains the most faced with various anthropogenic stresses (population growth, urban and industrial development, overgrazing, etc.).

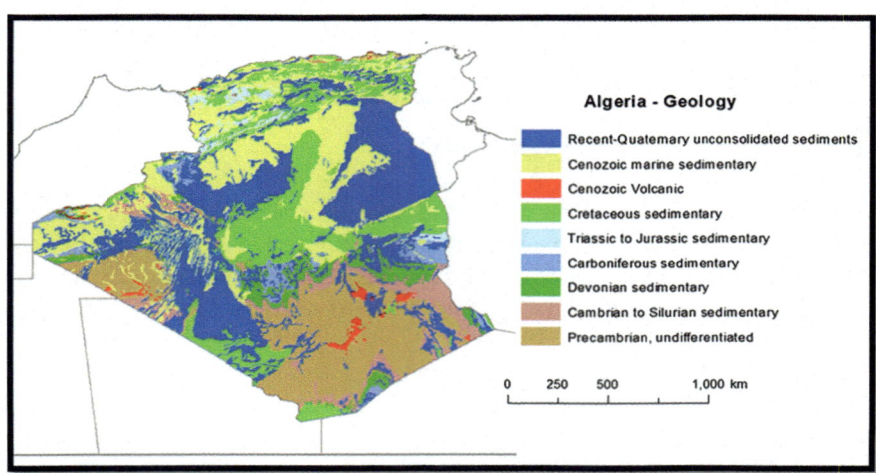

Fig. 2 Geology of Algeria at 1:5 million – scale [10]

- The Highlands (9% of the land) constitute an enclosed area between the Sahara desert and Tell Atlas, forming elevated plains under semiarid climatic conditions.
- The Sahara, which accounts for 87% of the territory's total region, is an outside region marked by severe aridity owing to rainfall scarcity (less than 100 mm).

The geomorphological structures identify basin slopes consisting of marly geological formations represented by impermeable soil, accentuated reliefs, and lack of vegetable cover. This describes the significance of ground erosion procedures (silting dams).

Aquifer types of Algeria are shown in Fig. 3. Algeria is defined by a semiarid to an arid environment where there is a continuous danger of drought and aridity "even in humid areas where the average annual precipitation seems high" [12]. The part of the land receiving more than 400 mm of rain is restricted to a strip of land 150 km broad in the coastal region [13].

4 The Scarce Soil and Water Resources

4.1 Soil Resources

Algeria is a large region of 2,381,000 km^2, but the agricultural area (UAA) is very restricted: almost 8 million ha (3.3% of the country region) for 40 million ha used for farming. There is restricted and declining fertile land appropriate for farming. Indeed, at the time of independence (1962), the UAA accounted for 0.82 ha per capita; it was only 0.26 ha in 2004 and could fall to 0.18 ha in 2010. Nearly 6% of UAA (450,000 ha) are irrigated, more than three-quarters of which are assigned to fruit and vegetable plants. These indicators indicate that in terms of amount and quality, the potential of agricultural property in Algeria is comparatively small. The soils are shallow and poorly irrigated and worn by steep geomorphological constructions and suffered through the centuries from the process of degradation driven by the combined impact of natural and anthropogenic processes leading to a significant loss of their natural fertility limit affecting agricultural production. Water erosion (rainfall exceeding 400 mm) primarily impacts northern Algerian soil and threatens almost 12 million ha in mountain regions. Land degradation creates important fertility loss and adversely impacts the productivity of agriculture. Since the early 1970s, human intervention has taken on more alarming proportions, including the coastal areas of the nation that contains Algeria's most fertile soil. Rapid urbanizations as the anarchic expansion of the industrial structure and development of desertification are all factors in Algerian land degradation.

Due to erosion, pollution, and salinization, soils in Algeria are subject to serious degradation. Therefore, increasing aridity will amplify the soil and ecosystem degradation phenomenon, causing desertification and fragile regions such as steppes. The area of land possibly impacted by water erosion is estimated at 4 million ha, 53% of which is deemed arable land. Regarding salinity-affected soils, statistics indicate that this phenomenon affects more than 50% of irrigated soil. With regard to

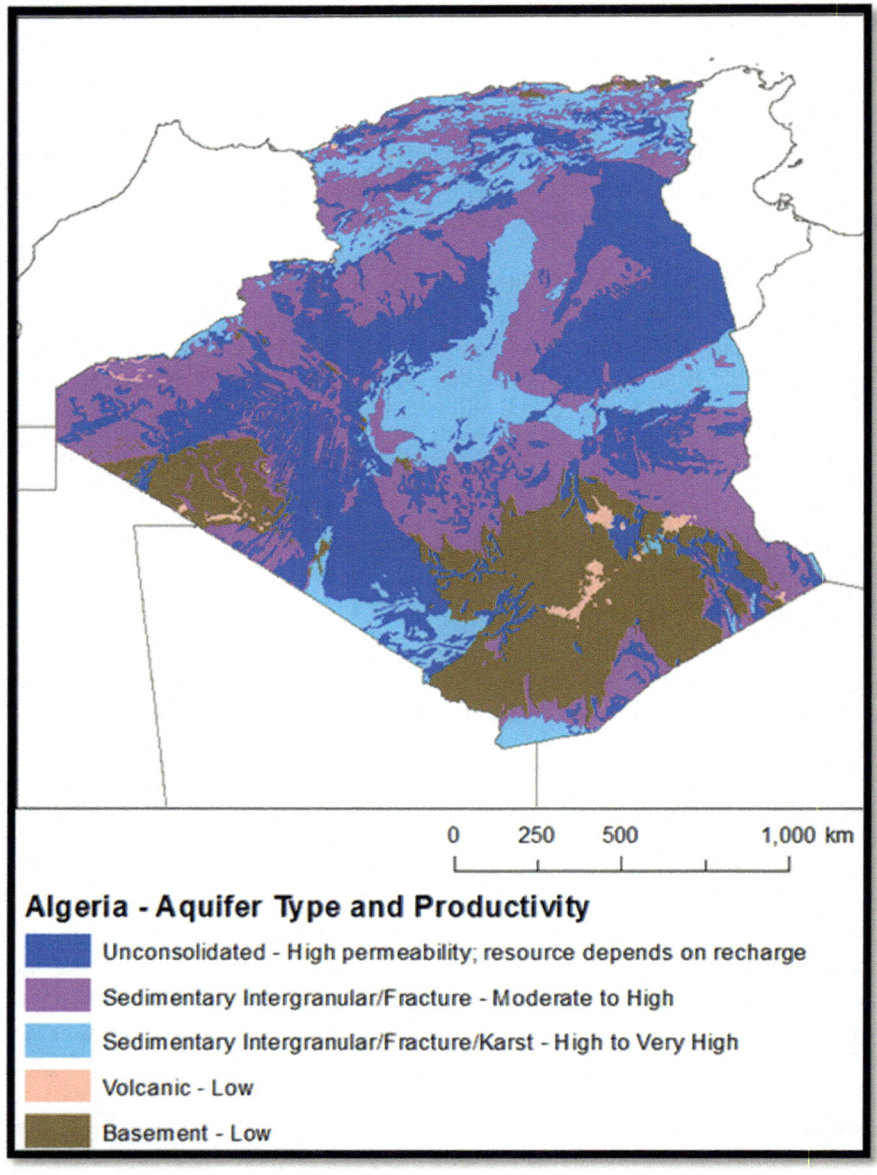

Fig. 3 Aquifer types of Algeria [11]

desertification, it should be observed that this phenomenon is linked to the harshness of the accentuated climate due to orographic circumstances and geopedology, as well as the impacts of anthropogenic stress. Therefore, in terms of rehabilitation, rationalization, conservation, and sustainable management, the problem of physical natural resources now occurs. Agricultural research has to provide appropriate responses to these issues.

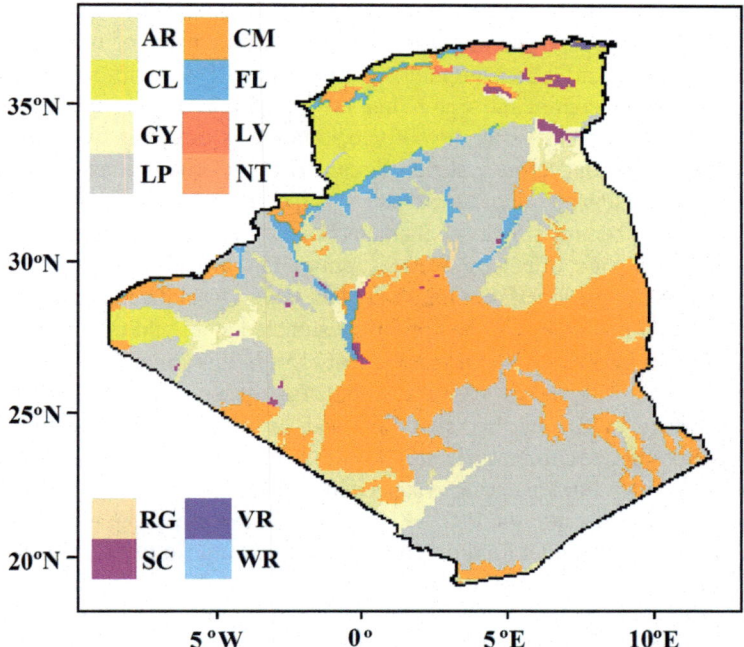

Fig. 4 Soil map of Algeria (European Soil Portal) [14]

Stony leptosols dominate soil in Algeria's mountainous Atlas area (Fig. 4). Soils are usually rich in calcium carbonate (calcisols) in the drier region south of the Atlas. Many of these soils are appropriate for agriculture, but the main limitation on crop development is the availability of water. Soils are better created along the wetter coastal region, resulting in more vegetation including luvisols and cambisols. Vertisols are found in the eastern coastal region of Algeria, supporting comprehensive cereal cultivation and grazing http://earthwise.bgs.ac.uk/index.php/Hydrogeology_of_Algeria. South of the Atlas, the arid region is described by poorly formed leptosols containing little organic matter. Arenosol regions indicate big sand dune regions. Fluvisols are discovered in the valleys of the river. Rivers are usually ephemeral to the south of the Atlas, but the valleys are often extensively cultivated in the wetter northern region, where rivers are more or less perennial.

4.2 Water Resources

Algeria is Africa's second most water-scarce country, after Libya. There are no large rivers in Algeria. The Cheliff River in the coastal plain is the only significant stream, providing some water for irrigation. From 2015, Algeria will face a water shortage scenario that will primarily impact all industries of the economy and agriculture. Water in Algeria is a resource that is increasingly restricted, vulnerable, and fragile. Algeria faces the issue of water quality as well as insufficient water availability. Every year, approximately 600 million m³ of untreated waste-water is discharged, affecting soil and water resources. This factor is now regarded as Algeria's most significant water and environmental management challenge. Excellent spatiotemporal variability of precipitation is ongoing stress for natural ecosystems and rainfed crop systems. Recent rises in droughts and temperatures frequency and intensity, due to climate change, result in even higher aridity. The drought has been a structural characteristic of the Algerian climate since the 1990s. The drought's effect on water supplies has already led in a worsening water resource deficit, low dam filling rate, and lower groundwater reserves.

Natural water resources will be presently estimated at around 600 m³/capita/year and in 2025 will be around 500 m³/capita/year [15]. In this context, Algeria is below the "poverty line" of 1,000 m³/capita/year and slightly above the 500 m³/capita/year "target shortage" [15]. In this situation, and given the arid environment and recurring droughts, securing agricultural output includes irrigation growth, which inevitably includes implementing water economy-focused policies. Such a policy component is inserted into the PREAR. The water economy program is introduced to protect 70% of Algeria's food requirements by moving from 900,000 ha in 2008 to 1.6 million ha in 2014, with an incompressible water requirement of 12 billion m³. Dealing with these requirements involves enhancing the productivity of current water resources, mobilizing fresh conventional and unconventional resources, and developing/disseminating water-saving irrigation methods as a factor in modernizing agriculture and enhancing the effectiveness of agricultural water [16].

There are 17 significant watersheds in Algeria for surface water (Fig. 5). Low rainfall implies that most rivers are ephemeral in Algeria's mountainous

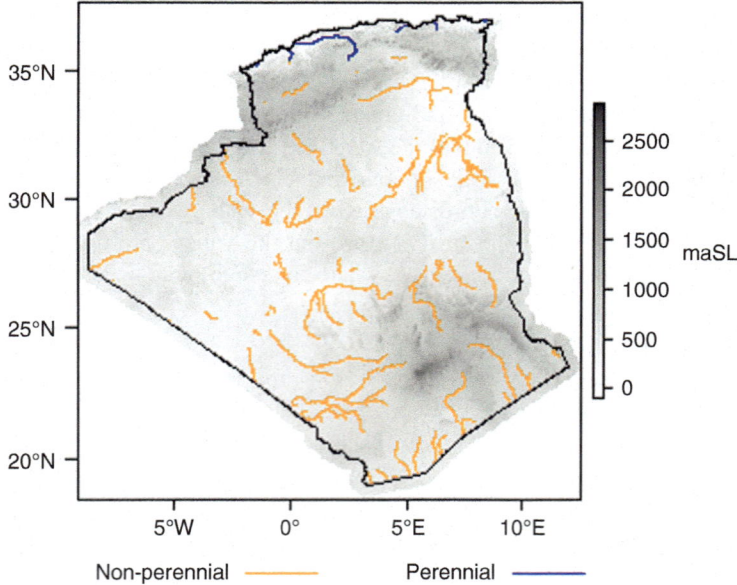

Fig. 5 Major surface water features of Algeria [11]

and desert areas, flowing only after major rainfall occurrences. Only the rivers that flow throughout the year in the northern coastal region are perennial. In the south, wadis (ephemeral rivers) drain into locked inner sinks – chotts or sebhkas – subject to elevated rates of evaporation. Chelif (or Cheliff) River is the longest river in Algeria, which runs for 700 km from its source in the Saharan Atlas to end where it meets the Mediterranean Sea http://earthwise.bgs.ac.uk/index.php/Hydrogeology_ of_Algeria.

Algeria is subject to physical circumstances and unfavorable hydroclimate due to its geographical place within the arid and semiarid region, exacerbated by periods of chronic drought. Observed climate and drought changes in North Africa for several centuries, particularly in Algeria, have exacerbated the country's adverse effect on water resources (see Fig. 6).

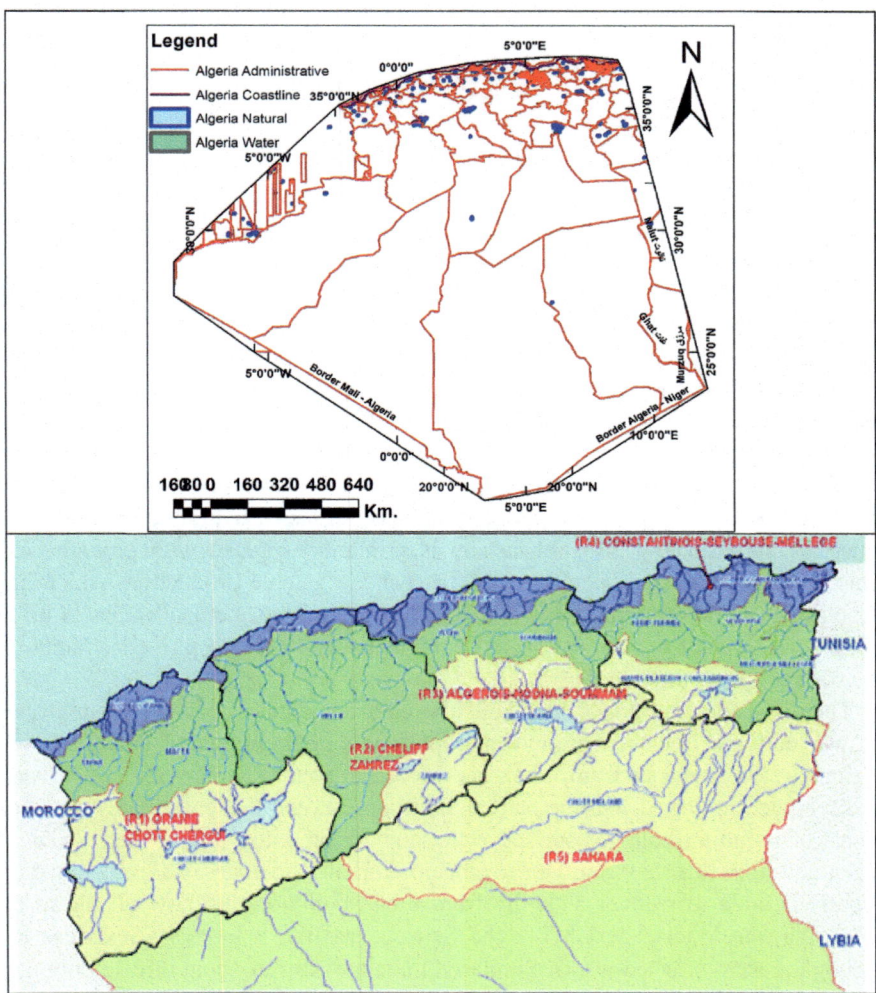

Fig. 6 Natural areas of surface water resources

In the coming decades, Algeria's hydrological year 2001–2002 was a year of hinge crisis (drought) for both water supply and a specific challenge strategy to adapt to a decline in renewable water resources. This crisis and water supply disturbance have highlighted the very random nature of water resources and contributed to a new awareness of the need to use nonconventional water resources (seawater desalination and sewage reuse by expanding and implementing a new water resource industry strategy). This development policy has two goals: to secure the population's supply of drinking water and to improve food security rates through possibilities provided by irrigated areas maintenance and extension.

The water sector is planning to move the water from certain dams in the coastal region to the Tell Atlas region, which will then be transported to the Highlands in turn, to guarantee the water needed. Then, desalination of seawater and water conservation should compensate for the deficit in the coastal area. This last option is a priority. A possible transfer of water from the Sahara (Albian aquifer) will also compensate for the deficit of the remaining Highlands region.

5 Agriculture and Environment

Despite the diversity of types of ownership, almost all productive farms are now managed privately. Cereal manufacturing dominates, but the high-value cropping of fruits and vegetables has grown in the latest years, approximately 8.3 million ha of the cultivated region. Although approximately 569,000 ha are estimated to be equipped for irrigation, in fact only 453,000 are irrigated. This, however, adds more than 40% of domestic agricultural manufacturing. Agriculture is the primary water consumer in the country taking nearly 4,000 MCM/year and is the industry that has the biggest effect on groundwater. Irrigation is split into significant irrigated fields, mostly built during the colonial era (GPI) and based on surface water, and the latest developments known as tiny and medium-sized hydraulic systems. The latter is approximately 363,000 ha and is primarily irrigated with groundwater. This industry adds significantly to the fruit and vegetable production, which has benefited from the National Development Fund's big investment subsidies of up to 80%. This level of investment stayed comparatively stable despite constraints on the budget during the oil crisis.

There are increasing worries about the environmental impact of groundwater growth and the effect of groundwater pollution in households, industries, and agriculture. Wastewater from urban and industrial sources, some 820 MCM/year, is discharged untreated into the natural environment, and this contributes to worry levels of pollution both for surface water and for alluvial groundwater, which is recharged from surface water. Due to enhanced urban discharges, wadis, who would usually be able to recover naturally from urban pollution, are now unable to do so. Tafna, the Macta, the Cheliff, the Sébaou, and the Soummam Seybouse are among the most affected wadis. Similarly, there are worries about diffuse pollution from intensive farming and irrigation using brackish water and untreated wastewater related to bad water management.

6 Environmental Impact of Pesticides

The growth of organic chemistry that started in the 1940s brought a fresh era of synthetic pesticides that became the most widely used spectrum of industrial chemicals in contemporary culture. More than 900 active ingredients [17, 18] enter in the composition of thousands of pesticides products, mainly used in agriculture, to control insects, diseases, weeds, and other pests. Even though the use of pesticides becomes an important instrument for increasing productivity, improving quality, protecting livestock, and fighting vector illnesses, there is now evidence that these products generate risks for man and his environment. Each year, pesticides contribute to an estimated 26 million human poisonings and 220,000 deaths worldwide [19]. Pesticides poisoning in Algeria came in second place in the causes of acute intoxication after drugs. While chronic impacts of exposure to pesticides on human health are less easily recognizable, wealth study has been carried out in this regard [20, 21] found a signification association between pesticides exposure and many health problems including cancer, neurological damage, reproductive and developmental hazards, immunotoxicity, and endocrine disruption. Despite their extensive implementation, their objectives are effectively reached by only a tiny quantity of the pesticides used.

6.1 Environmental Issues

In the twentieth century, the prodigious development of the chemical industry profoundly and irreversibly altered the patterns of manufacturing and consumption in both technologically and economically more developed areas than in less wealthy areas of the planet. Mass production and extensive use of chemicals in agriculture, in specific mineral fertilizers and phytosanitary products, have a drastic rise in crop yields to intensify agriculture. There is no doubt that pesticides and associated products are among the chemicals most used in our present setting. Pesticides (insecticides, raticides, fungicides, and herbicides) are toxicologically characterized chemical compounds whose first extensive use (DDT) goes back to World War II. Indeed, organochlorine pesticides (OCPs), deemed to be the most poisonous and persistent organic pollutant (POPs), have been widely used as contact insecticides throughout the globe and to a lesser extent as fungicides and acaricides. Besides their positive impacts on crop safety and safety, organochlorine pesticides are silently expressing their negative health impacts.

Characterization and rates of herbicides choice focus on two weed killers, widely used in Algeria [22]. Glyphosate is a soft organic acid from a natural amino acid analog in white powder that belongs to the chemical organophosphorus family. $C_3H_8NO_5P$ is a complete foliar systemic weed killer, endowed with a phosphonate grouping N-(phosphonomethyl)glycine. Glyphosate is heavily absorbed into the soil; microorganisms have degraded it, and it can be more or less persistent.

Acid 2.4-dichlorophenoxyacetic (also reported 2.4-D) is a chemical organochlorine weed killer of $C_8H_6C_{12}O_3$ fundamental formula. Colorless crystal or white powder with no odor, specific against weed but inactive on lawns and cereals, prevents fruit falling and acting as growth hormone on dying crops. It is one of water, soil, air, and rain contaminants that we also discover in the indoor atmosphere. We maintained 2.5 μg for glyphosate (easy) and 12.1 μg for 2.4-D [23].

Scientists have rapidly identified harmful impacts with the clear proof [24–26]. This has resulted in more or less severe laws designed to restrict their use or their complete prohibition. In this context, a whole arsenal of legality has been created to align it with the obligations of our country and the global conventions to which it has acceded [27].

During the 1960s and 1970s, the misuse of pesticides in Algeria's agricultural industry produced big unused stocks across the domestic territory that are stored anarchically and diffusely without taking into consideration the hazards to people's health and the environment, especially water resources. This threat of degradation of the quality of these resources and groundwater, is more vulnerable [28].

The impacts of pesticides on nontarget species are the consequences of pesticides. Pesticides are chemical preparations that are used to kill pests of fungi or animals. More than 98% of sprayed insecticides and 95% of herbicides achieve a destination other than their target species because they are sprayed or distributed throughout the entire agricultural sector [29]. Runoff can bring pesticides into aquatic environments, while wind can bring them to other fields, pasture regions, human settlements, and undeveloped regions, which may affect other species. Other issues arise from bad practices in manufacturing, transportation, and storage. Repeated use improves the strength of pests over time, while its impacts on other species can promote the revival of the pest [30]. Each class of pesticides or pesticides has a particular set of environmental issues. Such undesirable impacts have resulted in many pesticides being banned, while other uses have been restricted and/or decreased by legislation. In general, pesticides have become less persistent and more species-specific over time, decreasing their footprint on the environment. Furthermore, the quantities of pesticides used per hectare have decreased by 99% in some instances. However, the worldwide spread of pesticide use, including the use of older/obsolete pesticides which in some jurisdictions have been banned, has risen generally [31, 32].

6.2   Pesticides Impacts on Biodiversity

Farmers have increasingly used chemical pesticides (defined here to include insecticides, nematocides, fungicides, and herbicides) since the mid-1900s to reduce crop losses from pests, illnesses, and weed competition. In the past, developing nations have used fewer pesticides, but the use of pesticides in these nations is anticipated to develop faster than in the developed globe. The main environmental impact of the pesticide is biodiversity due to leaching into soil and water.

Filtration of pesticides into soil and water is harmful to animal and human health, and impacts in Africa can be magnified. Pesticide application efficiency levels are even smaller than for fertilizer, with some estimating that the planned pest reaches less than 0.1% of the pesticides applied to plants. The rest accumulates in soils where it can filter into soil or surface water and prove to be poisonous to microorganisms, aquatic animals, and humans. Pesticides accumulated in soils can damage arthropods, earthworms, fungi, bacteria, protozoa, and other organisms that contribute to soil function and composition. Birds' exposure to pesticides can trigger reproductive failure or even kill them immediately in sufficiently elevated doses. Exposure to pesticides may also affect domesticated livestock. Pesticides may continue for lengthy periods once they join an ecosystem. For example, organochlorine insecticides such as DDT were found 20 years after their use was prohibited in surface waters in the United States. Furthermore, biomagnification of pesticides entering the food chain may occur, whereby accumulated levels in organism tissues are many times greater than in the surrounding setting.

6.3   Pesticides Impacts on Soils

Intensive use of pesticides in agricultural production can degrade and harm the community of microorganisms residing in the soil, especially when overuse or misuse of these chemicals. The complete effect of pesticides on soil microorganisms is not yet fully understood; many studies have discovered harmful impacts of pesticides on soil microorganisms and biochemical procedures, while others have discovered that microorganisms can degrade and assimilate residues of some pesticides [33]. Besides multiple environmental variables, the impact of pesticides on soil microorganisms is affected by the persistence, concentration, and toxicity of the pesticide applied. This complicated interaction of variables makes definitive conclusions about the interaction between pesticides and the soil ecosystem hard to draw. However, in particular, long-term application of pesticides can interfere with the biochemical procedures of nutrient cycling [33]. Several chemicals used in pesticides are constant soil contaminants, the effect of which can last for centuries and affect soil conservation adversely.

Pesticide use reduces overall soil biodiversity. Not using chemicals leads to greater soil quality, with the extra impact that more organic matter in the soil enables greater water retention [34]. This enables to boost farm returns in drought years, with organic farms yielding 20–40% greater than their standard counterparts [35]. A lesser content of organic matter in the soil leads to an increase in the amount of pesticide that leaves the scope because organic matter binds and helps break down pesticides [34].

Both degradation and sorption are factors that affect pesticide persistence in soil. Depending on the pesticide's chemical nature, these procedures directly regulate the transportation from land to water and, in turn, our food and air. Degradation, breaking down organic substances, includes interactions in the soil between microorganisms. Sorption impacts pesticide bioaccumulation, which is dependent on soil organic matter. Because of pH and mostly acidic composition, it has been shown that weak organic acids are weakly sorbed by soil. It has been shown that sorbed chemicals are less available to microorganisms. Aging mechanisms are poorly defined, but the residue of pesticides becomes more resistant to degradation and extraction as they lose biological activity [36].

Pesticides or their degradation products may be further converted or degraded by other microorganisms in natural settings or may eventually lead to complete degradation by the microbial consortium. However, constant xenobiotics such as pesticides and dead-end metabolism products will accumulate in the setting, become a component of the soil humus, or enter the food chain leading to biomagnification (Fig. 7).

Over the previous half-century, supplementing agricultural systems with synthetically derived nitrogen (N), phosphorus (P), potassium (K), calcium, magnesium, and micronutrients has made it possible for humans to dramatically

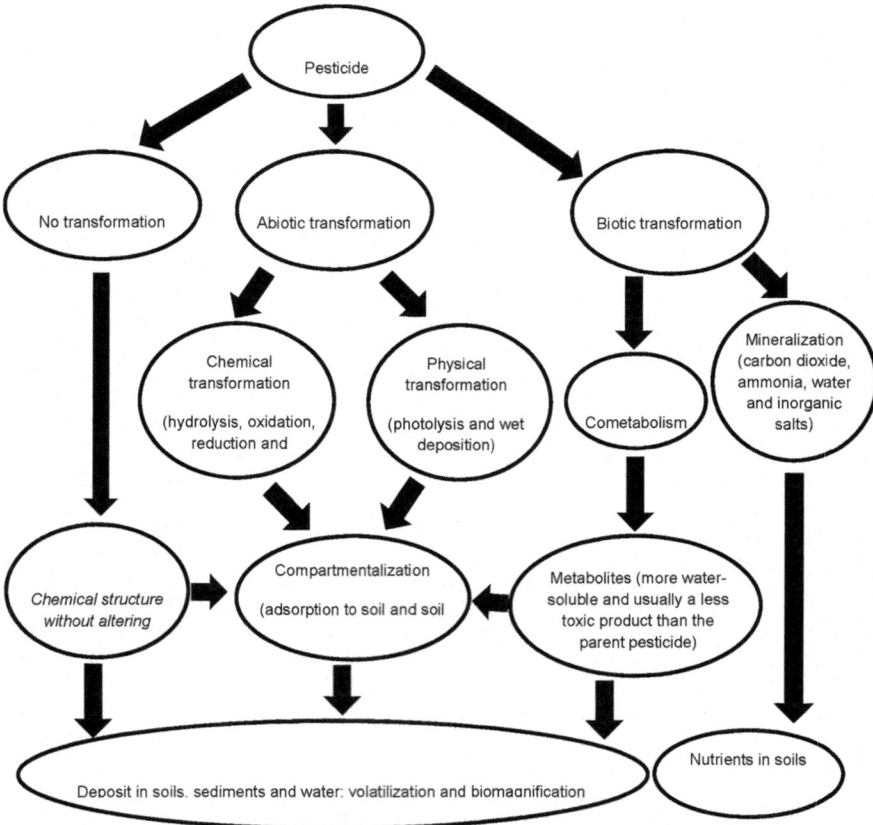

Fig. 7 Fate of pesticides in the environment [37]

boost per unit returns. However, rises in fertilizer use have affected soil fertility, water quality, air quality, and greenhouse gas emissions owing to inefficiencies in fertilizer implementation and plant uptake.

Leaching nitrates and ammonium-based fertilizers contribute to soil acidification. High fertilization rates of nitrogen can lead to soil acidification, a process that results in aluminum and manganese poisonous concentrations and reduces vital nutrient quantities. Acidification happens when ammonium undergoes nitrification to form nitrate in certain nitrogen fertilizers, and the nitrate then leaches into the soil. In the lack of nitrogen leaching, ammonium-based fertilizers may also lead directly to acidification. In advanced and developing countries, especially in East Asia, soil acidification is an issue [38]. A recent study of China's main crop production fields, for instance, discovered important acidification of all top soils mainly owing to elevated inputs of nitrogen fertilizers [39].

Nitrogen is a nutrient that is highly mobile and readily lost from agricultural soils. The average effectiveness of absorption of fertilizers is only 30–50%, which means that soils can accumulate big amounts of unabsorbed nitrogen and other nutrients. These nutrients can leak into aquatic ecosystems in a number of ways [40]. Excessive rainfall or irrigation may result in leaching of accumulated soil nitrates under the rooting area of a crop and entering groundwater. Nitrates can also flow through layers of the soil into ecosystems of surface water. Leaching nitrogen and other fertilizer nutrients into fresh and saltwater settings can contribute to eutrophication (overabundant concentrations of nutrients), leading to enhanced blooms of algae and depletion of oxygen. In these fields, "dead zones" may grow, with reduced oxygen concentrations in dramatically reducing fish populations and diversity of species. The Baltic Sea, the Black Sea, India's west coast, and the Gulf of Mexico's Mississippi River outlet comprise important dead areas created by eutrophication. Polluted water consumption can also have a negative effect on human health. Even after the leakage of nitrogen is slowed or eliminated, it may take decades to restore contaminated water bodies [38].

7 Arresting Emerging Pollution Problems

Urban and industrial pollution, as well as the absence of access to safe water and sanitation, pose growing public health threats. In general, antipollution measures were aimed at only one medium, such as water or air, often resulting in another medium's fast decay. Integrated pollution control should tackle the effect on all media and concentrate on pollution prevention waste decrease and recycling. To promote clean industries, governments should:

1. Encourage private investment in clean industries by removing tariff obstacles and nontariff obstacles to clean process technology.
2. Apply the polluter-pay principle to guarantee that companies pay pollution expenses and create licensing processes to guarantee that steps are taken to control pollution.

To clean up "hot spots" where pollution is endangering human health or ecosystems, governments should:

- Improve the quality and extend the coverage of environmentally secure water and sanitation facilities by improving the operation and maintenance of current facilities, rehabilitating nonoperational treatment plants, and promoting investment in suitable low-cost techniques.
- Identify and rank hot spots through environmental audits and then create low-cost mitigation plans on which to base compliance contracts with polluting companies on phased emission cuts.
- Ensure that the expenses of cleanup are borne by the financially feasible government and private companies. It is necessary to shut down highly polluting public undertakings that are not financially viable and cannot be restructured.
- Reduce air pollution from transportation and power sources by phasing out leaded petrol and high-sulfur fuels, setting emission requirements for new cars and imposing gradual ownership charges on used cars and accelerating the growth of natural gas. Several nations are starting to address emerging pollution issues. In choosing industrial fields, Algeria, Morocco, and Saudi Arabia adopt an integrated strategy to pollution control.

8 Improving the Management of Scarce Natural Resources

Environmental management is a fairly fresh area in the region. Therefore, it is a top priority to build institutional ability in environmental policymaking. Capacity building will involve well-focused help to enhance organizational and technical capacity and enhance surveillance and enforcement of norms of environmental quality. It will also enable that environmental hazards are publicized and the public be involved in setting priorities and making choices. Increased environmental awareness can stimulate government intervention and readiness to pay for environmental services. Algeria, Egypt, Morocco, and Saudi Arabia are restructuring their environmental institutions. In order to define priorities, risk assessment has been implemented in Algeria, Egypt, and Tunisia. Several nations have made environmental impact assessment compulsory for fresh development initiatives, including members of the Gulf Cooperation Council. Public access to data on the environment has enhanced, but much more effort is needed to obtain true government involvement in environmental management.

The region's water scarcity and arable land are a basic constraint to its future economic growth. While historically abundant, energy will also become scarce if present patterns of consumption continue. The issue needs that development be reoriented away from mining the natural resource base by enhancing management through the three measures as follows:

1. Adapt to improve water scarcity by raising water prices, promoting the preservation and mobilizing investment economic resources, and enhancing conflict mediation organizations, and introducing integrated water resource planning and management.
2. Eliminate energy, fertilizer, pesticide, and other agrochemicals subsidies gradually.
3. Intensify attempts to promote the implementation of established techniques for the effective use and conservation of water, soil, and power through data campaigns, incentives such as environmental taxes and pollution fees, and the removal of trade barriers to new technologies.

9 The Future of Pollution Prevention Using Green Chemistry

Pollution prevention may be important, but not the best solution? As awareness of the importance of pollution prevention among scientists and the industry community grows, there is a need to reduce the sources of pollution by changing the design of preparation methods and production processes so that they do not produce waste in the first place. Even the absence of waste is not sufficient to protect the environment, but chemical products must be designed so that hazardous substances do not need to be used during production. Excluding the use of such materials, the impact on environmental gains would be twofold because of these new design processes.

This chapter shows the urgent need to develop new branches of chemistry that are less dangerous to human health and the environment. Therefore, we must pursue the goals of green chemistry. Robert F. Kennedy has a beautiful saying: "Some see things as they are and ask why? However, I see things as they should be and ask why not?" One of the most philosophical reasons why this must be done is that we can do so. Because they know how to trade and transform chemical compounds and the risks that can occur. Thus they have the ability to reduce or eliminate risks for themselves and society. Another reason for the intensive application of green chemistry by society is that this chemistry is based on basic molecular science as a way to solve environmental problems and does not address problems with bandage or patchwork to reduce risks.

The early pioneers of this science have established some principles on which to base and outline their future directions:

1. It is best to prevent waste from being formed or disposed of.
2. Preparation methods should be designed so that most reactors are integrated into the final product.
3. Manufacturing methods must be designed so that reactive substances are less toxic or are not at all dangerous to human health and environmental safety.
4. The chemical product must have the highest functional efficiency and the lowest toxicity.

5. Reactions are preferred without the use of additional materials such as solvents or separation materials, and, if necessary, they should not be dangerous.
6. Due to its cost and environmental impact, energy needs to be considered – so its use is minimal, and it is preferable to design interactions at normal temperature.
7. Raw materials with indent materials should be renewable rather than depletion of nonrenewable materials.
8. Products must be designed so that they do not become stable after they function and must be biodegradable in the environment into simple, harmless materials.
9. Safe chemicals should be selected in terms of type and composition so that chemical incidents can minimize the risk of gas, explosion, or fire.

Human health and the environment have certainly been affected by chemicals and their various stages of manufacture. Green chemistry became responsible for finding suitable solutions to all old manufacturing problems by finding alternative solutions to all previous negatives. Green chemistry provides agricultural products, herbs, molasses, and natural raw materials as alternative stock for the preparation of many compounds.

10   Conclusion and Recommendations

Algeria, with its arid and semiarid climate, is highly vulnerable to climate change with desertification as a major concern. Overall, temperature and evaporative demand are expected to increase for Algeria. The change in future rainfall is however more debated. In Algeria, even if global climate change is not taken into consideration, water shortages are a significant acuity issue in many parts of the nation. Agricultural research in Algeria will face many difficulties mainly determined by the momentum produced by economic reforms, which are themselves conditioned by two main limitations:

1. Climate changes and its effects such as growing drought and flooding and water scarcity. From now on, the last will be a significant determinant in defining all the elements of Algeria's food safety policy.
2. The globalization of the economy causes powerful entropy on the markets and excellent instability of the prices of agricultural products due to both political and climatic hazards. Also the replaceability of the products contributes to this.

Naturally, the current situation in Algeria is exacerbated by several constraints:

1. An imbalance between requirements and accessible resources: over the previous two decades, population growth and economic and social development have resulted in a significant rise in demand for drinking, industrial, and agricultural water. Water need expressed by various users is mobilized well above water resources: this generates assignment disputes and sometimes involves challenging decisions.

2. A geographic imbalance between requirements and resources: the elevated concentration of water requirements in the coastal strip (60%) requires reassignment by moving rather costly economic resources to balance deficits in inland regions, including the entire Highlands region.
3. Groundwater and surface water pollution: domestic, industrial, and agricultural waste far exceeds the ability of sewage systems, significantly reducing the quantity of treated water that can be used.
4. Risk of sustainable development rupture: In relation to pollution, severe issues arose in groundwater evaluated samples that exceed natural resource renewal boundaries and need to tap into nonrenewable reserves.

Besides, the weakness of our resources is overused by the poor spatial and temporal distribution of these resources.

Algeria suffers from the soil erosion and siltation of dams and losses due to outdated distribution and poor management.

Among the biggest challenges that will face Algeria to the Horizon 2023, we can mention:

- The necessity to ensure sustainable management of natural resources and ecosystems.
- The necessity to ensure food security of the nation and citizens.
- The resolution of the throbbing questions of employment through the development of a productive and competitive economy.
- The establishment of the foundations of effective governance of both the economy and society.

For future studies, a number of convergence points have been recognized by analyzing the strategic directions of studies in Algeria and their comparison with those of the European Union, based on the significant problems for Horizon 2023. These relate to:

- The research planification based on major challenges (food security, climate change, and water economy).
- Priority research themes (food security, sustainable agriculture, fight against climate change, efficient use of natural resources; inclusive, innovative, and secure societies).
- Integrating SMEs in the process of research and innovation (industrial primacy pillar).
- Willingness to mobilize industry stakeholders and engineering sciences, most directly concerned and most likely to integrate scientific knowledge in an innovation perspective.
- Necessity to develop innovation in a direction favorable to smart, sustainable, and inclusive growth.
- A greater role for social sciences in the development of research to address all societal challenges.

References

1. United Nations, Department of Economic and Social Affairs, Population Division (2017) World population prospects: the 2017 revision. United Nations, New York
2. FAO (2005) Agricultural extension and training needs of farmers in the small island countries: a case study from Samoa. M.K. Qamar and S.S. Lameta. FAO, Rome
3. Malabo MP (2018) Water-wise: smart irrigation strategies for Africa. International Food Policy Research Institute (IFPRI) and Malabo Montpellier Panel, Dakar
4. Kabata-Pendias A, Pendias H (2001) Trace elements in soils and plants. CRC Press, New York
5. Huang SS, Liao QL, Hua M (2007) Survey of heavy metal pollution and assessment of agricultural soil in Yangzhong district, Jiangsu Province, China. Chemosphere 67 (11):2148–2155
6. N'guessan YM, Probst MJL, Bur T, Probst A (2009) Trace elements in stream bed sediments from agricultural catchments (Gascogne region, S-W France): where do they come from? Sci Total Environ 407(8):2939–2952
7. Anikwe MAN, Okonkwo CI, Mbah CN (2003) Yield of soybean. Tropicultura 5:22–27
8. Kanissery RG, Gerald-Sims K (2011) Biostimulation for the enhanced degradation of herbicides in soil. Appl Environ Soil Sci 2011: 10 p
9. Yamamoto Y, Sukchan S (2017) Land suitability analysis concerning water resource and soil property. Analysis concerning water resource JIRCAS Working Report No. 30, 2017 (Report No. 30)
10. Persits F, Ahlbrandt T, Tuttle M, Charpentier R, Brownfield M, Takahashi K (2002) Map showing geology, oil and gas fields and geologic provinces of Africa. USGS Open File report 97-470 A. Ver 2.0
11. Africa Water Atlas (2019) Africa Groundwater Atlas Country Hydrogeology maps. British Geological Survey. http://earthwise.bgs.ac.uk/index.php/Hydrogeology_of_Algeria. Accessed 10 Jan 2019
12. Perennes J (1993) Water and men in the Maghreb. Karthala Edition, Paris
13. MATE (2002) National Plan of actions for the environment and the durable development, p 39
14. Eusoils (2012) Data collection on contaminated sites. http://eusoils.jrc.ec.europa.eu/library/ data/eionet/2011_Contaminated_Sites.htm, 2011
15. CEDARE (2014) Algeria water sector m&e rapid assessment report. Monitoring & evaluation for water in North Africa (MEWINA) project. Water resources management program. CEDARE, Heliopolis
16. MADR (2012) Rapport d'audition MADR. Volet hydraulique. Direction des zones arides et semi-arides, p 27
17. Ware G, Whitacre D (2004) The pesticide book, 6th edn. Meister Pro Information Resources, Willoughby
18. Pimentel D, Pimentel MH (2008) Food, energy and society, 3rd edn. CRC Press: Taylor and Francis Group, LLC, Boca Raton, p 402
19. Gusdorf J (2019) Ecological living. Taylor & Francis, Milton Park
20. Krieger R, Doull J, Hemmen J, Hodgson E, Maibach H, Reiter L, Ritter L, Ross J, Slikker W, Vega H (2010) Hayes handbook of pesticide toxicology, 3rd edn. Academic Press, Cambridge, p 1
21. WHO (2008) Pesticides: children's health and the environment. WHO, Geneva
22. Gauvrit C (1996) Efficacité et sélectivité des herbicides. INRA, Paris, p 158
23. Saiyed H, Dewan A, Bhatnagar V, Shenoy U, Shenoy R (2003) Effect of endosulfan on male reproductive development. Environ Health Perspect 111:1958–1962
24. Forté A, Colacino J, Polemi K, Guytingco A, Peraino NJ, Jindaphong S, Kaviya T, Westrick J, Neitzel R, Nambunmee K (2019) Pesticide exposure and adverse health effects associated with farm work in Northern Thailand. bioRxiv:549618
25. Zikankuba V, Mwanyika G, Ntwenya J, James A (2019) Pesticide regulations and their malpractice implications on food and environment safety. Cogent Food Agric 5:1601544

26. Mobarak YM, Al-Asmari MA (2011) Endosulfan impacts on the developing Chick embryos: morphological, morphometric and skeletal changes. Int J Zool Res 7(2):107–127
27. UNEP (2001) Stockholm convention on persistent organic pollutants. United Nations, New York
28. PNM (2006) National Implementation Plan (NIP) ALGERIA. Stockholm Convention POP's Project – Algeria GF/ALG/02/001
29. Miller G (2004) Sustaining the Earth: an integrated approach. Thomson/Brooks/Cole, Pacific Grove, pp 211–216
30. Damalas CA, Eleftherohorinos IG (2011) Pesticide exposure, safety issues, and risk assessment indicators. Int J Environ Res Public Health 8(12):1402–1419
31. Lamberth C Jeanmart S, Luksch T, Plant A (2013) Current challenges and trends in the discovery of agrochemicals. Sci Technol 341(6147):742–746
32. Tosi S, Cecilia C, Umberto V, Giancarlo Q, Giovanni G (2018) A 3-year survey of Italian honey bee-collected pollen reveals widespread contamination by agricultural pesticides. Sci Total Environ 615:208–218
33. Hussain S, Siddique T, Saleem M, Arshad M, Khalid A (2009) Impact of pesticides on soil microbial diversity, enzymes, and biochemical reactions. Adv Agronomy 102:159–200
34. Kellogg R, Nehring R, Grube A, Goss D, Plotkin S (2000) Environmental indicators of pesticide leaching and runoff from farm fields archived June 18, 2002, at the Wayback machine. United States Department of Agriculture Natural Resources Conservation Service, Washington, D.C.
35. Lotter D, Seidel R, Liebhardt W (2003) The performance of organic and conventional cropping systems in an extreme climate year. Am J Altern Agric 18(3):146–154
36. Arias-Estévez M, López-Periago E, Martínez-Carballo E, Simal-Gándara J, Juan-Carlos M, García-Río L (2008) The mobility and degradation of pesticides in soils and the pollution of groundwater resources. Agric Ecosyst Environ 123(4):247–260
37. Ortiz-Hernández ML, Sánchez-Salinas E, Dantán-González E, Castrejón-Godínez ML (2013) Pesticide biodegradation: mechanisms, genetics and strategies to enhance the process. Life Sci:251–287
38. FAO, Arias-Estevez, Lopez-Periago, Martinez-Carballo, Simal-Gandara, Mejuto & Garcia-Rio (2008), p 248 2003: p 348 65
39. Guo J, Liu X, Zhang Y, Shen J, Han W, Zhang W, Christie P, Goulding K, Vitousek P, Zhang F (2010) Significant acidification in major Chinese croplands. Science 327(5968):1008–1010
40. Millennium Ecosystem Assessment (2005) Guide to the millennium assessment reports, p 767

Part III
Climate Change Impact and Hydrogeological Investigations

Analysis of Flood Characteristics in the Context of Climate Variability in Northern Algeria: Case of Cheliff Watershed

Abdelkader Sadeuk Ben Abbes, Mohamed Meddi, Mohamed Renima, and Abdelkader Bouderbala

Contents

Abstract Extreme hydrological events, such as floods and droughts, are some of the natural disasters that occur in several parts of the world. They are regarded as being the most costly natural risks in terms of the disastrous consequences in human lives and in property damages. The main objective of the present study is to estimate flood events in Cheliff watershed in giving return periods at the gauged stations, which is

A. Sadeuk Ben Abbes (✉) and A. Bouderbala
Department of Earth Sciences, University of Khemis Miliana, Ain Defla, Algeria
e-mail: a.sadekbenabbes@univ-dbkm.dz

M. Meddi
Water Engineering and Environment Laboratory, Higher National School of Hydraulics, Blida, Algeria

M. Renima
Laboratory of Chemistry Vegetable-Water-Energy, University of Hassib Benbouali, Chlef, Algeria

Abdelazim M. Negm, Abdelkader Bouderbala, Haroun Chenchouni, and
Damià Barceló (eds.), *Water Resources in Algeria - Part I: Assessment of Surface and Groundwater Resources*, Hdb Env Chem (2020) 97: 95–110,
DOI 10.1007/698_2020_525, © Springer Nature Switzerland AG 2020,
Published online: 16 June 2020

located in a semiarid region in the northwest of Algeria. The choice of this area is due
to the significant floods observed in this watershed, which are occurring mainly from
1960 and 2006. A study is carried out to know the temporal variability and the place
occupied by peak flows for both annual and monthly levels, which can finally
determine the peak output of flood. We will try to understand the evolution of
average and extreme flows according to rainfalls that are temporally associated
with them. Frequency analysis is performed on different series of observed annual
and monthly average discharges, including classical statistical tools as well as recent
techniques. The obtained results show that the annual maximum approach is more
appropriate in this case. This study also indicates the importance of continuous data
monitoring in these stations.

Keywords Algeria, Annual maximum, Cheliff watershed, Floods, Peak flows

1 Introduction

During floods, the maximum levels of water, the maximum speed or the maximum
flow, and the duration of the overflow can cause problems which are ever dangerous.
For this reason, it is very important to do studies flow, not a single flow, but for many
return periods and many durations, so many data flows.

Floods are one of the fundamental features of the regime of a watercourse.
Unfortunately we do not have a long time series of flood data to draw global
conclusions. Many studies on the genesis and the danger of the flood have been
carried out for a few years in the world [1–7]. The prone Mediterranean countries
which have been exposed to this type of phenomena that we can cite are France,
Spain, and Italy [8], Morocco [9, 10], and Algeria [11].

Flooding is a common environmental hazard and a leading cause of natural
disaster fatalities and economic damages worldwide [12–15].

It has been estimated that floods caused approximately 6.8 million deaths
throughout the twentieth century and more than 500,000 fatalities between 1980
and 2009 [16]. In recent decades, we have also seen a steady increase in frequency
and economic losses that come from flooding events [17, 18].

That raises the question of what factors contribute to the change of flood impacts,
with a possible increase of exogenous flood hazard (e.g., extreme precipitation or
high streamflow events) or increased vulnerability of populations and assets to flood
hazards [17–19].

There are several definitions of the word flood. It refers to an event characterized
by a rapid increase in flow [1, 20, 21]. Several parameters are used to describe a flood
and peak flow. It is the maximum flow rate of the flood for a given period; it is
defined as an instantaneous flow, which is difficult to estimate from one series of data
flow records with long time steps (few hours a day).

The study of the Algerian watercourse floods remains a quasi-unknown field due
to the very limited specific indications about data which are given in the Algerian
hydrological directories [22–25].

In this work, we will study some characteristics of flows in some stations to characterize the phenomenon of floods. We are interested in the annual floods that marked the basin and observed at the different stations between 1960 and 2006, to know their temporal variability and the place occupied by peak flows for average flows at both annual and monthly levels. Finally, this study gives the methods to determine the peak of the flood and to understand the evolution of average and extreme flows according to the rains that are temporally associated with them.

2 Materials and Methods

2.1 Study Area

The Cheliff Basin numbered 01 according to the nomenclature adopted by NAHR (National Agency of Hydraulic Resources) corresponds to an intra-mountain basin. It is located in northern Algeria (see Fig. 1). It is situated between the two Tellian Atlas chains which are parallel to the Mediterranean coast. It lies between meridians $0°$ $12'$ and $3°$ $87'$ and latitudes between $33°$ $91'$ and $36°$ $58'$ North. This basin covers three sub-basins: the Cheliff upstream of Boughezoul, Upper and Middle Cheliff, and Lower Cheliff and Mina. They occupy an area of about 43,750 km^2, covering 77% of the total area of Cheliff-Zahrez watershed basin. Cheliff Basin is limited in the North by the Mediterranean Sea, in the South by the mountains of Ouarsenis, in the East by the Algiers basin, and in the West by the Oranian basin. This basin is crossed by the longest wadi in Algeria, which is Cheliff wadi, with a length of 750 km, in the end, the Mediterranean Sea, is near Mostaganem City.

The Cheliff wadi is a notable exception among the large North African wadis; it is the only one that drains a portion of the highlands and one of those with the longest watercourse and the significant flow. It is formed initially by Nahr Ouassel and Nahr Touil, which takes its initial source from the Atlas Sahara in Djebel Amour near to Aflou, and crosses a distance over 750 km, before joining the Mediterranean Sea, near Mostaganem City.

2.2 Data Collection and Analysis

The data set used in this study is provided by the National Agency for Hydraulics Resources (ANRH) of Blida. It consists of the daily average flows data collected at the gauge stations, based on daily hydrometric data for the available periods. We conducted an inventory of all annual flows (daily and instantaneous peak flow values for the year). The entire time series is represented in Fig. 2.

It was considered more methodologically sound to start this part by studying the flow to show the magnitude of the flood phenomenon in the basin and then to approach the rainfall analysis as an explanatory factor of the flow.

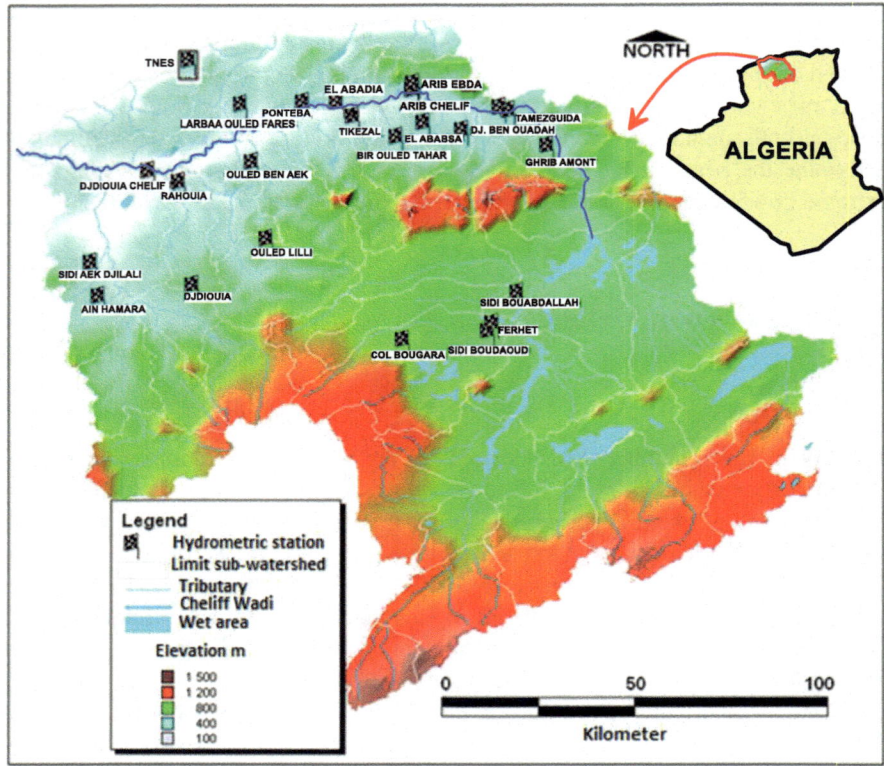

Fig. 1 Elevation map showing locations of hydrometric stations studied at the Cheliff watershed in Algeria

To detect exceptional floods and their ratio with average flows, we are based on the observed annual and monthly values of flow, and the recurrent values of the flows.

3   Results

3.1   Annual Maximum Flows (Variability and Duration of Recurrence)

3.1.1   Ababsa Station (011715)

Interannual Fluctuations

Figure 2 shows the interannual variations in mean annual, maximum daily, and maximum annual flows for Ababsa station. It shows a clear irregularity of the mean annual flow rates.

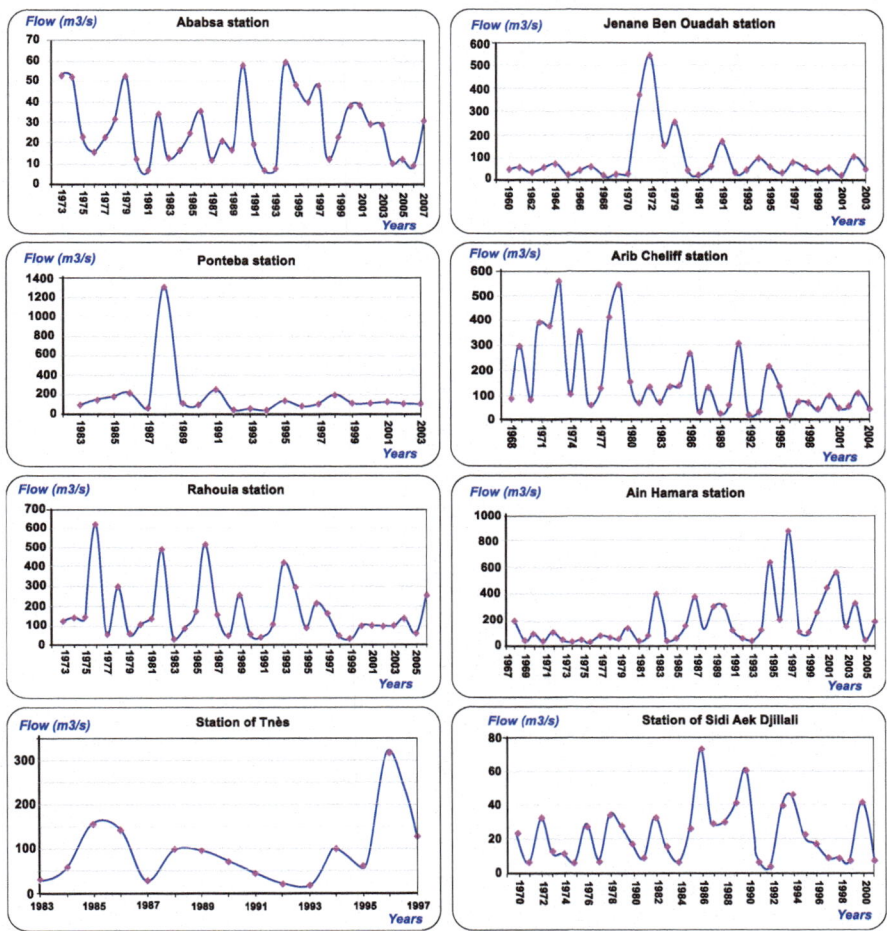

Fig. 2 Variations of annual flows in different stations in Cheliff watershed in Algeria

In the interannual flow of 35 values for the period 1973–2007, only 12 values exceed the average flow (27 m³/s). The year 1994 recorded the highest value (59 m³/s) of average flow, as well as the year 1990 also recorded an important value (58 m³/s). Statistical characteristics of the observation series are given in Table 1.

Recurrence Times

The study of the interannual variability of the normal average flows is always supplemented by the frequential study. It has a goal to estimate the limit values reached or exceeded during a given period; this requires to seek the physical law, which is the most appropriate adjustment to the annual mean flow distribution, and then to estimate the parameters needed (mean and standard deviation), as well as the reduced variable used to calculate the quantiles.

Table 1 Statistical characteristics of different stations in the Cheliff watershed in Algeria (ANRH)

Stations	Code	N	Min	Max	Mean	SD	Median	CV	Skewness	Kurtosis
El Ababsa	011715	35	5	59	27	16	23	0.6	0.55	2
Djenane Ben Ouadah	011514	31	10	543	99.6	129	54	1.3	2.45	6.99
Ponteba	012203	21	18	1,300	160	267	96	1.67	4.26	15.8
Arib Cheliff	011702	37	10	562	154	149	99	0.966	1.37	3.52
Rahouia	012701	34	19	609	160	147	109	0.919	1.66	4.51
Ain Hamara	013302	35	8	878	177	198	109	1.12	2	6.15
Sidi A. Djilali	013401	32	3	73	22.1	17.4	18.5	0.788	1.13	3.6
Tenes	020207	15	18	313	89.3	75.8	70	0.849	1.94	5.27

N number of observations, *SD* standard deviation, *CV* coefficient of variation

To do this, we proceed with the arrangement of flow rates in ascending order (or descending order) by assigning each variable its rank in the normal series, counted from 1, and then calculate the corresponding experimental frequencies according to the formula:

$$f = (i - 0.5)/n.$$

where (*i*: rank, *n*: number of years of observation).

By performing the rate-frequency correlation, it appears that the average annual flows of the Ababsa station (1973/2007) adjust to the log-normal distribution.

The theoretical values estimated by this appropriate adjustment of Ababsa flow rates are acceptable until a return period of 20 years (Fig. 3). Beyond this period, the estimated quantiles are very much exceeded.

3.1.2 Djenane Ben Ouadah Station (Code 011514)

Interannual Fluctuations

In Fig. 2, we note that before the 1970s, the flow rates are very low, but after 1970s we note that the power of the floods becomes very important until 1991 when we noticed the decrease of these flows. We observed an important flow in the years 1973 (543 m^3/s), 1979 (250 m^3/s), and 1991 (180 m^3/s) which took 20 years. After 1991, the floods became less important as it is illustrated in Fig. 2 and the flow rates are in general less than 100 m^3/s, and the very remarkable peaks are observed during the years 1992, 1994, and 2002. It is noted that the station is installed in a river regularized by certain hydraulic works; if not, the floods became three times more than that recorded at this station.

Recurrence Times

Recurrence times are obtained in the same way as the previous station. The results are shown in Table 1.

In Fig. 2, we can see that the extreme flows are greater than 500 m^3/s when the return period is 100 years, whereas the flows are above 300 m^3/s when the return period exceeds 20 years.

3.1.3 Ponteba Station Flows (Code 012203)

Interannual Fluctuations

The variation of the annual flows of the Ponteba station is represented in Fig. 2. According to Fig. 2, we note that the flows become more important from the year

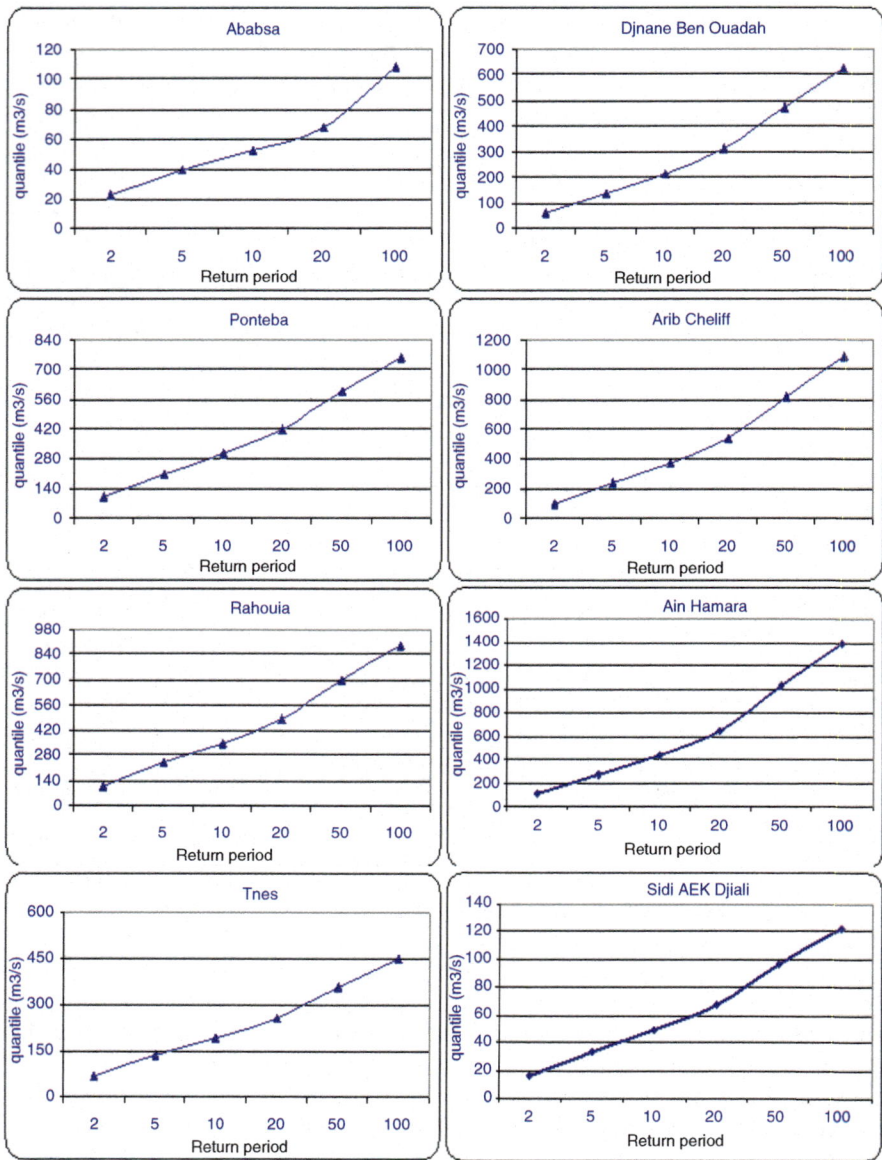

Fig. 3 Theoretical values of the quantiles according to their return period at different stations studied in the Cheliff Watershed

1988, where the maximum value is 1,300 m³/s. After the year 1989, we recorded some remarkable peaks as these in years 1991 (230 m³/s), 1995 (119 m³/s), 1998 (186.75 m³/s), and 2001 (111 m³/s).

Recurrence Times

Figure 3 shows that the extreme values of max flow are greater than 500 m^3/s for return periods 50 and 100 years, while the average quantiles (200 m^3/s) will be recorded every 5 years.

3.1.4   Arib Cheliff Station

Interannual Fluctuations

Figure 2 represents the variation of the annual maximum flow rates of the Arib Cheliff station; the analysis of this figure shows three distinct periods: The first one extended from 1968 to 1980; it is characterized by important flow rates, with a maximum of 562.2 m^3/s which was recorded during the year 1973. The second period was in the period from 1980 to 1986 when the flows become important in months. Then we notice an increase in these flows from 1986 to 1994. The rest of the series is marked by a remarkable decrease of flows and became less than 100 m^3/s.

It is noted that this station provides a good representation of the Upper Cheliff's hydrological regime, as it is installed at the outlet of the basin and on the wadi Cheliff.

Recurrence Times

Frequency analysis of the observation series of Arib Cheliff station (Fig. 3) shows the return periods of extreme flows. It flows above 800 m^3/s in every 50 years, while the important flows more than 1,000 m^3/s have a chance every 100 years.

3.1.5   Rahouia Station (Code 012701)

Interannual Fluctuations

The flood hydrograph in Fig. 2 shows that the highest flow values are observed during the period 1976–1994 with maximum flows of 609 m^3/s, 478 m^3/s, 504 m^3/s, and 404 m^3/s recorded in 1976, 1982, 1986, and 1993, respectively (Table 1). After that 1994, the flows became less important (<250 m^3/s).

Recurrence Times

For the observation series of the Rahouia station (Fig. 3), the frequency analysis makes it possible to exit the return periods of the extreme flows, such as the flow rates higher than 400 m³/s which they are every 20 years, whereas the flows above 800 m³/s have a chance over 100 years.

3.1.6 Ain Hamara Station (013302)

Interannual Fluctuations

From the hydrograph of the floods observed in the Ain Hamara station (Fig. 2), it can be seen that these flows became more important from the year 1993 with a maximum of 878m³/s which was recorded during the year 1998 (Table 1).

Recurrence Times

According to Fig. 3, the flows above 500 m³/s are exceeded every 20 years, while flow rates of 1,000 m³/s can return every 100 years.

3.1.7 Station of Ténès (020207)

Interannual Fluctuations

Flows at the Ténès station became powerful in the years 1995–1996, when the maximum was observed during 1996 with a value of 313 m³/s (see Fig. 2).

Recurrence Times

From this graph (Fig. 3), we note that the quantiles higher than 400 m³/s have the chance to reproduce once every 100 years, while the maximum average value which corresponds to a quantile of 313 m³/s can have a chance on 50 years.

3.1.8 Station of Sidi Aek Djilali (013401)

Interannual Fluctuations

The Sidi Aek Djilali station was characterized by two distinct periods, the first from 1969 to 1984 when flows were very low, but the rest of the series was characterized by a power floods, where we observed flow rates exceeded 50 m³/s, with a maximum of 73 m³/s recorded during the year 1986 (Fig. 2).

Recurrence Times

In this station (Fig. 3), the maximum flow recorded is 30 m^3/s in the return period 5 years and 120 m^3/s for the return period of 100 years.

3.2 Monthly Variations of Maximum Flows

The analysis of the monthly flows makes it possible to highlight the regimes of the rivers and their interannual or interseasonal variations.

To achieve the objective, we will try to highlight the most abundant months in the surface flow by its maximum values (peak flows), in order to understand the maximum flow regime.

3.2.1 Ababsa Station (011715) (Harreza Watershed)

In fact, the wet period from February to May (spring) is the most abundant in maximum average flow, it is followed by another period from September to November (autumn), the average regime of Harreza basin is essentially rain-fed, with a maximum in October 58.82 m^3/s, followed by February (52.5 m^3/s).

3.2.2 Djenane Ben Ouadah Station (011514)

The area controlled by this station is characterized by a very wet period, compared to others, where the flow exceeds 300 m^3/s during January, February, and March. The maximum is recorded during March with a value of 543.24 m^3/s.

3.2.3 Arib Cheliff Station (011702)

According to the graph in Fig. 4, the wettest period is located between September and April, with a maximum of 562.2 m^3/s recorded in March.

3.2.4 Ponteba Station Flows (012203)

This station is characterized by two peaks; the first one characterizes the month of December, with a maximum flow of 1,300 m^3/s, and the second corresponds to March with a flow rate of 823.3 m^3/s.

3.2.5   Rahouia Station (012701)

At this station, the maximum flows are observed in the autumn period, of which the month of September presents the maximum value (609 m^3/s).

3.2.6   Ain Hamara Station (013302)

Figure 4 shows that the wettest month is in August, but this is still an exception, since the flow at this month is not much better than the average, except in the year of 1996 when the value recorded was 878 m^3/s.

3.2.7   Station of Tnès (020207)

In this graph (Fig. 4), it is noted that 5 out of 12 months have flow rates above average; the maximum recorded during April was 313.4 m^3/s.

3.2.8   Station of Sidi AEK Djilali (013401)

At Sidi Aek Djilali station, the maximum flows recorded during July was 35.78 m^3/s and in March was 35.41 m^3/s.

4   Discussions

Furthermore, several studies showed that the flows are low in the Upper plain of Cheliff, due to low rainfall, strong evapotranspiration, and high permeability of lithological formations [26–29]. On the other hand, the flows are relatively high in the Middle and Lower Cheliff, which combine the precipitation and low permeability of geological outcrops.

The flow trends of Cheliff Basin depended on decreases in precipitation and on the increase of temperatures. The current study highlights an irregularity of flow trends throughout the Cheliff Basin. The estimation of extreme quantiles for different return periods should take into consideration the recording period and the right tail of the distribution. The formally gauged record represents a relatively small sample of a much larger population of flood events. Thus, the extrapolation for long return periods is less accurate. In these stations, only the following return periods were considered for the estimation of quantiles: 2, 5, 10, 20, 50, and 100 years.

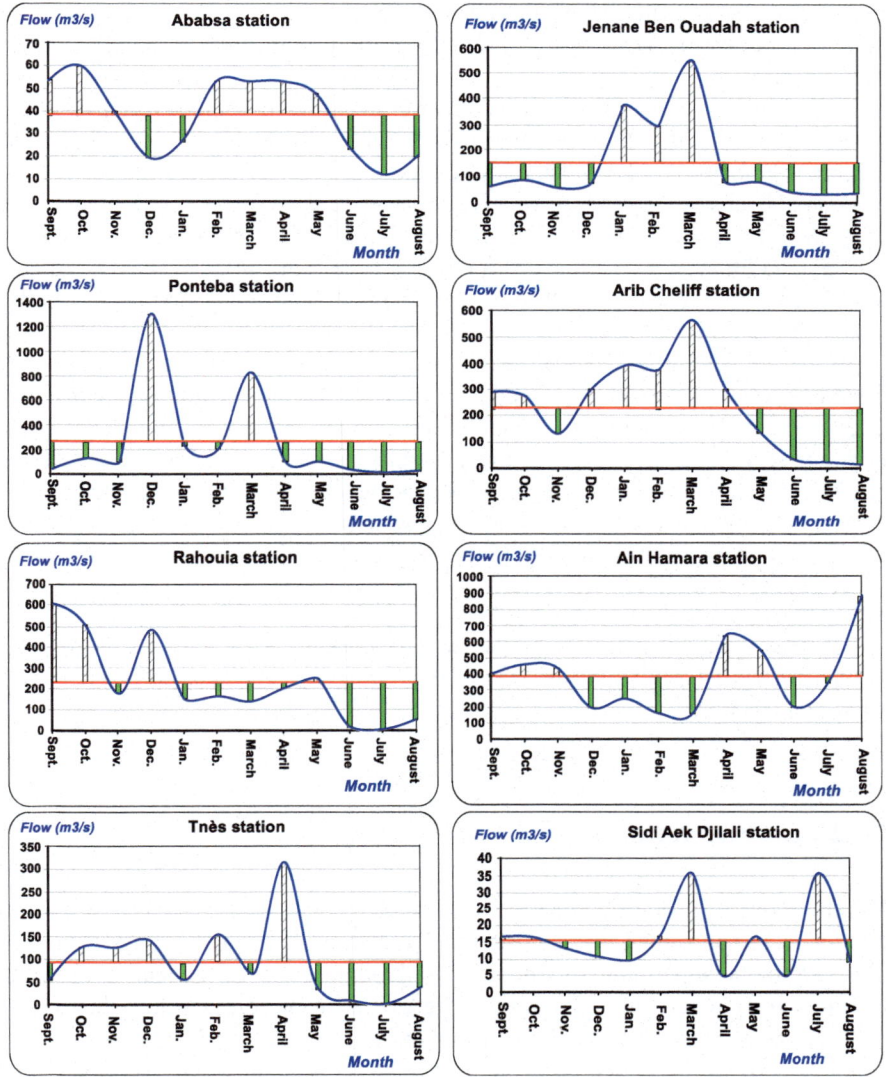

Fig. 4 Monthly fluctuations of maximum flows recorded at different stations across the Cheliff Watershed, Algeria

5 Conclusion

The study of the Algerian Wadis floods remains a quasi-unknown field as only some very specific indications are given in the Algerian hydrological directories. Floods are one of the basic features of a stream regime. The present study, which is the first one carried out in northern oust of Algeria, is based on the flow data recorded at the gauging stations of Cheliff watershed, which are available and considered in the present study.

The statistical study of the recorded flows of the different stations localized in the watershed of Cheliff shows the highest values of flows which were recorded in the station of Ponteba(1,300 m³/s) and in the station of Ain Hamara (878 m³/s).

Also, the most remarkable events were observed during the 1970s for the stations of Arib Cheliff, Rahouia, and Djenane Ben Ouadah, the 1980s for the station of Ponteba, and finally during the 1990s the stations of Ain Amara, Sidi Aek Djilali, and El Ababsa. We report here that the autumn floods characterize the 1990s and the winter floods for the 1980s.

In general, the different hydrometric stations show that there are two distinct periods in terms of duration and flood power. The first is the most important in terms of duration; it characterizes the spring season, with few floods sometimes. The other period is very short. It corresponds to the autumn season. It is characterized by flash floods. The flow is very abundant in March and October in second place. The analysis of rainfall and the monthly flows showed a relatively temporal concordance, where the rainiest months are usually the most abundant in flow.

In this region of Upper and Middle Cheliff, the average maximum flow reaches 400 m³/s, if the rainfall is around 500 mm.

The fact of precipitation falls almost on the whole basin, so the flows are directly influenced by certain parameters like the state of the soil, the vegetation cover, and the air temperature.

6   Recommendations

Flood forecasting is defined as a process of predicting the magnitude and duration of flooding based on known characteristics of a watershed, to prevent damage to human life and to the environment. The use of flood forecasting models is of great importance to watercourse managers. The existence of foreseeable natural risk, over a region, must lead the decision-makers to prohibit or admit, under certain conditions, certain modes of occupation or land use. Also provide, if necessary, the implementation of collective safeguarding and protection measures. The preventive measures to be adopted aim to ensure, on the one hand, better control of the flood hazard, on the other hand, a limitation of the vulnerability of the people, the goods, and the exposed activities.

References

1. Requena AI, Ouarda TB, Chebana F (2017) Flood frequency analysis at ungauged sites based on regionally estimated streamflows. J Hydrometeorol 18(9):2521–2539

2. Zischg AP, Hofer P, Mosimann M, Röthlisberger V, Ramirez JA, Keiler M, Weingartner R (2018) Flood risk (d) evolution: disentangling key drivers of flood risk change with a retro-model experiment. Sci Total Environ 639:195–207
3. Mediero L, Jiménez-Álvarez A, Garrote L (2010) Design flood hydrographs from the relationship between flood peak and volume. Hydrol Earth Syst Sci 14(12):2495–2505
4. Hirpa FA, Pappenberger F, Arnal L, Baugh CA, Cloke HL, Dutra E, Stephens E et al (2018) Global flood forecasting for averting disasters worldwide. In: Global flood hazard: applications in modeling, mapping, and forecasting. Wiley, Hoboken, pp 205–228
5. Vukmirović V, Vukmirović N (2017) Stochastic analysis of flood series. Hydrol Sci J 62 (11):1721–1735. https://doi.org/10.1080/02626667.2017.1342825
6. Guha-Sapir D, Below R, Hoyois P (2016) EM-DAT: the CRED/OFDA international disaster database. Université catholique de Louvain, Brussels. http://www.emdat.be
7. Guha-Sapir D, Hoyois P, Wallemacq P, Below R (2016) Annual disaster statistical review 2016: the numbers and trends. CRED, Brussels, p 91
8. Llasat MC (2004) Recent and historical Mediterranean floods (Spain, France, Italy), consequences-lessons-projects. La Houille Blanche 6:37–41. translated from French
9. Saidi ME, Daoudi L, Aresmouk ME, Fniguire F, Boukrim S (2010) The floods of the wadi Ourika (High Atlas, Morocco): extreme events in a semi-arid mountain context. Comunicações Geologicast 97:113–128. In French
10. Boumenni H, Bachnou A, Alaa NE (2017) The rainfall-runoff model GR4J optimization of parameter by genetic algorithms and Gauss-Newton method: application for the watershed Ourika (High Atlas, Morocco). Arab J Geosci 10(15):343
11. Sadeuk Ben Abbes A, Meddi M (2016) Study of propagation and floods routing in north-western region of Algeria. Int J Hydrol Sci Technol 6(2):118–142
12. Jonkman SN (2005) Global perspectives on losses of human life caused by floods. Nat Hazards 34:151–175
13. Kousky C (2014) Informing climate adaptation: a review of the economic costs of natural disasters. Energy Econ 46:576–592
14. Miller S, Muir-Wood R, Boissonnade A (2008) An exploration of trends in normalized weather-related catastrophe losses. In: Diaz HF, Murnane RJ (eds) Climate extremes and society. Cambridge University Press, Cambridge, pp 225–247
15. Smith K (2013) Environmental hazards: assessing risks and reducing disasters.6th edn. Routledge, New York
16. Doocy S, Daniels A, Murray S, Kirsch TD (2013) The human impact of floods: a historical review of events 1980–2009 and systematic review. PLoS Curr 16:5. https://doi.org/10.1371/currents.dis.f4deb457904936b07c09daa98ee8171a
17. Do HX, Westra S, Leonard M (2017) A global-scale investigation of trends in annual maximum streamflow. J Hydrol 552:28–43
18. Tanoue M, Hirabayashi Y, Ikeuchi H (2016) Global-scale river flood vulnerability in the last 50 years. Sci Rep 6:36021
19. IPCC (2012) Field CB, Barros V, Stocker TF, Qin D, Dokken DJ, Ebi KL, Mastrandrea MD, Mach KJ, Plattner G-K, Allen SK, Tignor M, Midgley PM (eds) Managing the risks of extreme events and disasters to advance climate change adaptation. A special report of working groups I and II of the Intergovernmental panel on climate change. Cambridge University Press, Cambridge
20. CTGREF, Météorologie (1979) Floods and sanitation; 1 to 10 day rainfall analysis on 300 metropolitan stations. CTREF Antony, National Meteorology, Office of Water. (Translated from French)
21. Touazi M, Laborde JP, Bhiry N (2004) Modelling rainfall-discharge at a mean inter-yearly scale in Northern Algeria. J Hydrol 296:179–191
22. Bouasria S, Khalladi M, Khaldi A (2010) Dynamic slowdown of floods in a watershed in western Algeria: case of Oued Mekerra (Sidi Bel Abbes). Eur J Sci Res 43:172–182. In French

23. Medejerab A (2009) The catastrophic floods of October 2008 in Ghardaïa Algeria. Geographia Technica (NS): 311–316 (In French)
24. Korichi K, Hazzab A (2012) Hydrodynamic investigation and numerical simulation of intermittent and ephemeral flows in semi-arid regions: Wadi Mekerra, Algeria. J Hydrol Hydromech 60(2):125–142
25. Maref N, Seddini A (2018) Modeling of flood generation in semi-arid catchment using a spatially distributed model: case of study Wadi Mekerra catchment (Northwest Algeria). Arab J Geosci 11(6):116
26. Theilen-Willige B (2014) Contribution of remote sensing and GIS methods to the detection of areas exposed to flooding in the coastal zones of Northern-Algeria. ICT-DM'2014-Conference, Algiers, Mar 24–25, p 5
27. Meddi M, Talia A (2013) Runoff evolution in Macta basin (Northwest of Algeria). Arab J Geosci 6(1):35–41
28. Meddi M, Sadeuk Ben Abbes A (2014) Statistical analysis and forecast of flood flows in the Oued Mekerra watershed (western Algeria). Revue Nature & Technologie C-Sciences de l'Environnement 10:21–31. In French
29. Ketrouci K, Meddi M, Abdesselam B (2012) Study of extreme floods in Algeria: case of the Tafna watershed. Science et Changements Planétaires/Sécheresse 23(4):297–305. In French

Assessing the Climate Change Impact on Water Resources and Adaptation Strategies in Algerian Cheliff Basin

Yamina Elmeddahi and Ragab Ragab

Contents

Abstract   The effect of climate change on the water resources of the Cheliff basin in Algeria was evaluated with a particular focus on the significant factors affecting the water reserves. The Cheliff basin, which is one of the largest basins in the north of Algeria, is affected by water scarcity due to the extension of industrial and agricultural activities with the population growth, on the one hand, and to a decline in water resources caused by extreme droughts, on the other hand. The results of the current climate change assessment revealed a downward trend in the precipitation ranging from

Y. Elmeddahi (✉)
Vegetable Chemistry-Water-Energy Research Laboratory, Faculty of Civil Engineering and Architecture, University Hassiba Ben Bouali of Chlef, Ouled Fares, Algeria
e-mail: y.elmeddah@univ-chlef.dz; elmeddahi-a@hotmail.fr

R. Ragab
Centre of Ecology and Hydrology, CEH, Wallingford, UK

Abdelazim M. Negm, Abdelkader Bouderbala, Haroun Chenchouni, and
Damià Barceló (eds.), *Water Resources in Algeria - Part I: Assessment of Surface and Groundwater Resources*, Hdb Env Chem (2020) 97: 111–134,
DOI 10.1007/698_2019_398, © Springer Nature Switzerland AG 2019,
Published online: 28 September 2019

14 to 54% and a reduction in streamflows that exceeds 40% with a break observed at the beginning of the 1980s. According to different emission scenarios, several general circulation models (GCMs) predict an increase in temperature of +0.9°C to +5°C on average at the end of the twenty-first century, with a decrease in average rainfall of 10–30%. A conceptual model predicted a flow deficit ranging from 10 to 48% at different periods and in different scenarios.

This study found that the problem of water scarcity was exacerbated by poor management of available water resources and by the significant increase in the population, which exceeded five million in 2010. Intensive use of water for irrigation and economic development has put additional pressure on the limited water resources. All these facts call for a proper "fit for purpose" integrated water management policy for the whole country.

Keywords Cheliff basin, Climate change, Trend, Water resource

1 Introduction

Over the past three decades of the twentieth century, many scientists have argued that climate change could lead to major changes threatening the very existence of humans on the planet.

Recently, considerable research has been led to address the impact of climate change on the environment and particularly on water resources [1–10]. During the last century, the average temperature across the Mediterranean basin and the North Africa region has increased by 0.56–0.92°C [9–17]. According to different emission scenarios [4, 14, 17–22], several general circulation models (GCM's) predict an increase in temperature from 0.7 to more than 4°C by the end of the twenty-first century. Algeria, in particular, has experienced a decrease in average annual precipitation. Rainfall is predicted to further reduce during the coming century [9, 14, 23–25].

This situation is particularly noticeable in regions subject to a semiarid climate regime. This is particularly the case in the Cheliff basin, which is one of the largest basins in northern Algeria. It is affected by water shortage due to the expansion of industrial and agricultural activities with population growth, on the one hand, and the reduction of water resources caused by extreme droughts, on the other hand. Recent studies over the past few years have revealed a declining rainfall trend in most Algerian regions [17, 26, 27] and a significant decrease in flow to dams and significant groundwater level drop and high vulnerability to groundwater pollution [28].

Precipitation is the fundamental driver of the Earth's hydrological system. As a result, any change in intensity, frequency, and timing of precipitation will have a direct impact on water resources and therefore on economic development. The severe droughts of the early 1980s can testify to the magnitude of the effects that climate extremes can have on the national economy. The floods in recent decades in the north give a further idea of the damage that can be caused.

At present, the major concern of the country is to predict, with scientifically accepted margins of uncertainty, the potential impacts of climate change predicted by the IPCC on water resources.

In this context, it is necessary to put in place sound adaptation strategies aimed at minimizing the negative impacts that climate change would bring in the future. Decision-makers thus need objective assessments of the vulnerability of different socioeconomic sectors through the integration of climate information at the national and local levels. This information concerns, in particular, observed trends and projected future scenarios. This chapter aims to assess the direct and indirect impacts of climate change on water resources by identifying major trends in precipitation over time, annual flows, and the significant contribution of the dam in the study area.

2 Materials and Methods

2.1 Description of Case Study Area

The Cheliff basin, one of largest basins in Algeria, is located in the northwest of Algeria and lies between 34° and 36° N in latitude and between 0° 12′ and 3° 87′ E in longitude. The watershed covers an area of 43,750 km^2, being bordered to the north by the coastal-Dahra basin, in the south by the Zahrez basin, in the east by the Algiers basin, and in the west by the Oran basin. It consists of three subregions, the upstream Boughezoul basin, the Upper and Middle Cheliff basin, and the Lower Cheliff basin and Mina. The study area is characterized by the heterogeneity of large natural units, to which is added a significant latitudinal extension and diversified geography (such as plains, mountains, and Tellian plateaus). The basin is 20–70 km from the Mediterranean Sea. All these factors determine the region's climate, which ranges from semi-humid to semiarid. The climate of the case study region itself is the semiarid Mediterranean with warm summers and cold winters. The rainfall has a wide interval of variability with a trend of decline from north to south and from east to west. The mean annual rainfall ranges from 300 mm in the high plains to 600 mm in the coastal watershed, except in the Zaccar massif where about 800 mm have been recorded.

Average annual temperatures decrease gradually from north to south with a minimum 14.2°C and a maximum 18.7°C. Average monthly temperatures follow the same pattern, but the decline is faster in the cold season than in the hot season, because of the particularly harsh effect of continentally winter and the regulating influence of the sea in summer. The hot season, months during which average monthly temperatures are higher than the annual average, extends from May to October, while the cold season lasts from November to April. The maximum temperatures of 27–28°C are reached in August or July, and the minimum temperatures are in January and February (3–10°C).

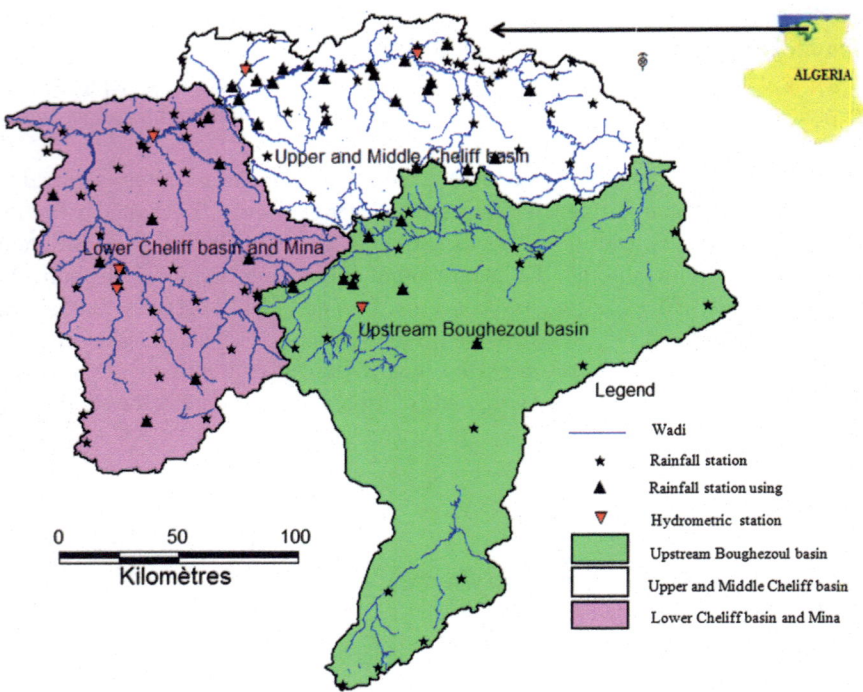

Fig. 1 Location of the study area including the sub-basins

The Cheliff Wadi is 750 km long and is the longest "river" in Algeria. Wadi refers to a dry (short-lived) riverbed that only contains water during periods of heavy rain. The Cheliff Wadi source is in the Saharan Atlas, with 70% of its annual runoff occurring from December to April. The large variability of hydrological regimes can be observed in the large range of variation of mean annual flow which varies from 0.06 to 15.36 m^3 s^{-1}. Extreme values of the flows are of great interest to hydrologists as they constitute a major concern due to their scale, aggressiveness, and negative impact on societies and economic and social development.

Hydrologic data monitoring stations of the three sub-basins of the Cheliff bowl are illustrated in Fig. 1. Monthly rainfall data and monthly flow data over the study area were obtained from the National Meteorological Office (NMO) and the National Agency of Hydraulic Resources (NAHR).

In this study, a total of 42 hydro-climatic variables were selected for the study. The stations considered are those with a long series of observation and few gaps. Data of surface water was obtained from the Ministry of Water Resources (MWE).

2.2 Methodology

The assessment of the rainfall variability was performed using the nonparametric test of Mann-Kendall (MK). This test is to confirm whether there is a positive or negative trend for a certain confidence level. The magnitude of linear tendency was calculated using Sen's estimator. The step change (break date) is detected by the Pettitt test. GR2M conceptual model has been applied to simulate the monthly streamflow. The simulation was performed with seasonal and annual rainfall data recorded at 36 stations covering a period of 38–105 years, while seasonal and annual flow data of six gauging stations covered a period of 26–46 years.

2.2.1 Mann-Kendall Test

The nonparametric test of Mann-Kendall [29, 30] was used for determining monotonic trends in hydrometeorological and other non-normal distribution series [31–33]. This test allows for testing of the correlation between the ranks of a time series and their time sequence [34]. It measures the degree of significance of the trend at a level of 0.05.

Let $(x_1 \ldots x_n)$ be a sample of independent values from a random variable X of which is to assess the stationarity. The null hypothesis H_0 is the hypothesis of stationarity of the series (no trend). The alternative hypothesis H_1 corresponds to the non-stationarity of the series. The Mann-Kendall S statistic is defined as:

$$S = \sum_{i=1}^{n-1} \sum_{k=i+1}^{n} \text{sgn}\,(X_k - X_i) \tag{1}$$

with:

$$\text{sgn}\,(X_k - X_i) = \begin{cases} +1 \text{ si } (X_k - X_i) > 0 \\ 0 \text{ si } (X_k - X_i) = 0 \\ -1 \text{ si } (X_k - X_i) < 0 \end{cases} \tag{2}$$

where X_k and X_i are of the time series and n is the length of the data sequence. The variance of S and test statistic Z is given by:

$$\text{Var}(S) = \frac{n\,(n-1)(2n+5) - \sum\limits_{p=1}^{q} t_p\bigl(t_p - 1\bigr)\,\bigl(2t_p + 5\bigr)}{18} \tag{3}$$

$$Z = \begin{cases} \dfrac{(S-1)}{\sqrt{\mathrm{Var}(S)}} \text{ si } S > 0 \\ \qquad 0 \text{ si } S = 0 \\ \dfrac{(S+1)}{\sqrt{\mathrm{Var}(S)}} \text{ si } S < 0 \end{cases} \qquad (4)$$

where q is the number of tied groups and t_p is the size of the pth tied group.

The null hypothesis is accepted or rejected at α depending on whether $\alpha_1 > \alpha$ or $\alpha_1 < \alpha$. Generally, the 0.05 level is largely used. In this study, the analysis of precipitation variability, 0.05 and 0.01 levels, was employed. When the statistical value of z is positive, the trend is increasing, while a negative value indicates a declining trend.

2.2.2   Sen's Median Slope Estimator

When the hypothesis of no trend is rejected by Mann-Kendall test, the Theil-Sen's slope [35] allows determining the magnitude of the linear trend. The slope is estimated for N pairs of data points, Q, as:

$$Q_i = \frac{X_j - X_k}{J - k}, \quad \text{For } i = 1 \dots n \qquad (5)$$

where X_j and X_k are the data values at time j and k, respectively, with $j > k$. The median of the slope between all pairs of data of Q slope estimates is Sen's linear slope.

2.2.3   Pettitt Test

The Pettitt [36] approach derived from the Mann-Whitney test makes it possible to test the significance of the breaks in the rainfall and hydrometric series considered from a null hypothesis [36]. This test is known for its robustness [37].

The implementation of the test requires that for any time t ranging from $1 - n$, the series (x_i), $i = 1 - t$ and $t + 1 - n$ belong to the same population. The statistic test is the maximum absolute value of the variable $U_{t,n}$ defined by:

$$U_{t,n} = \sum_{i=1}^{t} \sum_{j=t+1}^{n} D_{ij} \qquad (6)$$

where

$$D_{i,j} = \text{sgn}\left(x_i - x_j\right) = \begin{cases} 1 \text{ si } x_i > x_j \\ 0 \text{ si } x_i = x_j \\ -1 \text{ si } x_i < x_j \end{cases} \quad (7)$$

Pettitt proposes to test the null hypothesis using the K_n statistic defined by the absolute maximum of $U_{(t,n)}$. From the rank theory, it shows that if k denotes the value of K_n taken on the studied series, under the null hypothesis, the probability of exceeding the value k is given approximately by the relation (8) [37–39]:

$$\text{Prob}(K_n > k) \approx 2 \exp\left[\frac{-6k^2}{(n^3 + n^2)}\right] \quad (8)$$

On the other hand, for a given risk α of the first kind, if the estimated probability of exceedance is less than α, the null hypothesis, H_0, is rejected, and the estimated value of the break date is given by the time t defining the maximum absolute value of the variable $U_{t,n}$.

2.2.4   Conceptual Model GR2M

The GR2M model consists of a production reservoir that governs the production function, characterized by its maximum capacity, and a reservoir (gravity water) that governs the transfer function [40, 41]. This monthly model of water balance was based on two main optimizable parameters: $X1$ (mm), which represents the maximum capacity of the soil moisture reservoir, and $X2$ ($0 < X2 < 1$), the underground exchange coefficient. Rainfall (P) and potential evaporation (E), which is calculated by the Thornthwaite method, are both the main input parameters of the model which are modulated in the same portion. The outputs' results represent the streamflow in mm.

3   Results and Discussion

3.1   Precipitation

3.1.1   Spatiotemporal Variation in Rainfall

Rainfall distribution is heterogeneous across the Cheliff basin and is characterized by a decreasing average annual rainfall from north to south, with a change in altitude of Tellian chains that show the important role of the altitude. Rainfall varies from 778.8 mm in the north to 249.1 mm in the south. Calculating the standard deviation and coefficient of variation (CV %) for each station shows that for all selected stations the annual coefficient of variation is between 17.0 and 60.3%. The coefficient of variation of annual rainfall is generally increasing from north to south of the study area. This spatial variability is aggravated by heavy rains in association with northwesterly directional winds for the northwest and southwest regions. The variability is also due to the heavy winter and spring rains that affect the mountainous regions where the altitude exceeds 1,000 m as is the case of the Dahra mountains compared to the Lower Chellif plain and the western part of the basin (the basin of Mina).

Precipitation rarely exceeds 250 mm/year in upstream Boughezoul basin. This region is particularly semiarid to arid, and the effect of the latitude is very important.

3.1.2   Evolution of Rainfall

As reflected in Fig. 2, the representative stations of three characteristic regions of the study area show an excess of rainfall until the late 1970s and early 1980s. However, downward episodes are recorded during this period not exceeding 3 consecutive years (periods from 1930 to 1933, 1941 to 1944, and 1968 to 1970).

From the 1980s to the present day, this region experienced one of the most rain-deficit periods, both in intensity and persistence.

3.1.3   Annual Precipitation Trends

The results of the statistical analysis obtained are presented in Table 1. The analysis of trend by MK test shows that annual rainfall is trending downward across the basin.

Some of these trends are statistically significant. Several of the trends are significant at the 5% level, and two of the trends are significant at the 10% level. The value of the Sen's slope at the 0.05 level of significance reflects the magnitude of this trend. The decrease in annual rainfall varies from -0.75 mm/year in the southeast to -4.58 mm/year in the west with significant decreasing trends in the annual mean precipitation z-value: 5.22. The season's rainfall pattern and trends were very similar to the annual ones.

Significant decreasing trends were observed in seasonal average rainfall in winter for the southeast (at z-value -1.78 and -0.58 mm/year) to the west (at z-value -4.77 and -2.18 mm/year). The significant decreasing trends of average seasonal rainfall had the highest value of decreasing slope (i.e., -3.34 mm/year) in the west of the study area for Oued Lili station in spring.

However, the results of the two tests (Spearman and Mann-Kendall) applied to the series of annual rainfall in western Algeria, including some stations in our study area, revealed a downward trend in most of the studied stations [42], which is in agreement with the results obtained in this study. The results of a study conducted [26] in the northwest of Algeria showed a general downward trend in annual rainfall, which is consistent with the one obtained in this study.

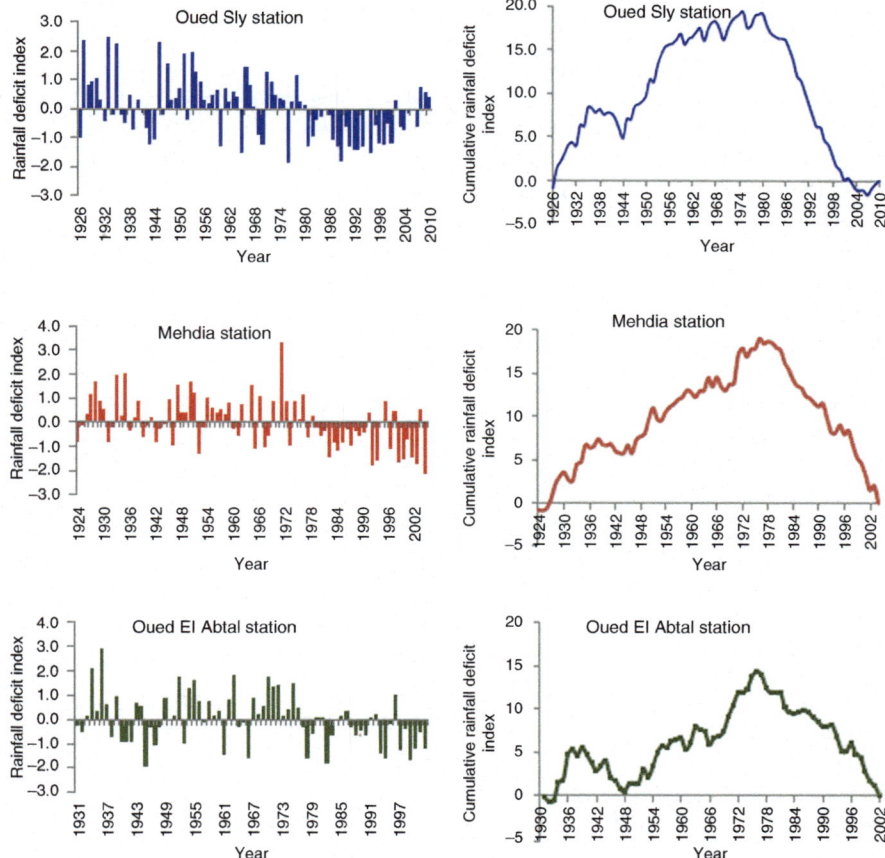

Fig. 2 Evolution of the rainfall deficit and accumulated deficit in the three stations of the study area

Analysis of precipitation series showed the importance of identifying local trends that differ from national or global trends. The results indicate that for the periods analyzed, there is a decrease in rainfall across the basin studied. Significant trends are appearing in the west and southeast parts of the region.

3.1.4 Break Detection

The analysis of long series makes it possible to detect possible changes in the rainfall regime. In the Cheliff basin, several significant trends have been identified for rainfall parameters. Several studies [17, 27, 43, 44] indicated a break during the 1970s for almost all the stations studied. The results obtained are presented in Fig. 3.

Table 1 Statistic Z of Mann-Kendall test for mean annual rainfall, winter (DJF) mean rainfall, spring (MAM) rainfall, and Sen's slope

Station	Years of record	MK Z-value (p ann)	Sen's slope (mm/year)	Z-value (DJF)	Sen's slope (mm/year) (DJF)	Z-value (MAM)	Sen's slope (mm/year) (MAM)
ONM Chlef	75	−1.35	−0.78	−1.49	−0.58	+0.09	+0.02
Chlef DDA	43	−1.08	−1.88	−0.87	−0.95	−0.72	−0.48
Essoula	43	−1.33	−1.46	−0.85	−0.58	−1.46	−0.63
O. Fares	43	−1.50	−2.86	−1.37	−1.39	−1.83^a	−1.30
Ponteba Bge	44	−1.60	−2.64	−1.60	−1.40	−1.57	−1.20
El Abadia	43	−1.79^a	−2.90	−0.91	−0.96	−1.96^a	−1.34
Khmesti	98	−1.04	−0.75	−1.78^a	−0.58	−1.60	−0.37
Merdja Amel	41	−1.57	−2.20	−2.11^b	−1.51	−1.48	−0.86
Rouina Mines	43	−1.60	−1.88	−1.06	−0.79	−1.98^b	−1.11
Arib Ebda	43	−1.77^a	−2.98	−0.70	−0.90	−1.77^a	−1.51
Harreza	43	−1.23	−1.93	−1.14	−1.01	−1.33	−0.93
O. Lili	38	−2.50^b	−3.76	−2.50^b	−2.02	−3.36^b	−3.34
Sidi Medjahed	95	−4.42^b	−3.62	−3.43^b	−1.72	−2.31^b	−0.83
El Ababsa	43	−1.37	−1.64	−0.95	−0.91	−1.64	−0.88
Ain Defla	101	−2.62^b	−1.26	−1.33	−0.37	−1.08	−0.22
Ain Boucif	84	−3.76^b	−2.97	−2.43^b	−0.99	−3.90^b	−1.12
Ammi Moussa	90	−2.73^b	−1.34	−2.25^b	−0.61	−0.85	−0.20
Derrag	93	−2.10^b	−1.42	−1.33	−0.51	−0.88	−0.31
Fodda Bge	68	−3.40^b	−2.65	−2.66^b	−1.10	−0.95	−0.34
Frenda	75	−2.91^b	−1.87	−2.58^b	−0.91	−1.89^a	−0.31
Grib Bge	66	−3.31^b	−2.53	−2.24^b	−1.21	−1.84^a	−0.93
Mehdia	95	−3.81^b	−1.95	−3.51^b	−1.05	−2.39^b	−0.54
O. Sly	85	−3.54^b	−1.76	−2.15^b	0.66	−1.38	−0.32
Rosfa	54	−3.93^b	−3.39	−3.14^b	−1.56	−3.05^b	−1.07
O. El Abtal	72	−2.44^b	−1.38	−4.14^b	−1.26	+0.66	+0.18
Sidi AEK Djillali	38	−3.32^b	−2.12	−0.72	−0.37	−2.76^b	−1.23
Ain Amara	38	−2.45^b	−1.35	−1.12	−0.75	−3.06^b	−1.88
Rouina Mairie	85	−2.78^b	−1.30	−2.06^b	−0.61	−1.21	−0.26
B. Amir AEK	85	−2.72^b	−1.55	−1.03	−0.43	−1.40	−0.46
Tissemsilt	89	−1.97^b	0.79	−1.72^a	−0.43	−0.65	−0.16
El Touaibia	43	−1.27	−1.70	−0.58	−0.43	−1.75^a	−0.88
Kenenda Farm	79	−5.22^b	−4.58	−4.77^b	−2.18	−3.00^b	−1.21

(continued)

Table 1 (continued)

Station	Years of record	MK Z-value (p ann)	Sen's slope (mm/year)	Z-value (DJF)	Sen's slope (mm/year) (DJF)	Z-value (MAM)	Sen's slope (mm/year) (MAM)
ThneitEl Had	106	−1.76[a]	−0.79	−0.79	−0.24	−1.47	−0.37
Rechagha	74	−4.07[b]	−1.99	−2.97	−0.61	−2.23[b]	−0.52
Ksar Chellala	100	−3.43[b]	−1.03	−4.10[b]	−0.46	−1.89[a]	−0.35
Dahmoni Trumulet	79	−5.07[b]	−3.12	−4.21[b]	−1.62	−2.95[b]	−0.91

[a]Trend statistically significant at 10%
[b]Trend statistically significant at 5%

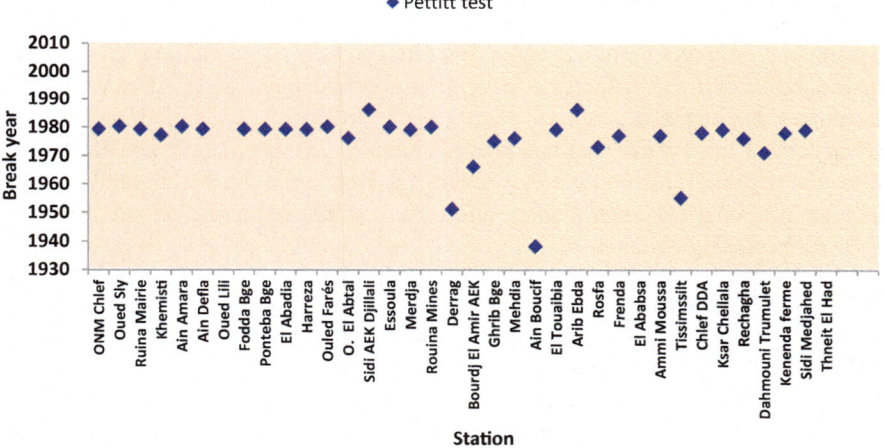

Fig. 3 Pettitt test in the study area

The analysis of these results shows that the majority of the series analyzed show breaks. In 73% of the stations, there is a break between 1970 and 1980. From one break to another, the average of rainfall decreases. These observations are consistent with the results of a number of studies, including [45], which are, for northern Algeria, most of the ruptures are between the beginning of the 1970s and end of the 1980s.

The study of the interannual variability of rainfall then allows to better quantify the deficit in the period after rupture compared to before rupture as shown in Fig. 4. Maximum reductions of around 54.8% were recorded at the Rosfa station in the south, and a minimum reduction rate of 14% was observed at the Theniet El Had station in the east.

A significant reduction, over 30%, was recorded in the west and south of the study area; the most of this region has less rainfall. Over the study's period, this

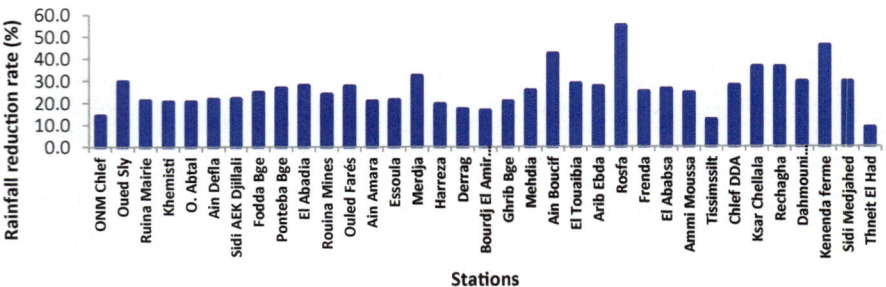

Fig. 4 Rainfall reduction rate in the study area

decline averages around 26% nationally; the end of the rainy season (February–April) shows more pronounced trends. The beginning of the rainy season shows, on the contrary, an upward trend but which remains low and statistically insignificant. Interannual rainfall variability increases as it approaches arid regions. The increase in variability follows the increase in longitude and the decrease in latitude. Altitude attenuates this increase.

Generally, there is a decrease in precipitation and a greater occurrence of droughts in recent decades. This break, in the sense of a decrease in the annual rainfall, gives an idea for thought to better manage an ever-dwindling water resource in the face of an ever-increasing demand.

3.2 Hydrometric Network and Presentation of Hydrometric Stations Selected

The Cheliff basin has few stations with the capacity to do hydrometric measurements. The archives reveal that the network is old; the first gauging station in the basin was established in 1925, but the number of hydrometric stations available and likely to provide sufficient good quality information for water resource management is insufficient. Cheliff basin, with an area of 43,750 km², has 43 gauging stations of which only 24 are operating hydrometric stations; this means 1 station for every 1,560 km².

For this study, six gauging stations were selected for having data over long periods with few data gaps.

3.2.1 Study of the Variability Hydrometric

The study of the hydrometric series conducted over a long period allows the assessment of the sensitivity of river flows to climate change. The average interannual discharge of the basin (Fig. 5) varies between 0 and 60.19 m³ s⁻¹ with

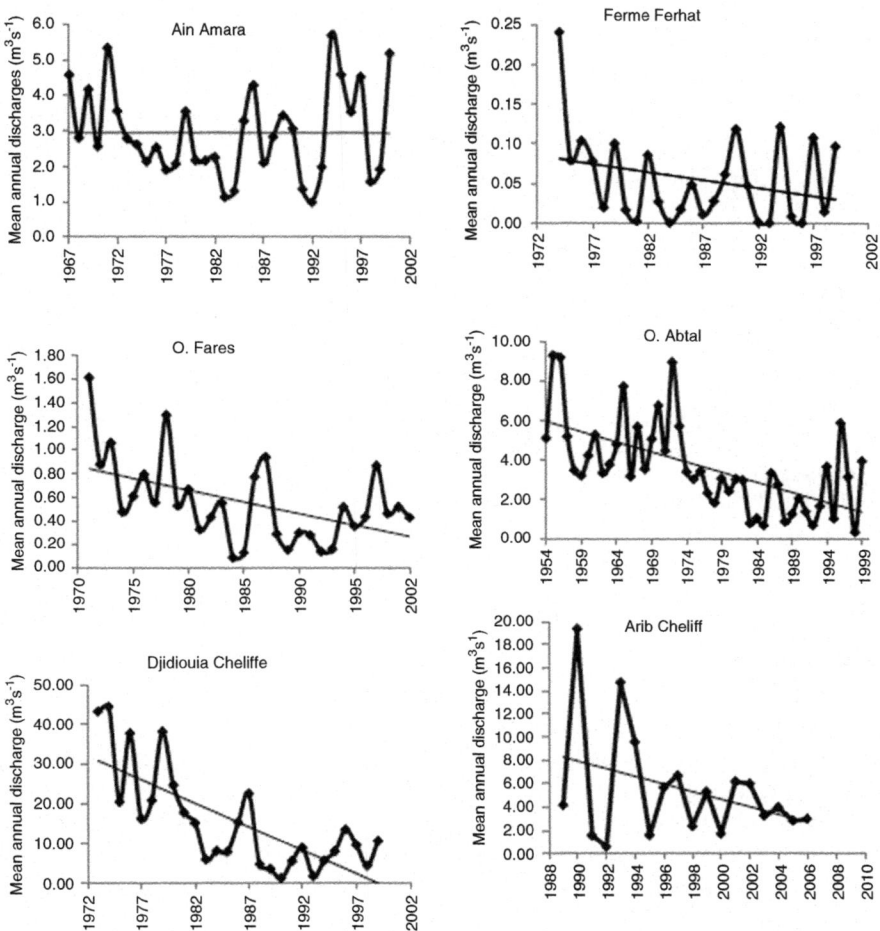

Fig. 5 Evolution of interannual reduced centered index discharges of six stations in the study area

a maximum of 952.20 m^3 s^{-1} in the eastern region and 0–1.15 m^3 s^{-1} with a maximum of 334.68 m^3 s^{-1} in the south of the region.

The high flows observed in the western part of the study area (the Ain Amara and O. El Abtal stations) are related to the importance of the Oued Mina inflow and the special morphometric characters. In this basin, altitude exceeds 1,200 m, which favors the generation of surface runoff. This gradual increase in inflow from east to west is consistent with climatic and physiographic data from the upstream Boughezoul Basin to Upper and Middle and Lower Cheliff to Mina. Flows are low in the high plains, due to low rainfall, high evapotranspiration, and high permeability of lithologic formations. On the other hand, they are relatively high in the Upper and Middle Cheliff, which combine abundant average precipitation and low permeability of geological outcrops. This suggests that the altitude parameter is an important factor.

Table 2 Statistic Z of Mann-Kendall test for mean annual streamflow, seasonal streamflow, and Sen's slope

Station	Z-value					Sen's slope
	ANN	Autumn (Sep–Nov)	Winter (Dec–Feb)	Spring (Mar–May)	Summer (Jun–Aug)	(Qann) $m^3s^{-1}y^{-1}$
O. El Abtal	-4.54^a	-2.94^a	-4.34^a	-3.32^a	-4.17^a	-0.103
Ain Amara	-0.67	$+0.25$	-2.92^a	-1.80^b	-3.48^a	-0.015
Arib Cheliff	-1.68	-3.07^a	-0.73	-0.99	-0.86	-0.32
Ferhat Farm	-1.30	$+0.49$	-3.76^a	-0.02^a	-0.02	-0.002
Djidiouia Cheliff	-3.61^a	-2.32^a	-3.44^a	-2.98^a	-2.44^a	-1.05
O. Fares	-2.94^a	2.23^a	-2.20^a	-3.02^a	-3.95^a	-0.03

[a]Trend statistically significant at 5%
[b]Trend statistically significant at 10%

3.3   Annual and Seasonal Flow Trends

Table 2 shows significant annual downward trends, which were observed in the western region (O. El Abtal station, Djidiouia Cheliff), with a maximum of -4.54 (O. El Abtal). Significant decreases in annual flow indicate a greater decreasing slope, respectively, for O. El Abtal and Djidiouia Cheliff (i.e., -0.103 and -1.18 m^3 s $^{-1}$). Same as the seasonal scale, a significant downward trend was observed in the west and south of the study area, with maximum z-values of -4.34 in winter (O. El Abtal station) and -3.42, -3.44, and -3.76 in summer and winter in Ain Amara, Djidiouia Cheliff, and Ferhat Farm, respectively.

3.4   Surface Water Resources

To meet a growing demand for water, the surface water storage capacity has been increased with the construction of new dams, going from 12 dams in 2000 to 15 dams at present with a capacity of 2.205 billion m^3 according to the Ministry of Water Resources [46].

There are 145 small dams in the Cheliff basin, but only 27 are in operation, and 118 are completely silted [47]. The available water resource corresponding to the capacity of the service reservoirs is 16 Hm^3 entirely intended for irrigation. There are seven canals and diversions from the dams in the study area. The derived volume is 13.4 Hm^3/year. Run-of-river sampling is estimated at an average of 57 Hm^3/year [46].

3.5   Impact of Climate Change on Surface Water Resources

The twentieth-century global warming in the study area has a negative impact on the precipitation cycle including all water resources. From 1968 to 2001, the interannual inflow of Cheliff basin was estimated at 1,025 Mm^3, consisting of 687 Mm^3 for Middle and Upper Cheliff, 242 Mm^3 for Lower Cheliff and Mina, and 96 Mm^3 in upstream Boughezoul. In 2009, precipitation decreased to less than 350 mm, and inflow decreased to 815 million m^3, so a rate of 20%. This decrease has resulted in a decreasing trend in runoff and consequently a reduction in dam capacity. To illustrate the decrease in storage of dams, three dams were chosen in the study area, Sidi Yakoub dam and Boughezoul dam located in the Middle and Upper Cheliff and Sidi M'Hamed Benaouda dam located in the Mina region (Figs. 6, 7, and 8).

The condition of the Sidi Yakoub dam filling ranged between 82% in 1994, 14% in 1998, and 65% in 2004. The contribution of runoff to surface water has consistently decreased due to the significant reduction in precipitation.

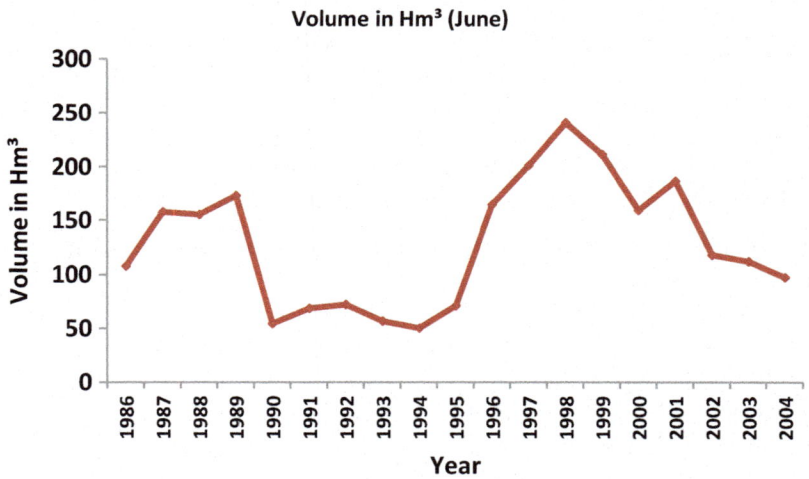

Fig. 6 Volume evolution in Sidi Yakoub dam (1986–2004)

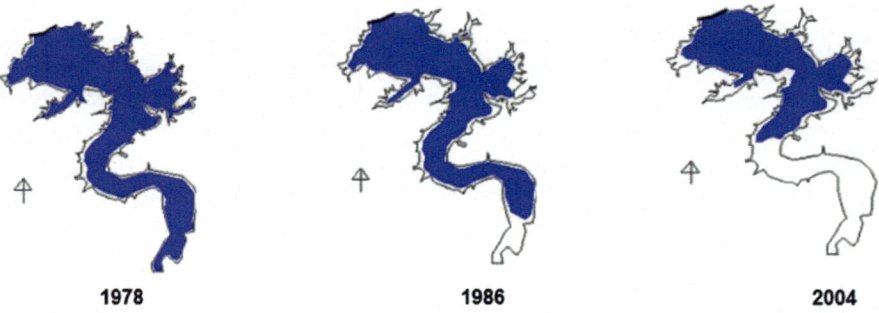

Fig. 7 Evolution of the capacity of Sidi M'Hamed Benaouda dam (Remini and Bensafia, 2011 in [48])

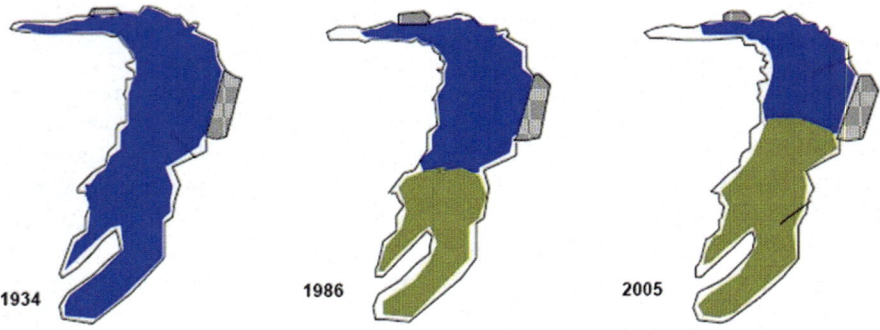

Fig. 8 Evolution of the capacity of Boughezoul dam [49]

Figures 7 and 8 shows the evolution of the capacity of the Sidi M'Hamed Benaouda dam and Boughezoul dam.

The reservoir of the Sidi M'Hamed Benaouda Dam had an initial total volume of 241 Hm^3, but this value was reduced over time to 226 Hm^3 in 1986 and 153 Hm^3 in 2004. The storage capacity of the reservoir is reduced by order of 4 Hm^3/year due to siltation. When it was built in 1934, the Boughezoul dam had a capacity of 55 million m^3. Its capacity decreased to 33 million m^3 in 1986 and 20 million m^3 in 2005. This development indicates the consequences of the decrease in precipitation and the severity of the siltation, which is mainly due to upstream erosion.

Other consequences of decreasing precipitation could be:

- Siltation of dams (concentrated precipitation and siltation accelerated by increased erosion)
- Unsteady river regime
- A decrease in the piezometric levels, inducing a decrease in the outflows of the natural outlets of groundwater
- The degradation of water quality
- Reduction in crop yields
- Increased water requirements for irrigated crops

It was observed that the problem of water scarcity was exacerbated by mismanagement of available water resources and a significant increase in the population that has exceeded five million in 2010 [44]. Intensive use of surface water for irrigation and water use for economic development has put additional pressure on limited water resources. Vital capacities are lost in urban water distribution systems, and networks of irrigation are dilapidated or badly maintained. All these failures reflect an inadequate mastery in the management of water resources policy of the country. Better water demand management that would control, reduce, and adjust consumption, while avoiding water losses and waste, is imperative [50].

3.6   Water Resources Management

Since independence 1962, the mobilization of water resources has been focused on groundwater resources. The rapid increase in water demand in the irrigation and industry sectors, as well as the progressive needs of the population, has led the public authorities to mobilize more and more the superficial resources. Thus, the efforts undertaken during the current decade have led to significant improvements.

Faced with these challenges, the country has introduced a new policy to more efficiently manage water resources via the building new dams. There are now 15 dams in the study area having a total capacity of 2.205 million m^3 according to the Ministry of Water Resources [46]. Other projects are in progress (i.e., five new dams) which will provide an additional volume of 1,608 Mm^3/year intended for the irrigation and domestic water supply. The use of nonconventional water resources is also being considered with the construction of small seawater desalinization plants in the year 2001. Two large stations with an estimated production capacity of 109.5 Hm^3 are also being built. Studies have also been launched on the reuse of treated wastewater for agricultural use.

3.7   Projected Climate Change and Adaptation on Surface Water Resources

Several sources can be used to obtain a consensual image of climate projections for Algeria and the study area at various time horizons. Climate change in Algeria was analyzed based on the UKHI and ECHAM3TR GCMs and the IPCC GIEC IS92a scenario. Other work has been done by [50] to assess the impacts of climate change on water resources, using HadCM3 low-resolution general circulation model (GCM) scenarios [51]. This involves the development of local-scale climate change scenarios using the statistical downscaling method performed by the LARS-WG stochastic generator [52, 53].

The seasonal and annual temperature changes simulated by the UKH1 and ECHAM3TR model for Algeria are qualitatively similar to those obtained by the HadCM3 model, although a little less important in terms of intensity. Projections by UKH1 model for the different horizons on the Cheliff basin would indicate a warming of the order of 0.6–1.1°C in 2020 and of 0.95–2.2°C for 2050. Similar trends were observed for the two scenarios HadCM3-A2 and HadCM3-B2. Overall, the simulations of temperature changes by different global and regional climate models are extremely homogeneous and still have the same order of magnitude of +2°C to +5°C on average at the end of the twenty-first century.

For precipitation (Fig. 9), the annual changes simulated by the UKH1 model show the decrease in rainier areas and the increase in less rainy areas compared to the

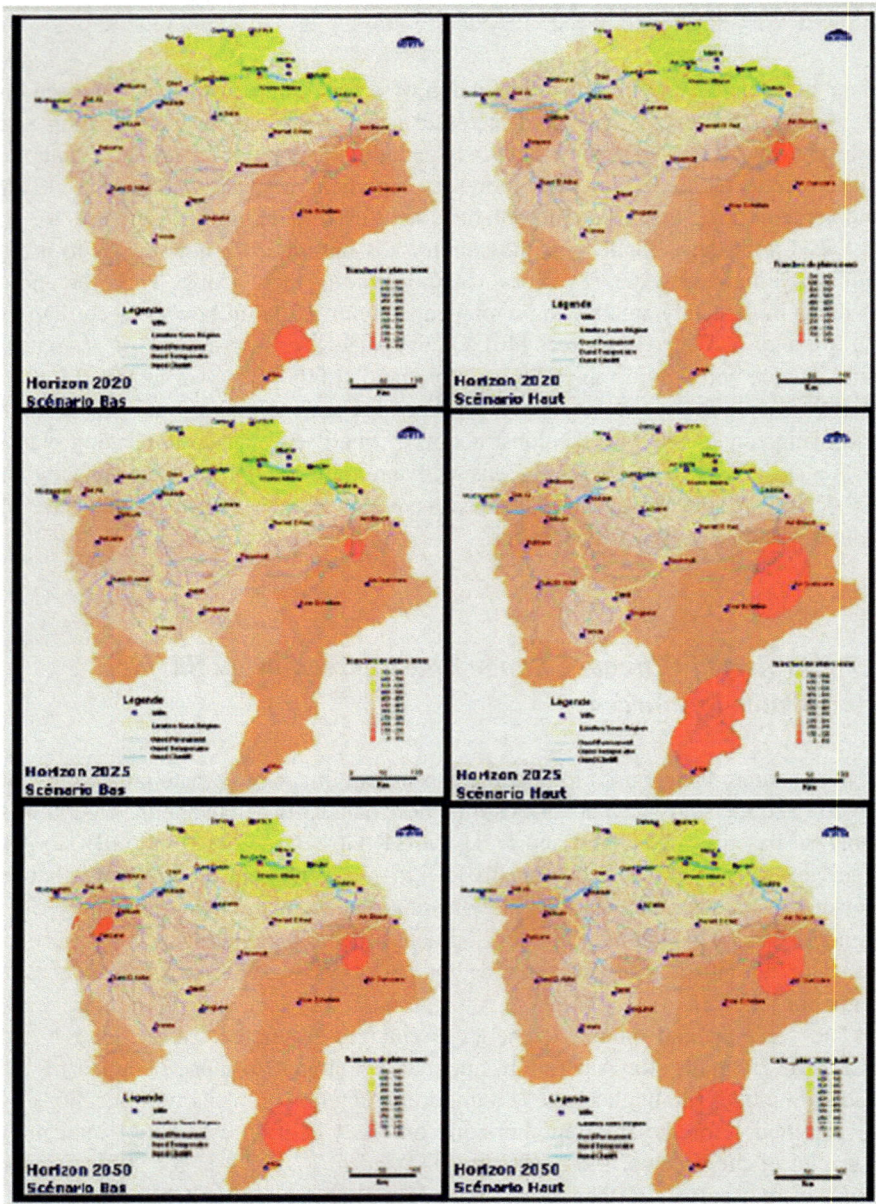

Fig. 9 Rainfall map of the Cheliff basin for 2020, 2025, and 2050 [54]

reference period (1961–1990). These different rainfall maps for the different horizons (2020, 2025, and 2050) reveal the same remarks about the spatial distribution of annual average rainfall but with values lower than the average relative to the reference period (1961–1990).

Table 3 Rate of precipitation variation (%) at horizons 2020, 2050, and 2080 compared to the reference period (1961–1990) in the study area

	A2	B2	A2	B2	A2	B2
Rate of change in precipitation (%)	2010–2039		2040–2069		2070–2099	
Annual	−10.35	−12.23	−20.78	−19.1	−33.65	−28.1
Autumn	−9	−12.1	−23.4	−16.7	−30.8	−27.3
Winter	−6.5	−9.9	−17.4	−15.1	−28.6	−24.9
Spring	−11.5	−8.1	−12.5	−17.9	−25.5	−20.3
Summer	−14.4	−18.8	−29.8	−26.7	−49.7	−39.9

The north region, whose rains were more than 450 mm for the period 1961–1990, becomes less important and would no longer exceed 400 mm for different scenarios and different horizons.

In the northeast region of the Cheliff basin and particularly the Zaccar massif, rainfall is becoming less important. A larger part of the basin will experience a reduction in precipitation; it includes the regions of Lower Cheliff and Mina and the upstream Boughezoul. This region will be between the isohyets 300 mm in the north and 100 mm in the south, where the semiarid to arid climate becomes more pronounced by occupying more surfaces faster.

The climate scenarios agree on a decrease in annual precipitation, on average, between 10.4 and 33.7% and between 12.2 and 28.1% for the A2 and B2 scenarios, respectively (Table 3). The decrease in annual precipitation is greater in summer and exceeds 40% by 2080. It should also be noted that the seasons would be more contrasting and the trend of decreasing precipitation would be stronger in autumn and winter for the A2 scenario. In general, the model predicts significant downward trends, suggesting that droughts are expected to increase substantially at the end of the twenty-first century in our study area.

This continuing decrease in rainfall at different horizons and in different scenarios confirms previous studies in the Mediterranean and Algeria, such as climate projections for rainfall in Algeria by 2071–2100 under three IPCC scenarios adapted from Giorgi and Lionello study [25] will indicate: 2011–2040 and 2041–2070, rainfall reductions range between 0 to 30% and from 1 to 40% for the scenarios B1 and A1B respectively and a decrease in the order of 25–40% by the period 2071–2100 for the A2 scenario.

The hydrologic model GR2M has been used to simulate flows over the next century. However, the results obtained must be treated with great caution because of some unavoidable difficulties related to both hydrological modeling and climate model uncertainty.

The two HadCM3 scenarios predict a decrease in flows for the three horizons (2020, 2050, and 2080) but with higher rates of variation (12, 30, and 45%) obtained with the A2 scenario. These results of the flow reduction for the A2 and B2 scenarios confirm the previous studies carried out in the Mediterranean basin, Algeria, and the Cheliff basin [18, 54, 55].

Also, an increase is expected in the frequency of droughts and a deficit of the surface water contribution of about 15% in the short term, resulting in a drop of groundwater of 4.4% in 2020 and 6.6% in 2050 [54]. So, the region is heading toward a much more severe water scarcity over the next few decades.

The sector most vulnerable to climate change is water resources and agriculture. In Algeria, adaptation initiatives are already in use and will have important consequences in several sectors. These actions must be integrated into a global adaptation policy of the country. The implementation of an adaptation plan:

- Water saving, it is a question of reducing the losses of water in the distribution and irrigation networks and of optimizing the consumption to adapt it to the needs of the different crops.
- Use of water-saving techniques, particularly in agriculture, such as drip irrigation and controlled suction, and the choice of crops that consume less water.
- Improvement of industrial water management methods (recycling, reuse).
- Use of unconventional waters.
- Protection of water resources.
- Mobilization of new water resources.

  - Collection and use of rainwater.

4 Conclusion

Climate change in the study area has had an adverse effect on the precipitation cycle, including all water resources. The reduction in the precipitation produced a downward trend in water inflow as shown by recording annual decreases from 1,025 to 815 Mm^3 between 1968–2001 and 2009. The consequences of water shortages are changes in the environmental balance which will consequently affect various human activities especially the available water supply for domestic and industrial consumption as well as for the agricultural economy.

Climatic scenarios agree on a decrease in annual precipitation, averaging between 10% and more than 30% at the end of the twenty-first century. These results reflect the availability of surface water resources, which will tend to decrease, with longer and more severe periods of low water. The Cheliff basin is considered particularly vulnerable to acute water scarcity in the coming years. A large water deficit due to population growth and an increase in water demand by different sectors of the economy are also expected (e.g., agriculture, industry).

Faced with these challenges, the country has implemented a new water resource management policy through the construction of new dams and use of unconventional water resources. It is hoped that the current investigation will help policy makers to make better decisions in developing water management strategies for the watersheds of Algeria.

5 Recommendations for Future Work

Future work on the impacts of climate change would benefit from integration of groundwater and surface water resources.

Better assessment of the effect of climate change on water resources at regional scale would require the application of the relevant regional models and the use of weather generator data for other future climate scenarios based on changes of different meteorological parameters (e.g., wind speed, radiation, relative humidity, number of cloudy days, etc.) in addition to temperature and rainfall. The results of these models may be used to simulate the runoff pattern by using a suitable hydrological model. The findings of such studies are important in preparation of regional water management plans.

Other benefits could be:

- Regionalization of parameters of hydrological models to assess hydrological behavior in ungauged basins
- Development of new drought forecasting tools based on new approaches

References

1. Arnell NW (1999) Climate change and global water resources. Glob Environ Chang 9:S31–S49
2. Vörösmarty CJ, Green P, Salisbury J (2000) Global water resources. Vulnerability from climate change and population growth. Science 289(5477):284–288
3. Muttiah RS, Wurbs RA (2002) Modeling the impacts of climate change on water supply reliabilities. J Water Int 27(3):407–419
4. Norrant C, Douguédroit A (2005) Monthly and daily precipitation trends in the Mediterranean (1950–2000). Theor Appl Climatol 83:89–106
5. Iglesias A, Garrote L, Diz A, Schlickenrieder J, Martin-Carrasco F (2011) Re-thinking water policy priorities in the Mediterranean region in view of climate change. Orig Environ Sci Policy 14(7):744–757
6. Alper B, Tayfur G, Gündüz O, Howard K-WF, Friedel MJ, Chambel A (2011) Climate change and its effects on water resources. Issues of national and global security. Springer, Berlin
7. Griffin MT, Montz BE, Arrigo JS (2013) Evaluating climate change induced water stress: a case study of the lower cape fear basin, NC. Appl Geogr 40:115–128
8. Bethoux JP, Gentili B, Tailliez D (1998) Warming and freshwater budget change in the Mediterranean since the 1940s, their possible relation to the greenhouse effect. Geophys Res Lett 25:1023–1026
9. Intergovernmental Panel on Climate Change IPCC (2001) Climate change, synthesis report (summary for policymakers). United Nations Environment Program, Geneva
10. Nicholson SE (2001) Climatic and environmental change in Africa during the last two centuries. Clim Res 17(1):123–144
11. Ragab R, Prudhomme C (2002) Climate change and water resources management in arid and semi-arid regions, prospective and challenges for the twenty-first century. Biosyst Eng 81 (1):3–34
12. Folland CK, Karl T, Salinger M (2002) Observed climate variability and change. Weather 57:269–278

13. Hansen J, Sato M, Ruedy R, Lo K, Lea DW, Medina-Elizade M (2006) Global temperature change. Proc Natl Acad Sci 103:14288–14293
14. Hasanean HM, Abdel Basset H (2006) Variability of summer temperature over Egypt. Int J Climatol 26:1619–1634
15. IPCC Intergovernmental Panel on Climate Change (2007) The physical science basis, summary for policymakers (contribution of WG I to the fourth assessment report of the IPCC). Cambridge University Press, Cambridge
16. Camuffo D, Bertolin C, Barriendos M, Dominguez-Castro F, Cocheo C, Enzi S, Sghedoni M, della Valle A, Garnier E, Alcoforado M-J, Xoplaki E, Luterbacher J, Diodato N, Maugeri M, Nunes MF, Rodriguez R (2010) 500-year temperature reconstruction in the Mediterranean basin by means of documentary data and instrumental observations. Clim Chang 10:1169–1199
17. Collins JM (2011) Temperature variability over Africa. J Clim 24:3649–3666
18. Elmeddahi Y, Issaadi A, Mahmoudi H, Tahar abbes M, Goossen M-FA (2014) Effect of climate change on water resources of the Algerian Middle Cheliff basin. Desalin Water Treat 52 (9–12):2073–2081
19. Hulme M, Doherty R, Ngara T, New M, Lister D (2001) African climate change: 1900–2100. Clim Res 17(2):145–168
20. Goubanova K, Li L (2007) Extremes in temperature and precipitation around the Mediterranean basin in an ensemble of future climate scenario simulations. Glob Planet Chang 57:27–42
21. Hertig E, Jacobeit J (2008) Downscaling future climate change: temperature scenarios for the Mediterranean area. Glob Planet Change 63:127–131
22. Marcos M, Tsimplis M (2008) Comparison of results of AOGCMs in the Mediterranean Sea during the twenty-first century. J Geophys Res 113:C12028. https://doi.org/10.1029/2008JC004820
23. Navarra A, Tubiana L (2013) Regional assessment of climate change in the Mediterranean volume 1: air, sea and precipitation and water. Springer, Berlin
24. IPCC (Intergovernmental Panel on Climate Change) (2013) Climate change 2013: the physical science basis. In: Stocker TF, Qin D, Plattner G-K, Tignor M, Allen SK, Boschung J, Nauels A, Xia Y, Bex V, Midgley PM (eds) Contribution of Working Group I to the fifth assessment report of the Intergovernmental Panel on Climate Change. Cambridge University Press, Cambridge
25. Giorgi F, Lionello P (2008) Climate change projections for the Mediterranean region. Glob Planet Chang 63:90–104
26. Meddi H (2013) Annual variability of precipitation of the north west of Algeria. APCBEE Proc 5:373–377
27. Elmeddahi Y, Mahmoudi H, Issaadi A, Goosen MFA, Ragab R (2016) Evaluating the effects of climate change and variability on water resources: a case study of the Cheliff Basin in Algeria. Am J Eng Appl Sci 9:835–845. https://doi.org/10.3844/ajeassp.2016.835.845
28. Meddi M, Boucefiane A (2013) Climate change impact on groundwater in Cheliff-Zahrez basin (Algeria). APCBEE Proc 5:446–450
29. Mann H-B (1945) Non-parametric tests against trend. Econometrica 13:245–259
30. Kendall M-G (1975) Rank correlation methods4th edn. Charles Griffin, London
31. Hirsch R-M, Helsel D-R, Cohn T-A, Gilroy E-J (1993) Statistical analysis of hydrologic data. Handb Hydrol 17:11–55
32. Yue S, Pilon P, Cavadias G (2002) Power of the Mann-Kendall and Spearman's rho tests for detecting monotonic trends in hydrological series. J Hydrol 259(1–4):254–271
33. Hamed K-H (2008) Trend detection in hydrologic data: the Mann-Kendall trend test under the scaling hypothesis. J Hydrol 349(3–4):350–363
34. Helsel DR, Hirsch RM (2002) Statistical methods in water resources techniques of water resources investigations. Book 4- Chapter A3. US Geological Survey, Reston
35. Sen PK (1968) Estimates of the regression coefficient based on Kendall's Tau. J Am Stat Assoc 63(324):1379–1389
36. Pettitt AN (1979) A non-parametric approach to the change-point problem. Appl Stat 28(2):126–135

37. Lubès H, Masson JM, Servat E, Paturel JE, Kouame B, Boyer JF (1994) Characterization of fluctuations in a time series by statistical test applications – bibliographic study. Report No 3 ICCARE programs. ORSTOM, Montpellier
38. Paturel JE, Servat E, Kouamé B, Lubès H, Masson JM, Boyer JF, Travaglio M, Marieu B (1996) Rainfall variability in humid Africa along the Gulf of Guinea. Integrated regional approach. First scientific workshop FRIEND AOC. PHIV. Technical documents in hydrology no 16, pp 1–31
39. Lubès-Niel H, Masson JM, Paturel JE, Servat E (1998) Climate variability and statistics. Simulation study of the power and robustness of some tests used to check the homogeneity of chronicles. J Water Sci 3:383–408
40. Perrin C, Michel C, Andréassian V (2003) Improvement of a parsimonious model for streamflow simulation. J Hydrol 279:275–289
41. Kouassi AM, N'guessan BTM, Kouamé KF, Kouamé KA, Okaingni JC, et Biémi J (2012) Application of the cross-simulation method to the analysis of trends in the rainfall-discharge relationship from the GR2M model, case of the N'Zi-Bandama watershed (Côte d'Ivoire). Compt Rendus Geosci 344(5):288–296
42. Hamlaoui-Moulai L, Mesbah M, Souag-Gamane D, Medjerab A (2013) Detecting hydroclimatic change using spatiotemporal analysis of rainfall time series in Western Algeria. Nat Hazards 65:1293–1311
43. Meddi H, Meddi M (2007) Spatial and temporal variability of rainfall in north west of Algeria. Geograph Tech 2:49–55
44. Meddi MM, Assani AA, Meddi H (2010) Temporal variability of annual rainfall in the Macta and Tafna catchments, northwestern Algeria. Water Resour Manag 24:3817–3833
45. Ministry of Water Resources (MWE). Note of synthesis, activity in 2012. www.mre.dz/
46. National Statistics Office (NOS) (2013) Series C – regional statistics and mapping environmental statistics. Statistics collections no 177
47. Toumi S (2013) Application of nuclear techniques and remote sensing to the study of water erosion in the Wadi Mina basin. PhD thesis, National School of Hydraulics
48. Remini B, Bensafia D, Nasroun T (2015) Impact of sediment transport of the Chellif River on silting of the Boughezoul reservoir (Algeria). J Water Land Dev 24(I–III):35–40
49. Benblidia M, Thivet G (2010) Water resources management: the limits of a supply-side policy. International Centre for Advanced Mediterranean Agronomic Studies-CIHEAM
50. Elmeddahi Y (2016) Climate change and its impacts on water resources, the case of the Cheliff basin. PhD thesis, University Hassiba Ben Bouali of Chlef
51. Semenov MA, Brooks RJ, Barrow EM, Richardson CW (1998) Comparison of the WGEN and LARS-WG stochastic weather generators for diverse climates. Clim Res 10:95–107
52. Wilks DS, Wilby RL (1999) The weather generation game – a review of stochastic weather models. Prog Phys Geogr 23:329–357
53. Pope VD, Gallani ML, Rowntree PR, Stratton RA (2000) The impact of new physical parametrizations in the Hadley Centre climate model – HadCM3. Clim Dyn 16:123–146
54. PNUD 00039149 (2010) Second national communication of Algeria on climate change
55. PNUD, Project ALG/98/G31 (2001) Development of the national climate change strategy and action plan

Assessment of Projected Precipitations and Temperatures Change Signals over Algeria Based on Regional Climate Model: RCA4 Simulations

Ayoub Zeroual, Ali A. Assani, Hind Meddi, Senna Bouabdelli, Sara Zeroual, and Ramdane Alkama

Contents

A. Zeroual (✉), H. Meddi, and S. Bouabdelli
Ecole Nationale Supérieure d'Hydraulique de Blida, GEE, Soumaâ, Algeria
e-mail: zeroualayoub34@yahoo.fr

A. A. Assani
Environmental Sciences Department, University of Quebec at Trois-Rivières, Trois-Rivières, Canada

S. Zeroual
VESDD Laboratory, University of M'sila, M'sila, Algeria

R. Alkama
European Commission, JRC, Directorate D-Sustainable Resources, Bio-Economy Unit, Ispra, Italy

Abdelazim M. Negm, Abdelkader Bouderbala, Haroun Chenchouni, and
Damià Barceló (eds.), *Water Resources in Algeria - Part I: Assessment
of Surface and Groundwater Resources*, Hdb Env Chem (2020) 97: 135–160,
DOI 10.1007/698_2020_526, © Springer Nature Switzerland AG 2020,
Published online: 16 June 2020

Abstract   Algeria is the largest African and Mediterranean country. It is located in the southern seashores of the Mediterranean Sea. Its climate conditions are ranging from relatively wet to very dry which makes it confronted to high levels of rainfall deficits. The future rainfall evolution may be critical for human activities since increased temperatures may further exacerbate droughts and water shortages. In this study, the regional climate simulations RCA4 are evaluated over historical period 1951–2005 and then used to examine the rainfall and temperature projections over the end of the twenty-first century under two Representative Concentration Pathway (RCP4.5 and RCP8.5) scenarios. The historical simulations are evaluated against observations coming from the recent data sets of Climatic Research Unit (CRU). The trends in precipitation and temperature over historical (1951–2005) and projected future scenarios (2006–2060 and 2045–2100) was depicted by the estimation of the shifts of the three main climate zones existing in Algeria (Köppen-Gieger classification): warm temperate climates (C), steppe climate (BS), and desert climate (BW). Comparative to the mean climate zone surface areas derived from observations (1951–2005), all model simulations predict an expansion of desert climate zone at the expense of the temperate and steppe climate zones. This shift seems to particularly increase by the end of twenty-first century (2045–2100) under RCP8.5 scenario.

Keywords   Climate change, Climate zone, Global warming, Precipitation and temperature, Regional climate simulations

1   Introduction

Given the ongoing aridification and/or sometimes very abrupt climate change, the current distribution of climate conditions at the global scale will be reorganized. Some climates will disappear completely, while others will appear in some regions [1]. This is particularly true over Mediterranean Basin due to its geographic location between dry (the Sahara) and wet regions of the Northern Europe [2, 3]. This basin is under the influence of the downward branch of the Hadley cell circulation in summer and of fluxes from the West in the Atlantic Ocean in winter [4]. It is a transition zone in which the competing influences of extratropical and tropical systems affect climate events modulated by their proximity to the Mediterranean Sea [5]. This region, which includes Algeria, has been considered as the region for which there is the widest consensus between projections and model types used by the International Panel on Climate Change (IPCC) about future decreases in total rainfall [6]. However, IPCC model resolution, which ranges from 100 to 200 km, does not allow a sufficient level of regional detail, and a higher resolution can be achieved by using regional climate models. Regional models provide a finer representation of sub-regions of the planet while using global models to describe conditions at the edges of these subregions, among other things. As such, the power of supercomputers is essentially used to enhance spatial resolution. In this chapter, we examine climate in Algeria in terms of monthly mean precipitation and temperature using future projections of regional climate model (RCM) – RCA4 at 0.44° resolution used as part of the CORDEX-Africa program (Coordinated Regional climate Downscaling Experiment) under the RCP4.5 and RCP8.5 forcing scenario. The RCA4 regional climate model of Rossby Centre (SMHI) used the boundary conditions of nine atmosphere-ocean general circulation models (AOGCMs) from the Coupled Model Intercomparison Project-Phase 5 (CMIP5) to drive an ensemble of RCM simulations for African domain [7, 8]. To do so, we analyze precipitation and temperature trends, as well as the average climate trend observed in Algeria over the period from 1951 to 2005. These trends are compared with those estimated from the outputs of nine regional climate model simulations. It is then possible to test how well these models reproduce average climate and trends (of temperature and precipitation) over historical time and into the future (2005–2060 and 2045–2100). Two scenarios are primarily used for future projections, namely, RCP4.5 and RCP8.5. Since the precipitations and temperatures are used to define climate zones, the future change of these two climate variables is also evaluated by the shift in surface area of each climate zone as defined in Köppen-Geiger classification [9]. The shift in surface area of each climate zone is computed after constructing the mean map of climate zones for each three studied periods (1951–2005, 2006–2060, and 2045–2100).

2 Observational Data

We selected monthly temperature and precipitation data measured on a 0.5° ×0.5° grid from the latest version of the Climatic Research Unit database (CRU Version TS.3.24) for the period from 1951 to 2005 of meteorological data based on stations measurements [10] with spatial coverage of Algeria. The University of East Anglia CRU database is one of the most widely used sources of climate data, providing temperature and precipitation data at a 0.5° resolution and temporal coverage from 1901 to 2015.

3 Regional Climate Model (RCM) Simulations

Regional climate model (RCM) simulations were used as part of the Coordinated Regional climate Downscaling EXperiment (CORDEX-Africa, http://www.cordex. org/). CORDEX is a numerical climate simulation coordinated experiment carried out jointly by several research centers to generate fine-scale climate data over

Table 1 RCA 4 Regional
climate models and their
AOGCM driving model

MCR	Driving model (AOGCM)
RCA4 (CanESM2)	CCCma (Canada)
RCA4 (CNRM-CM5)	CNRM-CERFACS (France)
RCA4 (CSIRO-MK3)	CSIRO (Australia)
RCA4 (IPSL-CM5A)	IPSL (France)
RCA4 (MIROC5)	MIROC (Japan)
RCA4 (HadGEM2-ES)	MOHC (UK)
RCA4 (MPI-ESM-LR)	MPI-M (Germany)
RCA4 (NorESM1-M)	NCC (Norway)
RCA4 (GFDL-ESM2M)	NOAA-GFDL (USA)

14 regional domains (South America, Central America, North America, Africa, Europe, South Asia, East Asia, Central Asia, Australasia, Antarctica, the Arctic region, the Mediterranean region (MED), the Middle East, and Southeast Asia) [11]. As part of CORDEX-Africa in its second phase, regional climate models including conditions at the edges of ten atmosphere-ocean (AO) general circulation models (GCM) from Phase 5 of the Coupled Model Intercomparison Project (CMIP5) produced new versions of the regional climate models for the African region. We used monthly precipitation and temperature simulations from the Rossby Centre by RCA4 Regional Climate Model for Africa. This database was obtained from the Swedish Meteorology and Hydrology Institute's (SMHI) regional model using boundary conditions of nine AOGCMs from the CMIP5 (see Table 1) [11, 12]. Simulated data as a whole ($0.44° \times 0.44°$) cover the period from 1951 to 2100, which is subdivided into two periods: the historical period (1951–2005) and the projection period (2006–2100). This latter period was forced by two Representative Concentration Pathway scenarios, RCP4.5 and RCP8.5, assuming boundary conditions from the AOGCMs [11].

3.1 Representative Concentration Pathways (RCPs)

Greenhouse gas (GHG) representative concentration pathways developed by the IPCC as part of Phase 5 of the Coupled Model Intercomparison Project (AR5) predict a somber future for humanity as a whole and a multitude of life forms [13]. RCPs are concentration evolution scenarios for GHG (carbon dioxide or CO_2, methane or CH_4, nitrous oxide or N_2O, etc.), aerosols, and chemically active gases in the atmosphere over the 2006–2100 period and extrapolated to 2300 [14]. All scenarios are assumed to be directly linked to CO_2 emissions; as carbon emissions increase, climate warms. Mankind is currently set on the worst-case pathway, RCP8.5, which should leads to a 2°C mean global warming by 2050 [14]. Four scenarios were selected from 300 published scenarios to cover the widest possible range of allowable future radiative forcing pathways. These four RCP scenarios are labeled according to their radiative forcing in 2100, namely, 2.6 W/m²,

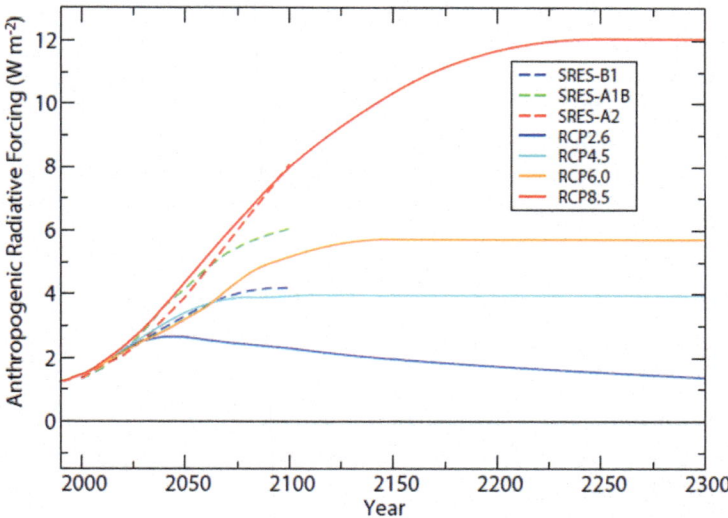

Fig. 1 Temporal evolution of anthropogenic radiative forcing between 2000 and 2300 for RCP scenarios (solid lines) and comparison with SRES scenarios used in AR4 (dashed lines). (From [14])

4.5 W/m^2, 6 W/m^2, and 8.5 W/m^2. Figure 1 shows the scenarios as well as the "SRES" scenarios previously used in the fourth Assessment Report (AR4) of 2007.

The most optimistic scenario, RCP2.6 (very low emissions $\approx$490 ppm CO_2 equiv), increases to 3 W/m^2 and then decreases to 2.6 W/m^2 by 2100. Scenarios RCP4.5 (low emissions $\approx$650 ppm CO_2 equiv.) and RCP6.0 (moderate emissions $\approx$850 ppm CO_2 equiv.) reach stable forcing values at 4.2 W/m^2 and 6.0 W/m^2, respectively, after 2100, while scenario RCP8.5 (very high emissions $\approx$1,370 ppm CO_2 equiv.) reaches 8.3 W/m^2 in 2100 along an ascending trajectory. These scenarios are not associated with any specific socioeconomic scenario, unlike the former SRES scenarios. Rather, they are projections that could arise from more than one underlying socioeconomic scenario to produce similar GHG emission and radiative forcing values.

4 Methodology

Once the observation datasets have been regridded into the same resolution of RCA4 simulation outputs (0.44° × 0.44°) using the first conservative remapping function of climate data operators (CDO) [15], we proceeded to the spatiotemporal variability analysis of precipitations and temperatures according to the two following stages:

- In the first stage, we produce mean climate maps, compute the mean surface area of each climate zone, and estimate linear trends over the historical period

(1951–2005) based on a precipitation and temperature dataset obtained from observation and simulation models.

- As a second step, we calculate the mean surface area of each climate zone and linear trends for the two future time periods (2006–2060 and 2045–2100) based on modeled monthly precipitation and temperature under the two RCP scenarios. Here, it must be recalled that the RCA4 simulations outputs in the future have been corrected using the quantile mapping (QM) bias correction algorithm [16, 17].

5   Results

5.1   Spatiotemporal Variability of Precipitation and Temperature in Algeria over the 1951–2005 Observation Period

As far as spatial variability is concerned, the mean interannual and monthly precipitation values derived from Climatic Research Unit (CRU TS3.24) data are presented in Figs. 2 and 3, respectively, and the spatial distribution of interannual mean and monthly mean temperatures are presented in Figs. 4 and 5, respectively.

We notice that the precipitation (Fig. 2) is characterized by a large spatial variation. The values of annual precipitation decrease when we move from North to South. The annual average precipitation is about 50 mm per year in the South, while it reaches 1,200 mm per year in the northeast of the country precisely at the stations of Jijel and El Kala. The annual average in coastal regions varies between 400 and 1,000 mm by increasing from the west to the east. The Eastern regions are thus better watered by rainfalls than western regions.

The examination of the monthly mean precipitation in Algeria over the study period (Fig. 3) allows us to divide the year into two marked seasons: the dry season, which extends from June till September, and the rainy season, which includes the eight remaining months of the year. Results show that the country seems that has a large rain distribution. Despite the rainy season begins in September, it is still disturbed by some periods of sirocco. The rainfall is well established in November, to increase until December, where the maximum is most frequently registered. Then, the monthly rainfall quantity decreases until June 1st, and the rain becomes rare or is reduced to some drops in some areas.

Figure 4 shows the annual temperature distribution over Algeria. We notice that the annual average temperatures of the country are divided according to three large geographical areas, from north to south as follows:

- The annual average temperature varies between 10 and 20°C, between the coast and the Tellian Atlas.
- Between the Tellian Atlas and Saharan Atlas, the annual average temperature varies between 20 and 22.5°C.

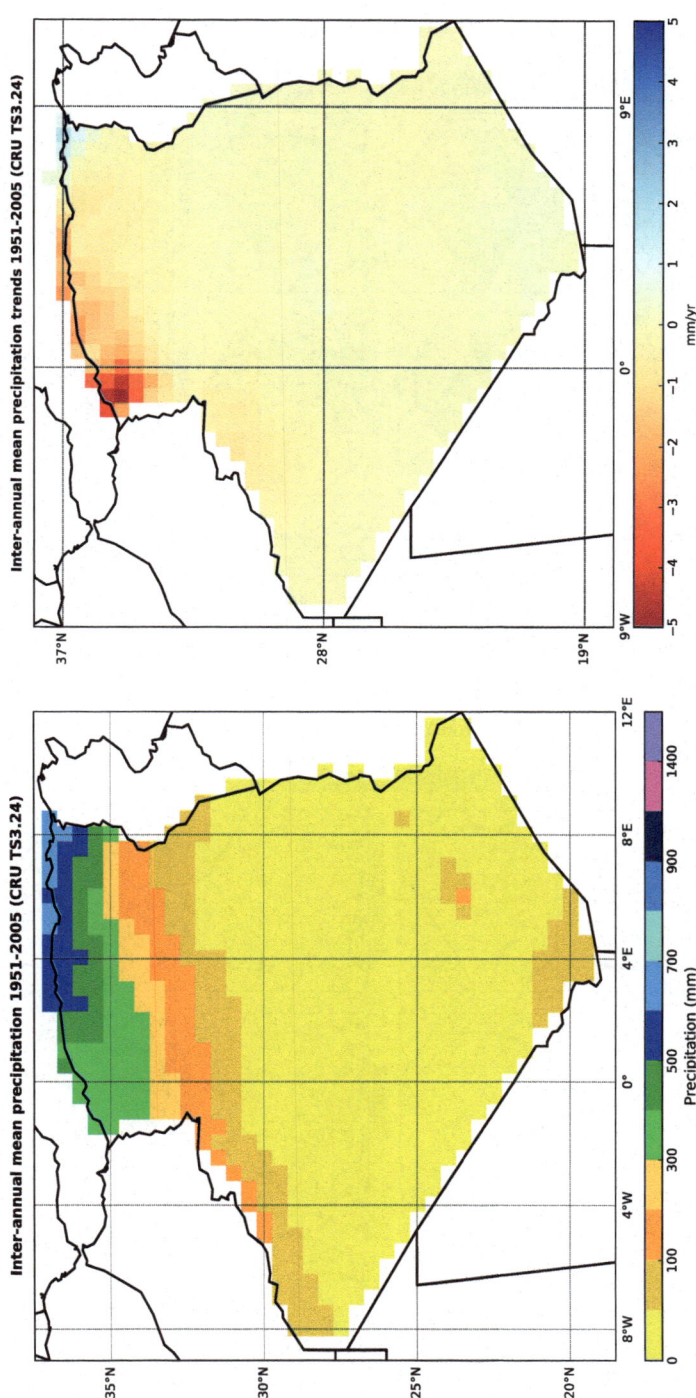

Fig. 2 Interannual mean precipitation in Algeria and related trends (1951–2005) (CRU TS3.24 at 0.5° × 0.5°)

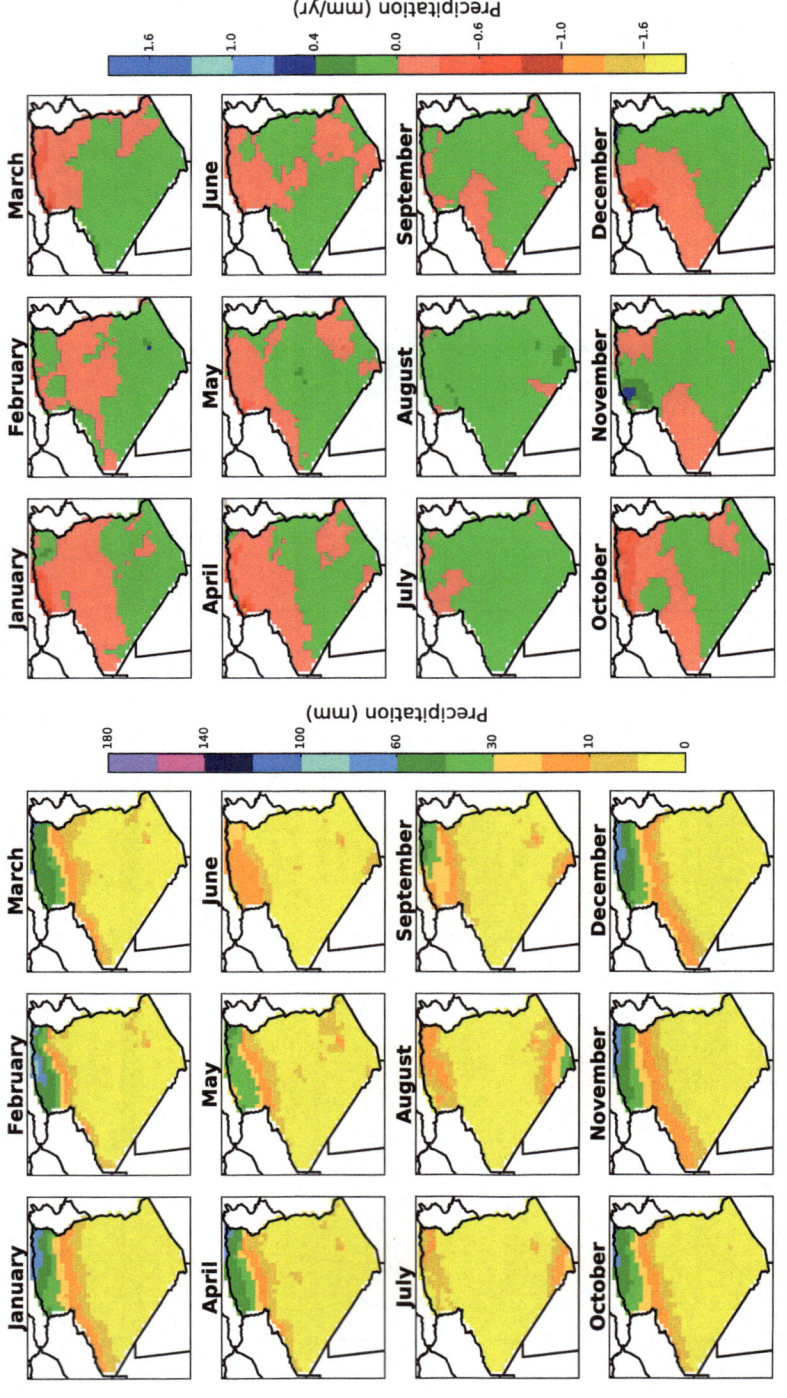

Fig. 3 Monthly mean precipitations in Algeria and related trends (1951–2005) (CRU TS3.24 at 0.5° × 0.5°)

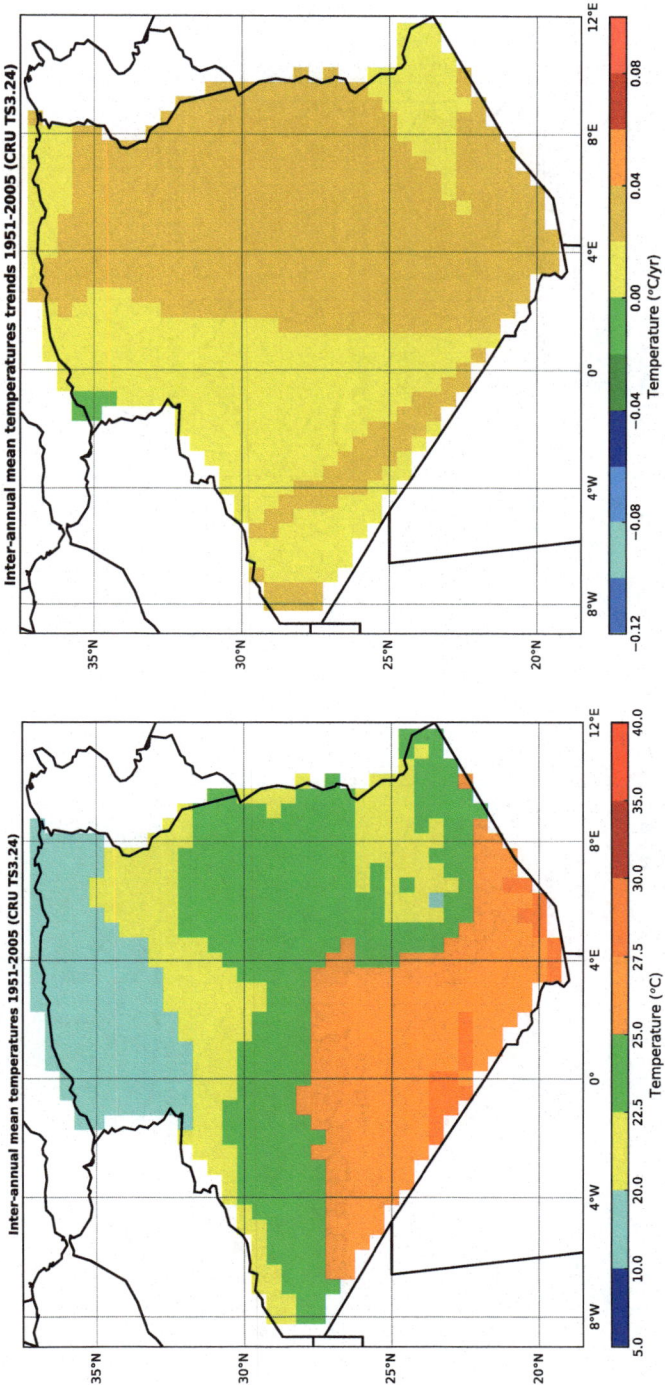

Fig. 4 Interannual mean temperatures in Algeria and related trends (1951–2005) (CRU TS3.24 at 0.5° × 0.5°)

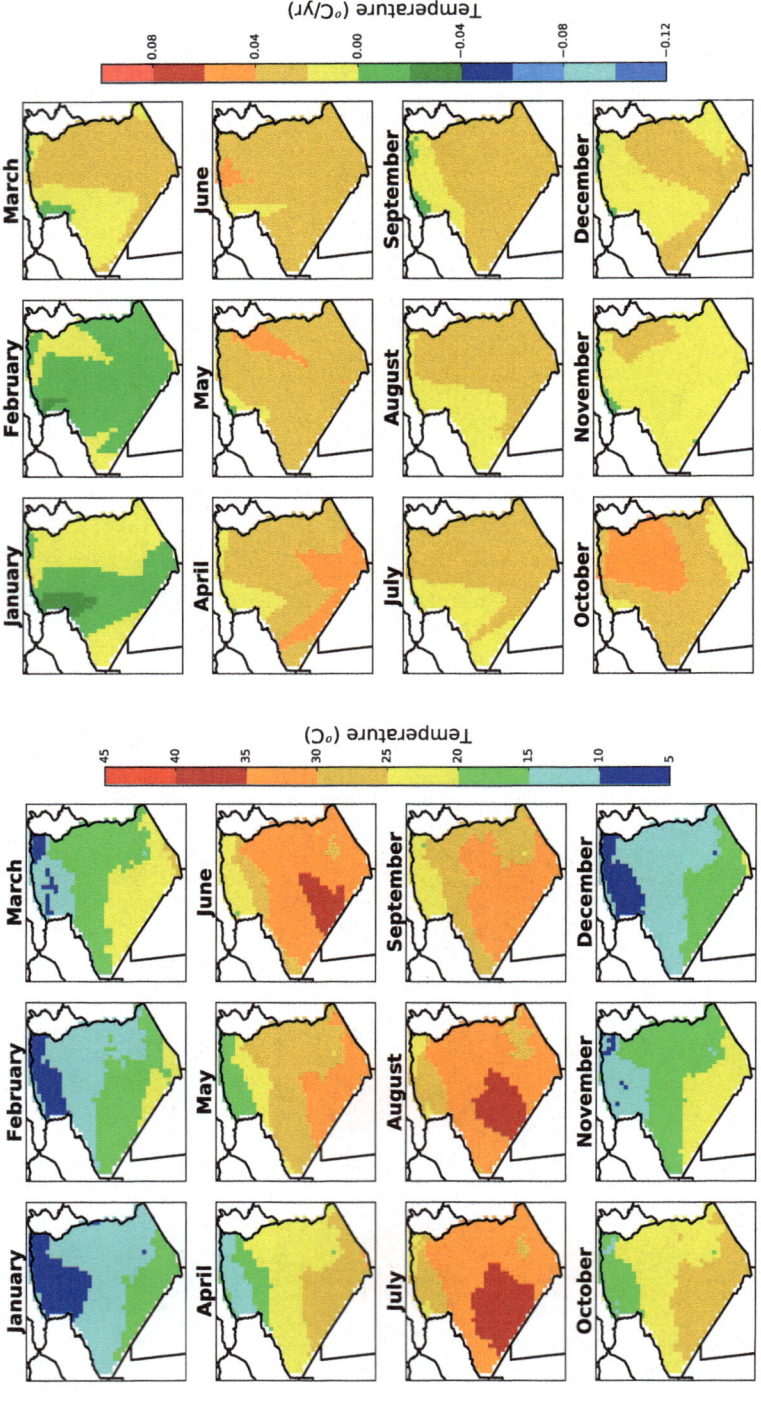

Fig. 5 Monthly mean temperatures in Algeria and related trends (1951–2005) (CRU TS3.24 at 0.5° × 0.5°)

• The annual average temperature in the Algerian Sahara varies between 22.5 and 30°C.

The monthly average temperatures in Algeria (Fig. 5), and concerning the winter temperatures, they vary between 5 and 10°C in the Western coastal cities and between 10 and 15°C in the eastern side, on plains and in the internal valleys. During the summer period, the coastal zone is refreshed by the marine winds; the average temperature varies between 25 and 35°C approximately.

However, the temperatures are particularly raised in the internal valleys and plains because of the enclosing and because of the exhibition to the Southern winds; they vary between 25 and 30°C in May, between 30 and 35°C in June, and between 30 and 40°C in the months of July and August. For the Southern part of the country, in desert, the temperature is ranged between 15 and 25°C in winter and reaches 35–40°C or even more in summer.

For temporal variability, results of the linear regression method applied to the interannual and monthly variability of precipitations and temperatures over the period from 1951 to 2005 are presented in the right sides of Figs. 2, 3, 4, and 5, respectively. The rates of increase and decrease were estimated from the slope of the linear regression line and are expressed as mm/year for precipitation and °C/year for temperature. Figure 2 (right side) shows the trend of interannual precipitation in Algeria during the period from 1951 to 2005.

This map shows that precipitations have decreased throughout the northern part of the country over this period. No change in precipitation trend is seen in the southern part of the country. The rate where annual rainfall decreases becomes smaller moving away from the coast and going from the West to the East. The rate of decrease of annual mean precipitation ranges from −1.5 mm/year in the North-west to −0.5 mm/year in the northeastern part of the country. Between the coast and the Tell Atlas, the rate of decrease is roughly −1 mm/year.

From the maps of monthly precipitation trends (Fig. 3, on the right side), it is possible to subdivide the rate of decrease in the northern part of the country into two distinct segments: a −0.3 mm/year decrease in monthly precipitation from January to June and from October to December and a +0.2 mm/year increase during the other 3 months (July, August, and September). In the southern part of the country, a 0.2 mm/year increase is observed for nearly all months.

Figure 4 (right side) shows the trend in annual temperatures in Algeria over the period from 1951 to 2005. On this map, an increase in temperature is seen over the whole country, from +0.02°C/year in the West to +0.04°C/year in the eastern part of the country. The maps of monthly mean temperature trends in Algeria over the period from 1951 to 2005 (Fig. 5, right side) show an increase ranging from +0.02°C/year to +0.04°C/year for the whole of the country except for the months of January and February and a −0.02°C/year decrease over nearly all of the country for these 2 months.

5.2 Spatial Variability of Climate Zone Surface Areas in Algeria Over the 1951–2005 Observation Period

The calculation detail overview and definition of all Köppen-Geiger classes applied here was briefly described in [18] or [19]. The Köppen climate zones are designated using a code of three letters. The first letter describes the main classes, in upper case, and indicates the vegetation group in the climate zone defined based on temperature and precipitation. The second letter, in lower case, accounts the distribution of precipitation at the annual scale. The third letter for temperature classes, also in lower case, reflects seasonal temperature variations. According to Zeroual et al. [20], the ten climate zones presented in the Table 2 have been observed in Algeria over the period 1951–2098. In this chapter, we are interested only in the surface areas of the main climatic zones, namely, warm temperate climates (C), steppe climate (BS), and desert climate (BW). Based on the Köppen-Geiger classification and the precipitation and temperature dataset taken from observations averaged over 1951–2005 period, we constructed a map of climate zones for this period and computed the surface extent of different climate zones. The surface area of each climate zone from 1951 to 2005 is presented in Table 3. Table 3 show that the desert climate (BW) extend over roughly 88.14% of the total surface area of the country, followed by the steppe climate (BS) zone, covering about 7.04%. The warm temperate climates (C) account for only about 4.83% of this area.

Table 2 Köppen -Geiger climate type observed in Algeria over the period 1951–2098

Type	Climate class	Third letter for temperature classification
B	Arid climates	
	BS: Steppe climate	h: Hot steppe/desert
	BW: Desert climate	k: Cold steppe/desert
C	Warm temperate climates	
	Cs: Warm temperate climate with dry summer	a: Hot summer
	Cw: Warm temperate climate with dry winter	b: Warm summer
	Cf: Warm temperate climate, fully humid	

Table 3 Surface area of each of the three main climatic zones observed over the period 1951–2005 (in % of total area)

	Climate zone in % of total area (1951–2005)		
Data	C	Bs	Bw
Climatic Research Unit (CRU)	4.83	7.04	88.14

5.3 Comparison of the Evolution of Observed and Simulated Precipitation and Temperatures During the Period from 1951 to 2005

Before examining the future evolution of precipitation and temperature, a comparison of climate (precipitation and temperature) observational data (CRU) from 1951 to 2005 and data from simulations from the nine climate models simulation of RCA4 (see Table 1) was carried out to test how well these models simulations reproduce observed climate and its evolution. Here, we should note that for this comparison, the monthly modeled precipitation and temperature were not bias corrected.

After building the maps of interannual precipitation and temperatures and monthly precipitation and temperatures and their trends over the 1951–2005 period derived using the nine climate models simulations, we notice that the climate model simulation RCA4-MIROC5 reproduces satisfactorily monthly mean precipitation as well as interannual and monthly mean temperatures, whereas climate model RCA4-NorESM1-M reproduces interannual mean precipitation.

As far as the temporal variability of annual and monthly precipitation and temperatures derived from climate models for the 1951–2005 period is concerned, the maps show that the rates of increase and decrease of monthly precipitation and monthly temperatures derived from climate model RCA4-NorESM1-M are in good agreement with observational data. However, observed annual precipitation trends are reproduced by the model RCA4-IPSL-CM5A and annual temperature trends are reproduced by the climate model RCA4-MPI-ESM-LR.

5.4 Spatial and Temporal Projected Variability of Precipitation and Temperature over the 2006–2060 and 2045–2100 Periods

Only results from the aforementioned models simulations, after bias correction for the two scenarios, are presented for each parameter (precipitation and temperature) and scale (annual and monthly).

As far as spatial variability is concerned, the distributions of interannual mean precipitation and temperature over the two projection periods (2006–2060 and 2045–2100) derived for the two scenarios are presented in Figs. 6 and 7.

Maps derived from the selected models show a decrease in interannual mean precipitation and an increase in interannual mean temperatures. Thus, for the two periods (2006–2060 and 2045–2100) and with both scenarios RCP4.5 and RCP8.5, the RCA4-NorESM1-M model predicts that mean precipitation will level off compared to the 1951–2005 time interval in the eastern and southern parts of the country and that they will decrease in the West. This decrease is greater for scenario RCP8.5

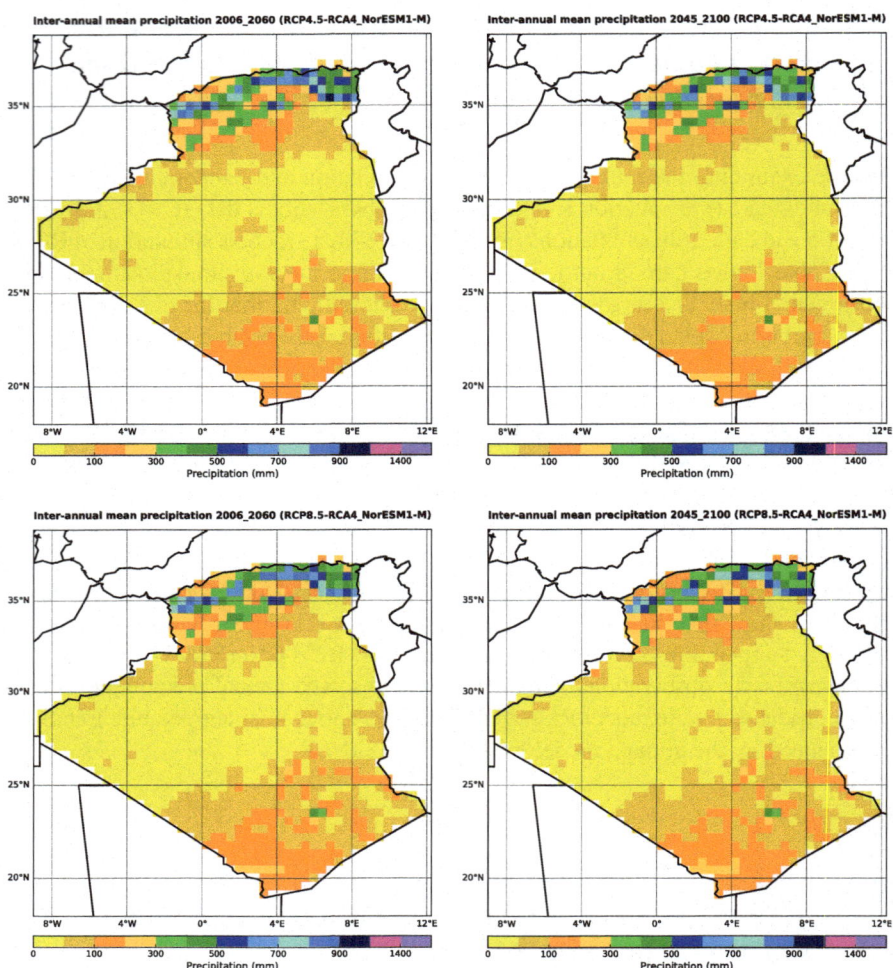

Fig. 6 Interannual mean precipitation in Algeria over the 2006–2060 and 2045–2100 time periods for the scenarios RCP4.5 and RCP8.5 (RCA4-NorESM1-M at a resolution of $0.44° \times 0.44°$

compared to the scenario RCP4.5, as well as for the 2045–2100 projection period compared to the 2006–2060 period.

The increase in interannual mean temperatures for the two projection periods with the two scenarios (see Fig. 7) predicted using model RCA4-MIROC5 simulation is obvious in the coastal region of the country especially during the second time interval (2045–2100) under RCP8.5scenario.

The monthly mean precipitation maps of Algeria generated using model RCA4-MIROC5 simulation for the 2006–2060 and 2045–2100 time intervals with the RCP4.5 and RCP8.5 scenarios (Fig. 8) show a slight decrease compared to the 1951–2005 reference time period in the western coastal region for all months. This

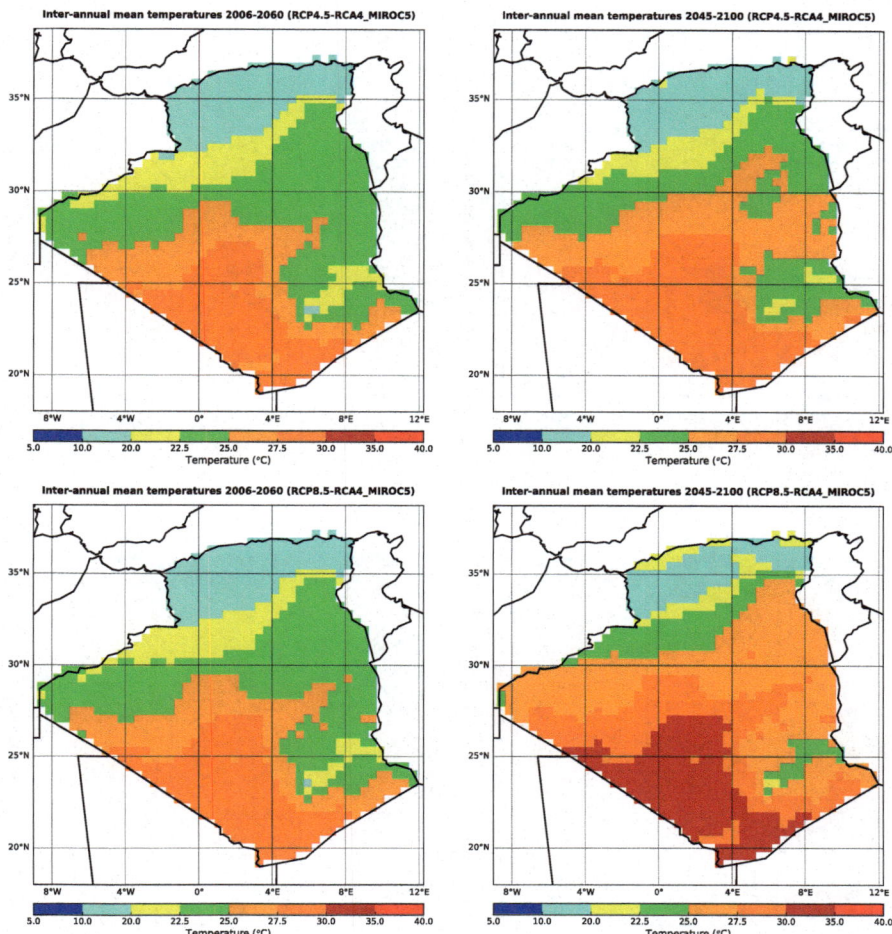

Fig. 7 Interannual temperatures in Algeria over the 2006–2060 and 2045–2100 time periods for the scenarios RCP4.5 and RCP8.5 (RCA4-MIROC5) at a resolution of 0.44° × 0.44°

decrease is marked for the months of November, December, and January during the second projection period (2045–2100) under RCP8.5 scenario.

The monthly mean temperature maps derived from model RCA4-MIROC5 simulation (Fig. 9) show an increase in temperature for all months. This increase is more marked for the second projection period (2045–2100) with the two scenarios RCP4.5 and RCP8.5 than for the first period (2006–2060).

As far as temporal variability over the two projection periods (2006–2060 and 2045–2100) is concerned, for both scenarios, the selected climate models predict a decrease in precipitation and an increase in temperature at the annual and monthly scales. The rates of decrease and increase are as follows:

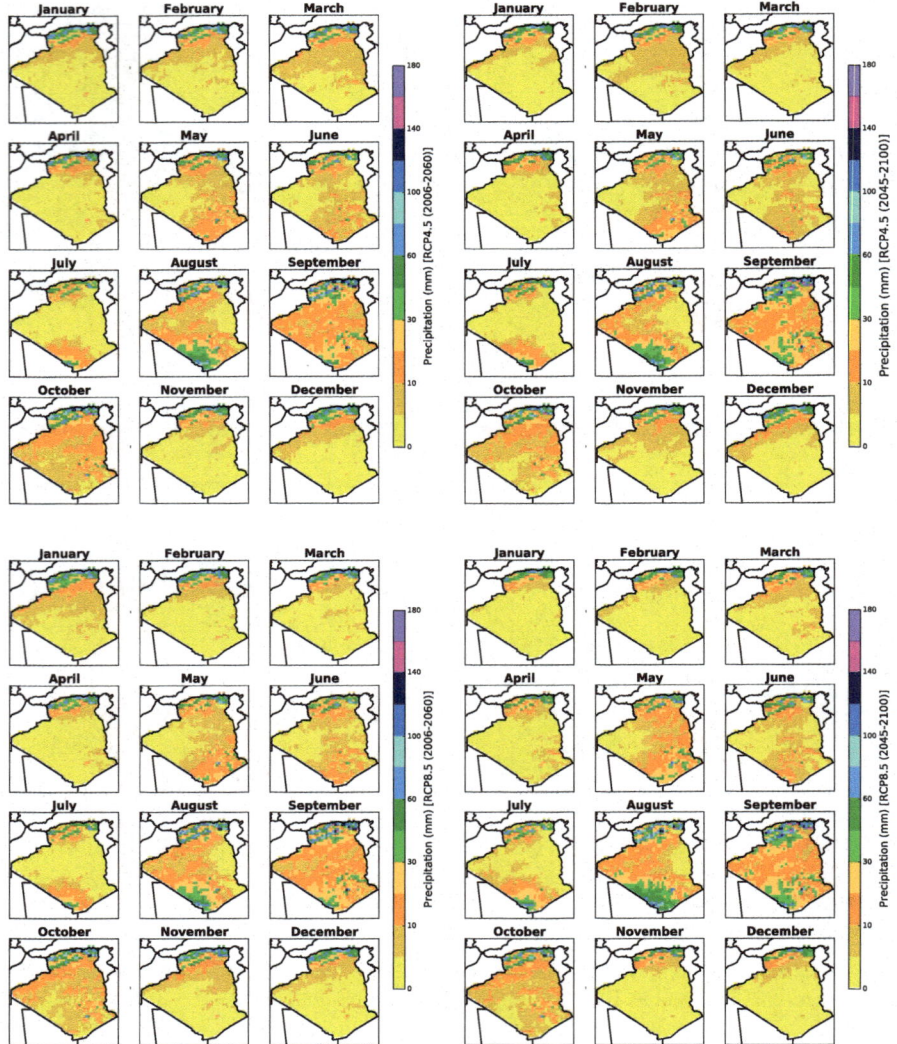

Fig. 8 Monthly precipitation in Algeria over the 2006–2060 and 2045–2100 time periods for the scenarios RCP4.5 and RCP8.5 (RCA4-MIROC5) at a resolution of 0.44° × 0.44°

1. According to climate model RCA4-IPSL-CM5A simulation, annual mean precipitation will decrease in the northern part of the country (Fig. 10) over the 2006–2060 time period at a rate of −0.5 to −1.5 mm/year with RCP4.5 scenario and of −1.5 to 2.5 mm/year with RCP8.5 scenario. During the second time period (2045–2100), the rate of decrease is nearly zero for scenario RCP4.5 and varies from 0.5 to −1.5 mm/year for scenario RCP8.5. No significant change in precipitation is seen for southern Algeria.

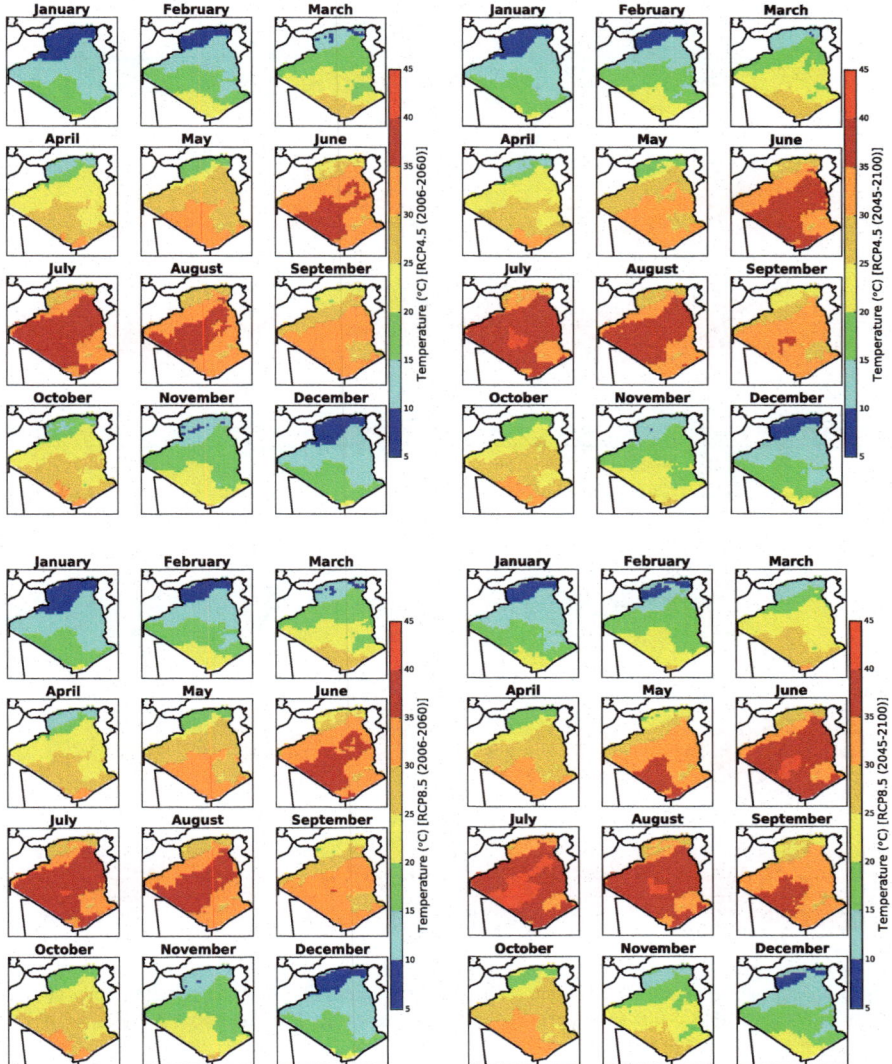

Fig. 9 Monthly temperatures in Algeria over the 2006–2060 and 2045–2100 time periods for the scenarios RCP4.5 and RCP8.5 (RCA4-MIROC5) at a resolution of 0.44° × 0.44°

2. According to climate model RCA4-MPI-ESM-LR simulation, mean annual temperatures will increase (Fig. 10) over the 2006–2060 time period at a rate of +0.02°C/year to +0.04°C/year in the North and of +0.04°C/year to +0.06°C/year in the South for the two scenarios. During the second time period (2045–2100), the rate of increase is +0.02°C/year over the whole country for the RCP4.5 scenario. For the RCP8.5 scenario, this rate ranges from +0.04°C/year to +0.06°C/year in the North and from +0.06°C/year to +0.08°C/year in the South.

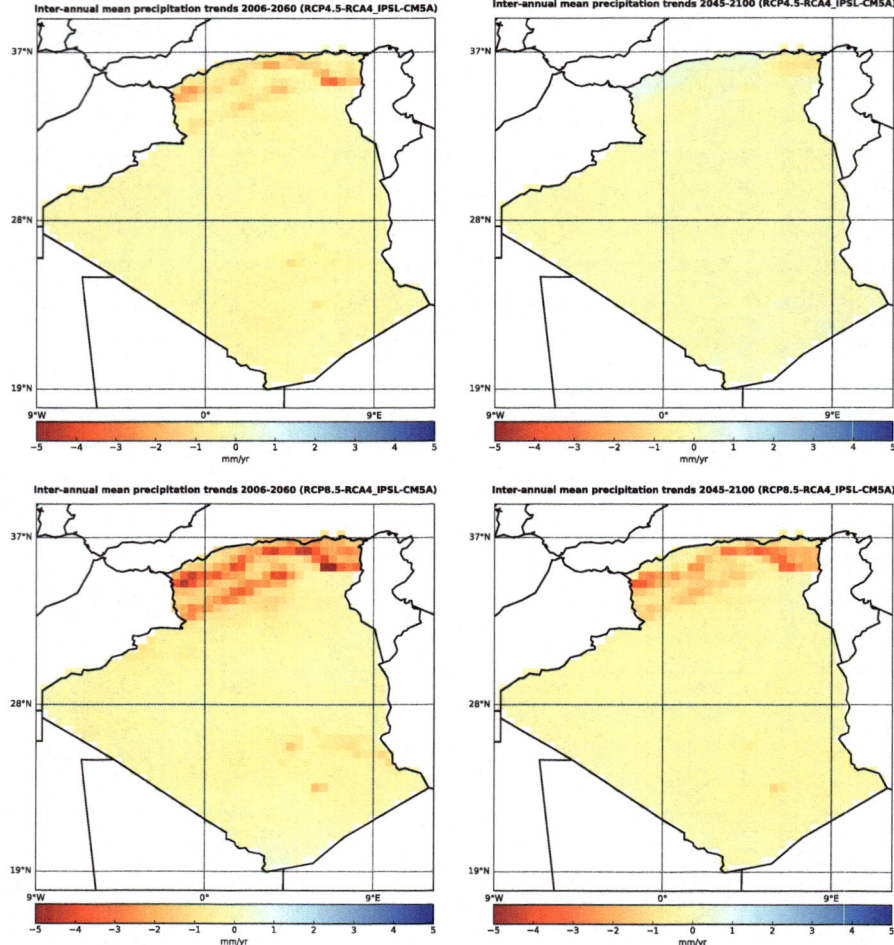

Fig. 10 Slopes of regression lines fitted to the temporal variability of annual precipitation over the 2006–2060 and 2045–2100 time periods for the scenarios RCP4.5 and RCP8.5 (RCA4-IPSL-CM5A) at a resolution of 0.44° × 0.44°

3. According to climate model RCA4-NorESM1-M simulation, monthly precipitation will decrease (Fig. 11) during the two periods with both scenarios. The rate of decrease is not synchronous for the different months, between regions, or between periods. This rate ranges from −0.3 mm/year to −0.8 mm/year, and it is higher for scenario RCP8.5 than for RCP4.5 for the six winter months (October to March).

4. According to climate model RCA4-MPI-ESM-LR simulation, annual mean temperatures will increase (Fig. 12) over the 2006–2060 time period at a rate of +0.02°C/year to +0.04°C/year in the North and of +0.04°C/year to +0.06°C/year in the South for the two scenarios. During the second time period (2045–2100),

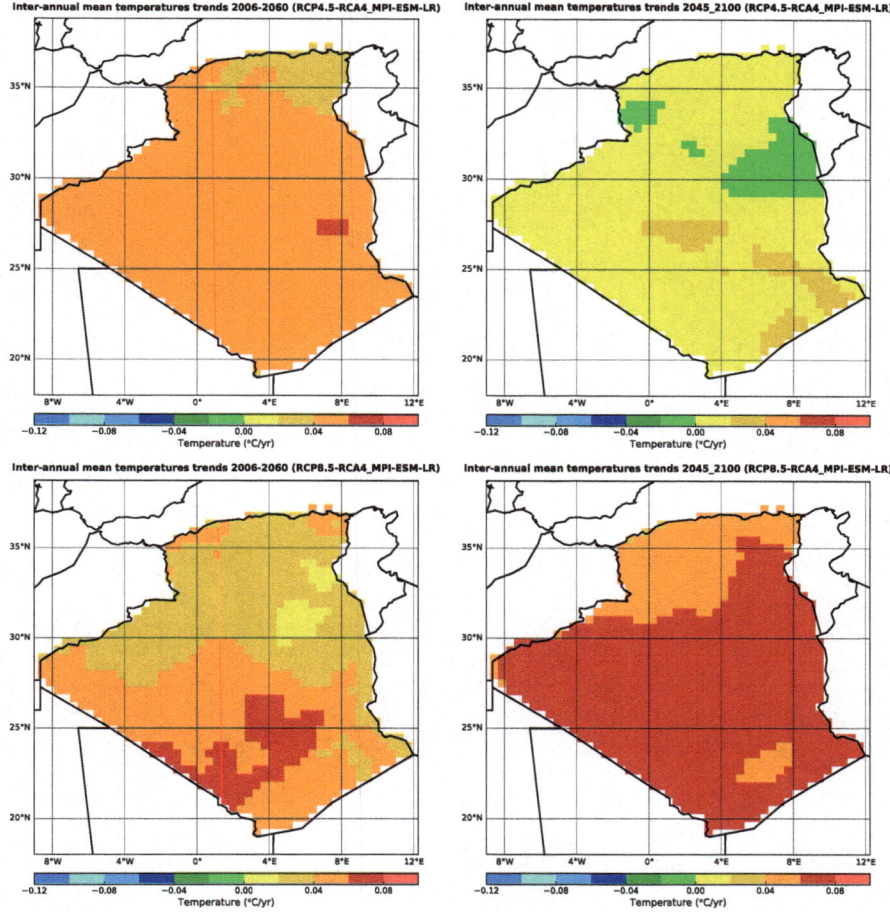

Fig. 11 Slopes of regression lines fitted to the temporal variability of annual temperatures over the 2006–2060 and 2045–2100 time periods for the scenarios RCP4.5 and RCP8.5 (RCA4-MPI-ESM-LR) at a resolution of 0.44° × 0.44°

the rate of temperature increase is +0.02°C/year over the whole country with scenario RCP4.5. For the RCP8.5 scenario, this rate is +0.04°C/year to +0.06°C/year in the North and +0.06°C/year to +0.08°C/year in the South.

5. According to climate model RCA4-NorESM1-M simulation, monthly mean temperatures will increase (Fig. 13) over the 2006–2060 time period at a rate of +0.04°C/year to +0.06°C/year for the months of February and May and of +0.02°C/year to +0.04°C/year for the other months for RCP4.5 scenario. For RCP8.5 scenario, over the 2006–2060 time period, the rate of increase will reach +0.08°C/year for the months of August and September in the northern part of the country and for the months of May and June in the South. For the other months, the rate of increase will range from +0.02°C/year to +0.06°C/year. During the

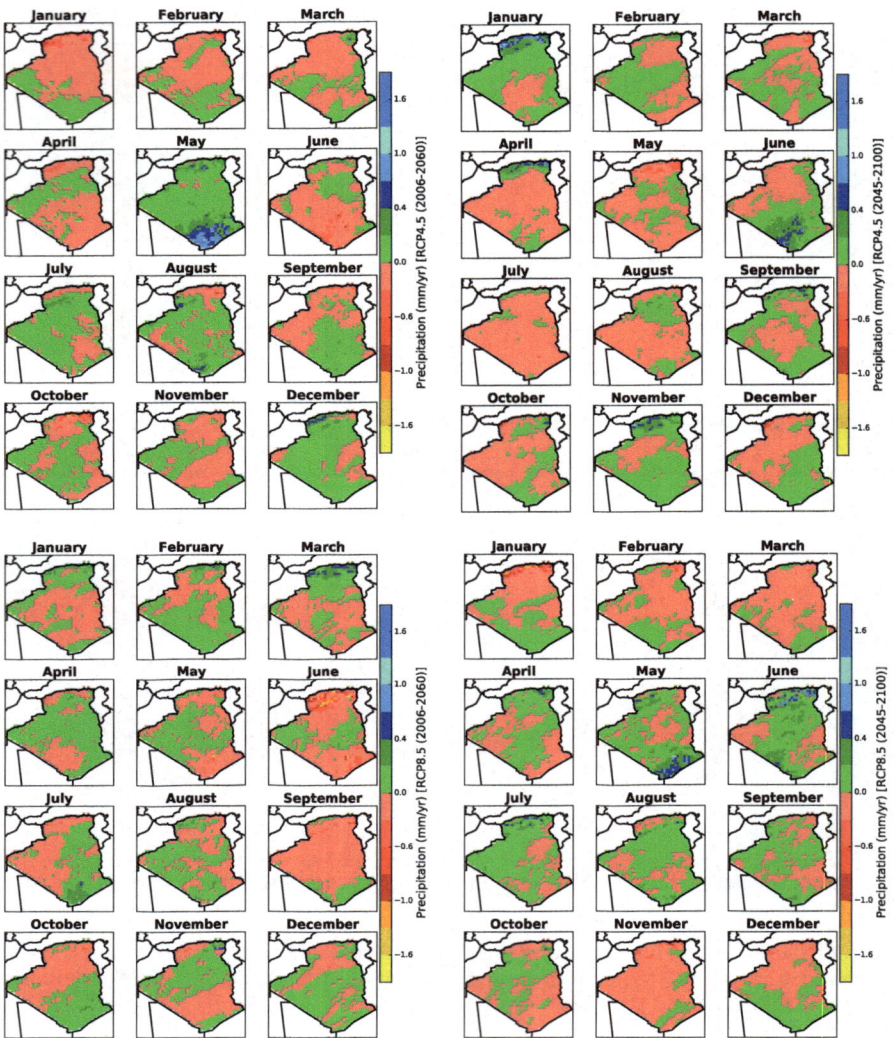

Fig. 12 Slopes of regression lines fitted to the temporal variability of monthly precipitation over the 2006–2060 and 2045–2100 time periods for the scenarios RCP4.5 and RCP8.5 (RCA4-NorESM1-M) at a resolution of 0.44° × 0.44°

second time period (2045–2100), the rate of increase will not exceed +0.04°C/year for RCP4.5 scenario for all months except for February and April, where a temperature decrease on the order of −0.04°C/year is predicted. For RCP8.5 scenario and during the second time period, the rate of increase ranges from +0.02°C/year to +0.04°C/year for the six winter months (October to March) and from +0.04°C/year to +0.1°C/year for the six summer months (April to September).

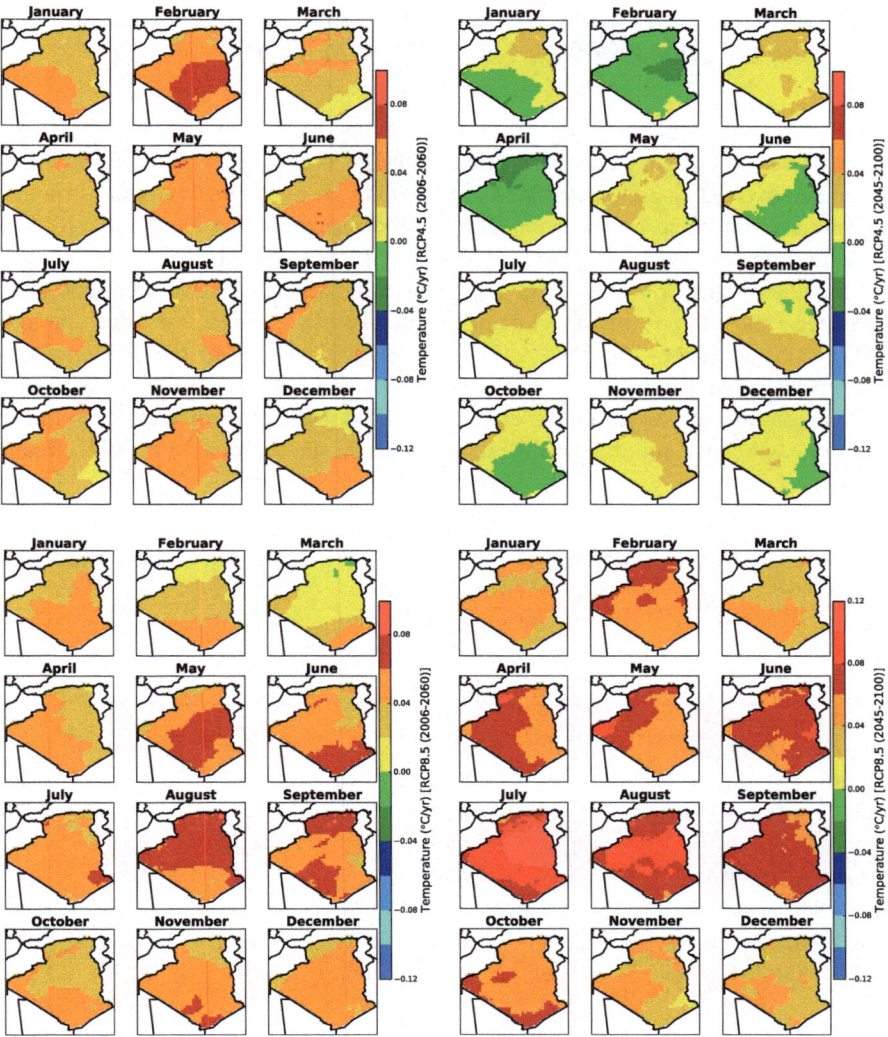

Fig. 13 Slopes of regression lines fitted to the temporal variability of monthly temperatures over the 2006–2060 and 2045–2100 time periods for the scenarios RCP4.5 and RCP8.5 (RCA4-NorESM1-M) at a resolution of 0.44° × 0.44°

5.5   Climate Zone Surface Areas and Their Projected Shifts for the 2006–2060 and 2045–2100 Periods

Once the modeled monthly precipitation and temperature of all models simulations were corrected using the quantile mapping (QM) bias correction algorithm and their averaged over 2006–2060 and 2045–2098 periods, we proceeded to the construction of climate zone map for each model simulation according to the Koppen-Geiger classification and computing the surface extent of each climate zones.

Results of shifts obtained from all models simulations (in % of total land area) between the three main climatic zones surface area of the periods 1951–2005 vs. 2006–2060 and 1951–2005 vs. 2045–2100 for the two scenarios are presented in Table 4. Projected shifts (in % of total land area) for all models simulations are similar for the desert climate (BW) and the warm temperate climate (C) zone, while we notice a considerable increase in surface area for the BW climate

Table 4 Shifts (in % of total land area) between the three main climatic zones surface area of the periods 1951–2005 vs. 2006–2060 and 1951–2005 vs. 2045–2100 for RCP 45 and RCP 85 scenarios

	2006–2060					
	RCP 45			RCP 85		
	C	BS	BW	C	BS	BW
CanESM (Canada)	−0.83	−0.52	1.33	−1.53	−0.08	1.6
CNRM-CM5 (France)	−2.46	−0.26	2.71	−1.12	0.07	1.04
CSIRO-MK3 (Australia)	−2.32	0.43	1.88	−2.01	0.44	1.54
IPSL-CM5A (France)	−2.99	−0.32	3.29	−2.71	−0.5	3.18
MIROC5 (Japan)	−1.66	0.42	1.23	−1.22	−0.34	1.54
HadGEM2-ES (UK)	−2.45	−0.09	2.52	−2.53	0.29	2.24
MPI-ESM-LR (Germany)	−2.44	−0.51	2.94	−2.19	0.21	1.97
NorESM1-M (Norway)	−2.04	0.24	1.79	−3.03	0.45	2.58
GFDL-ESM2M (USA)	−1.76	−0.23	1.98	−2.79	0.18	2.6
Mean	−2.106	−0.093	2.186	−2.126	0.080	2.032
	2045–2100					
	RCP 45			RCP 85		
CanESM (Canada)	−2.29	−0.12	2.41	−3	0.15	2.84
CNRM-CM5 (France)	−2.57	−0.33	2.9	−2.84	0.13	2.69
CSIRO-MK3 (Australia)	−3.19	0.63	2.55	−3.59	0.55	3.03
IPSL-CM5A (France)	−3.71	−0.53	4.23	−4.45	−3.27	7.71
MIROC5 (Japan)	−2.32	0.78	1.53	−3.18	0.24	2.93
HadGEM2-ES (UK)	−3.15	0.49	2.65	−3.9	−0.35	4.24
MPI-ESM-LR (Germany)	−3.42	−1.35	4.75	−3.94	−0.6	4.54
NorESM1-M (Norway)	−3.26	0.19	3.06	−3.9	0.21	3.69
GFDL-ESM2M (USA)	−2.45	−0.26	2.7	−3.55	−0.76	4.29
Mean	−2.929	−0.056	2.976	−3.594	−0.411	3.996

Note that 1% corresponds to an area of 23,820 km^2

zone and also a decrease in surface area of (C) zone. For steppe climate zone (BS), the projected shift was not found synchronous from one model simulation to another and from one scenario to another.

For (BW) and (C) climate zones, the shift of expansion and contraction between the main Koppen-Geiger climate class areas simulated by all models over the first projected period 2006–2060 is similar for both scenarios. However, the RCP8.5 scenario shows the largest shifts for the second projected period 2045–2100. In the observational period 1951–2006, a total of 88.14 % of the total land area of Algeria is covered by climates of type BW, followed by 7.04% BS climates and 4.83% C climates, respectively. Assuming an RCP8.5 scenario for the second projected period 2045–2100 (Table 4), the nine model projections result in increased ranges from +2.7% to +7.7% of BW climates zones area, as well as a decreased ranges from −4.45% to −2.84% of C climate zone area. The mean area of the nine RCM-RCA4 projections during the second period results in a small-decreased ranges from −0.05% for RCP4.5 to −0.4% for RCP8.5. Therefore, the desert zone expansion has been established at the expense of warm temperate zones.

6 Discussion and Conclusion

The spatial variability and temporal evolution of precipitation and temperature over Algeria were analyzed over 1951–2100 using a set of observational data and nine RCM RCA4 simulations from the CORDEX-Africa program. The analyses are done at annual and monthly time scales that lead to the three main findings:

1. Over the historical period 1951–2005, the long-term trends of precipitation and temperature are characterized by an increase in annual mean temperature of about +0.02°C/year in the western part of the country and of +0.04°C/year in the eastern part and in the same time a 0.5–1.5 mm/year decrease in annual mean precipitation in the northern part of the country. A range of 1–1.25°C of warming has been reported in the fifth IPCC report [14] over Algeria during 1901–2010. Similar magnitude of warming was also reported by Giorgi [21] and New et al. [22] over Mediterranean Basin at different time intervals from the twentieth century. The same trends were noted in various regions of the Mediterranean Basin, for instance, Zeroual et al. [23] in northern Algeria, Philandras et al. [24] for the eastern Mediterranean, and Driouech [25] for Morocco. In Lebanon, mean temperatures have changed little before 1970 and then increased substantially over the last 30 years [26]. Several studies have highlighted a decrease in precipitation in the southern part of the Mediterranean Basin over the second half of the twentieth century [21, 27, 28]. The fifth IPCC report also noted a decrease in precipitation in northern Algeria of about 2.5–5 mm/year per decade from 1951 to 2005 [29]. Raymond et al. [30] found similar results for total precipitation from September to April in northern Algeria.

2. As far as the future evolution (2005–2100) of the temporal variability of annual and monthly precipitation and temperature in Algeria is concerned, the study shows that all models project an increase in temperature and a decrease in precipitation during the 1945–2100 period especially under RCP8.5 scenario.
3. The current decrease in precipitation and increase in temperature and the anticipated shrinking of the surface area of the temperate climate zone will lead to numerous problems related, among other things, to food security and displacement of local populations in Algeria. These considerations must be included in future socioeconomic development plans. The rate at which such changes will occur in the future in Algeria's three climate zones deserves special attention.

Acknowledgment The authors acknowledge the use of temperature and precipitation gridded data from both the Climatic Research Unit (CRU) and the RCA4 regional climate model of Rossby Centre (SMHI).

Conflict of Interest The authors declare that they have no conflict of interest.

References

1. Williams JW, Jackson ST, Kutzbach JE (2007) Projected distributions of novel and disappearing climates by 2100 AD. Proc Natl Acad Sci U S A 104(14):5738–5742
2. Folland CK, Karl TR, Christy JR, Clarke RA, Gruza GV, Jouzel J, Mann ME, Oerlemans J, Salinger MJ, Wang S-W (2001) Chapter 2: observed climate variability and change. In: Climate change 2001: the scientific basis. Cambridge University Press, Cambridge, pp 99–182
3. Lionello P, Malanotte-Rizzoli P, Boscolo R, Alpert P, Artale V, Li L, Luterbacher J, May W, Trigo R, Tsimplis M, Ulbrich U, Xoplaki E (2006) The Mediterranean climate: an overview of the main characteristics and issues. Dev Earth Environ Sci 4(C):1–26
4. Bolle H-J (2003) Mediterranean climate: variability and trends. Springer, Berlin
5. Driouech F, Déqué M, Mokssit A (2008) Numerical simulation of the probability distribution function of precipitation over Morocco. Climate Dynam 32(7–8):1055–1063
6. Christensen JH, Hewitson B, Busuioc A, Chen A, Gao X, Held R et al (2007) Regional climate projections. Climate change, 2007: the physical science basis. Contribution of Working Group I to the fourth assessment report of the intergovernmental panel on climate change, vol 11. Cambridge University Press, Cambridge, pp 847–940
7. Giorgi F, Jones C, Asrar GR (2009) Addressing climate information needs at the regional level: the CORDEX framework. Bull World Meteorol Organ 58(3):175–183
8. Kalognomou EA, Lennard C, Shongwe M, Pinto I, Favre A, Kent M, Hewitson B, Dosio A, Nikulin G, Panitz HJ, Büchner M (2013) A diagnostic evaluation of precipitation in CORDEX models over Southern Africa. J Climate 26(23):9477–9506
9. Köppen W (1936) Das geographisca System der Klimate. In: Koppen W, Geiger G, Gebr C (eds) Handbuch der Klimatologie. Borntraeger, Berlin, pp 1–44
10. Tanarhte M, Hadjinicolaou P, Lelieveld J (2012) Intercomparison of temperature and precipitation data sets based on observations in the Mediterranean and the Middle East. J Geophys Res 117:D12
11. Jones C, Giorgi F, Asrar G (2011) The Coordinated Regional Downscaling Experiment: CORDEX–an international downscaling link to CMIP5. Clivar Exch 56:34–40
12. Nikulin G, Jones C, Giorgi F, Asrar G, Büchner M, Cerezo-Mota R, Christensen OB, Déqué M, Fernandez J, Hänsler A, Van Meijgaard E, Samuelsson P, Sylla MB, Sushama L (2012)

Precipitation climatology in an ensemble of CORDEX-Africa regional climate simulations. J Climate 25(18):6057–6078

13. IPCC Working Group 1 et al (2013) IPCC, 2013: Climate change (2013): the physical science basis. Contribution of Working Group I to the fifth assessment report of the intergovernmental panel on climate change, IPCC, vol AR5, p 1535

14. IPCC (2013) Working Group I contribution to the IPCC fifth assessment report, climate change 2013: the physical science basis, IPCC, vol AR5, March 2013, p 2014

15. Schulzweida U, Kornblueh L, Quast R (2007) CDO user's guide, Climate Data Operators

16. Sennikovs J, Bethers U (2009) Statistical downscaling method of regional climate model results for hydrological modelling. 18th World IMACS/MODSIM Congress, Cairns, Australia, 13–17 July

17. Nguyen H, Mehrotra R, Sharma A (2017) Can the variability in precipitation simulations across GCMs be reduced through sensible bias correction? Climate Dynam 49:1–19

18. Kottek M, Grieser J, Beck C, Rudolf B, Rubel F (2006) World map of the Köppen-Geiger climate classification updated. Meteorol Z 15(3):259–263

19. Peel BL, Finlayson BL, McMahon TA (2007) Updated world map of the Koppen-Geiger climate classification. Hydrol Earth Syst Sci 11:1633–1644

20. Zeroual A, Assani AA, Meddi M, Alkama R (2019) Assessment of climate change in Algeria from 1951 to 2098 using the Köppen–Geiger climate classification scheme. Climate Dynam 52 (1–2):227–243

21. Giorgi F (2002) Variability and trends of sub-continental scale surface climate in the twentieth century. Part II: AOGCM simulations. Climate Dynam 18(8):693–708

22. New M, Todd M, Hulme M, Jones P (2001) Precipitation measurements and trends in the twentieth century. Int J Climatol 21:1899–1922

23. Zeroual A, Assani A, Meddi M (2017) Combined analysis of temperature and rainfall variability as they relate to climate indices in Northern Algeria over the 1972–2013 period. Hydrol Res 48 (2):584–595

24. Philandras CM, Nastos PT, Kapsomenakis IN, Repapis CC (2015) Climatology of upper air temperature in the Eastern Mediterranean region. Atmos Res 152:29–42

25. Driouech F (2006) Étude des indices de changements climatiques sur le Maroc: températures et précipitations. Direction de la Meteorologie Nationale 'INFOMET', Casablanca

26. Ramadan HH, Beighley RE, Ramamurthy AS (2013) Temperature and precipitation trends in Lebanon's largest river: the Litani Basin. J Water Resour Plan Manag 139(1):86–95

27. Norrant C, Douguédroit A (2006) Monthly and daily precipitation trends in the Mediterranean (1950–2000). Theor Appl Climatol 83(1–4):89–106

28. Tramblay Y, El Adlouni S, Servat E (2013) Trends and variability in extreme precipitation indices over Maghreb countries. Nat Hazards Earth Syst Sci 13(12):3235–3248

29. Stocker T, Qin D, Plattner G, and Tignor M (2013) Climate change 2013: the physical science basis. Contribution of working group I to the fifth assessment report of the intergovernmental panel

30. Raymond F, Ullmann A, Camberlin P (2016) Précipitations intenses sur le Bassin Méditerranéen: quelles tendances entre 1950 et 2013? Cybergeo Eur J Geogr 760:8504

Comparison of Evolving Connectionist Systems (ECoS) and Neural Networks for Modelling Daily Pan Evaporation from Algerian Dam Reservoirs

Abderrazek Sebbar, Salim Heddam, Ozgur Kisi, Lakhdar Djemili,
and Larbi Houichi

Contents

A. Sebbar
Soil and Hydraulics Laboratory, Hydraulics Department, Faculty of Engineering Sciences,
University Badji-Mokhtar Annaba, Annaba, Algeria
e-mail: rsebbar@yahoo.fr

S. Heddam (✉)
Hydraulics Division, Agronomy Department, Faculty of Science, Laboratory of Research in
Biodiversity Interaction Ecosystem and Biotechnology, Skikda, Algeria
e-mail: heddamsalim@yahoo.fr

O. Kisi
School of Technology, Ilia State University, Tbilisi, Georgia
e-mail: ozgur.kisi@iliauni.edu.ge

L. Djemili
Research Laboratory of Natural Resources and Adjusting, Hydraulics Department, Faculty of
Engineering Sciences, University Badji-Mokhtar Annaba, Annaba, Algeria
e-mail: lakhdardjemili@gmail.com

L. Houichi
Department of Hydraulic, University of Batna 2, Batna, Algeria
e-mail: houichilarbi@yahoo.fr

Abdelazim M. Negm, Abdelkader Bouderbala, Haroun Chenchouni, and
Damià Barceló (eds.), *Water Resources in Algeria - Part I: Assessment
of Surface and Groundwater Resources*, Hdb Env Chem (2020) 97: 161–180,
DOI 10.1007/698_2020_527, © Springer Nature Switzerland AG 2020,
Published online: 18 June 2020

Abstract Evaporation (*EP*) from dams' reservoirs measured using pans is one of the most important methods adopted for quantifying the *loss* of *water* through *evaporation. Black box artificial intelligence techniques (AI) have been developed as alternative approaches for quantifying evaporation, and several kinds of models have been proposed worldwide. The present study uses the measurement of several climatic variables such as air temperature*, wind speed, and relative humidity to test the performances of new AI techniques called evolving connectionist systems (ECoS), applied for predicting daily evaporation from several dam reservoirs located in Algeria country. Two ECoS models, namely, (1) offline-based dynamic evolving neural-fuzzy inference systems named DENFIS_OF and (2) online-based dynamic evolving neural-fuzzy inference systems named DENFIS_ON, were applied and compared for predicting daily evaporation. The results using ECoS models were compared to multiple linear regression (MLR) and artificial neural network (ANN) models. From the results obtained, it is seen that the ECoS models could predict daily evaporation from dam reservoirs with better accuracy than the ANN and MLR models.

Keywords ANN, Dam reservoirs, DENFIS, ECoS, Evaporation, MLR, Modelling

1 Introduction

Exact quantification of the component of the water budget especially for surface water such as dam reservoirs is a challenge. It is necessary to have a clear knowledge of the water budget of a reservoir to meet to any possible demand for water and for a rational *water* planning and management. Among all the components of water budget, estimation of the water losses through evaporation is highly important and has an indirect relation with the other components. Water evaporation generally called as pan evaporation (EP) is measured using (1) empirical equations in which measurement of several weather variables are needed and (2) direct measurement using evaporimeter pan. According to the World Meteorological Organization (WMO), the Class A pan is *the most precise instrument for measuring* EP. Recently an alternative approach based on the use of artificial intelligence (AI) techniques has gained much popularity, and a number of researches worldwide have reported the successful use of AI models for modelling EP at many time steps.

Keskin et al. [1] compared two data-driven models for modelling daily EP in the Lake Eğirdir, Turkey. Using several climatic variables such as air and water temperatures (T_{mean} and TE), relative humidity (RH), solar radiation (SR), wind speed (W), air pressure (Pa), and sunshine hours (SH), the authors applied the adaptive neuro-fuzzy inference systems (ANFIS) and the fuzzy logic (FL) model and found that the ANFIS was better than the FL in modelling daily EP with a coefficient of determination (R^2) equal to 0.96 in validation phase. Rahimikhoob [2] compared the multilayer perceptron neural network (MLPNN) and the empirical Hargreaves equation (HG) for modelling daily EP in the semiarid region of Iran using only limited climatic variables, such as maximum and minimum air temperatures (T_{max} and T_{min}) and the extraterrestrial radiation (Ra). Dogan et al. [3] compared the ANFIS model and multiple linear regression (MLR) for modelling daily EP in the reservoir of Yuvacik dam, Turkey. By using four climatic variables (T_{mean}, RH, SR, and W), the authors demonstrated that the ANFIS provided better accuracy compared to the MLR model, with R^2 values equal to 0.96 and 0.90 for ANFIS and MLR, respectively. Sanikhani et al. [4] compared several models for modelling daily EP using T_{mean}, RH, W, and SR as input variables. The authors compared two ANFIS models; ANFIS with grid partitioning (ANFIS_G) and ANFIS with subtractive clustering (ANFIS_S), MLPNN, multivariate nonlinear regression (MNLR), and Stephens-Stewart (SS) and Penman (PN) models. According to the results obtained, two ANFIS models provided relatively similar results, and they were better than the MLPNN, MNLR, and SS and Penman models. Kim et al. [5] applied three kinds of artificial neural network models, namely, MLPNN, generalized regression neural networks (GRNN) model, and support vector machine (SVM), for modelling daily EP using climatic data from four weather stations: two weather stations in Republic of Korea and two weather stations in Iran.

Kisi et al. [6] investigated the capabilities of generalized ANFIS model called (GNF) for modelling daily EP using large data sets from the USA and one station from Iran. The authors have compared the results obtained using the GNF model with those obtained using PN, SS, and Griffiths methods. According to the results obtained, the authors demonstrated the robustness of the GNF model compared to the empirical models. Tabari et al. [7] proposed a new kind of model based on fuzzy logic, called coactive neuro-fuzzy inference system (CANFIS), for modelling daily EP in the semiarid region of Iran. By comparing several input combinations of climatic variables, the authors demonstrated that the MLPNN model is better than the CANFIS model. Wu et al. [8] compared the SVM and SS models for estimating daily EP in Taiwan. The authors demonstrated that by using the optimal input variables, the SVM model performed better compared to the SS model. Terzi [9] compared the gene expression programming (GEP) and ANFIS model in estimating daily EP in Turkey using the EP measured at the previous day as inputs and demonstrated that the GEP is superior to ANFIS model. Kim et al. [10] firstly applied cascade correlation neural networks (CCNN) model for estimating daily EP in two different regions; the inland and coastal in Republic of Korea. By comparing the CCNN model to the standard MLPNN using two different scenarios: local implementation and cross station scenarios, the authors demonstrated that the

CCNN provided better accuracy. In another study, Kim et al. [11] proposed a new model called Kohonen self-organizing feature maps – neural networks (K_ANN) for modelling daily EP in a dry climate region of Iran. The authors have proposed new strategies for modelling daily EP using the T_{mean}, SR, and SH measured at 3 previous days as inputs. Compared to the GEP, MLPNN, and MLR models, the authors reported that the K_ANN is slightly superior to the other three models. Using T_{max}, T_{min}, RH, W, and SH as input variables, Malik and Kumar [12] demonstrated that the MLPNN model is better than CANFIS and MLR in modelling daily EP in India.

Kisi et al. [13] compared three machine learning models, namely, classification and regression tree (CRT), chi-squared automatic interaction detector (CHAID) and MLPNN, for modelling daily EP using climatic variables from two weather stations in Turkey. The authors demonstrated that MLPNN is slightly superior to the two other models. Allawi and El-Shafie [14] investigated the ability of radial basis function neural network (RBF-NN) and ANFIS in predicting daily EP at Layang Reservoir, Malaysia. Some other important investigations can be found in the literature which highlights the importance of data-driven models in estimating pan evaporation [15–17]. Recently, Eray et al. [18] introduced two machine learning methods at the first time in the area of pan evaporation modelling: (1) evolving connectionist systems, the dynamic evolving neural-fuzzy inference systems named DENFIS and (2) multi-gene genetic programming (MGGP). Using five climatic variables: T_{max}, T_{min}, W, SH, and RH, collected at two climatic stations in Turkey, the authors demonstrated that the DENFIS performed better at one station, while the MGGP had better accuracy in other station. They used only one DENFIS method in their study.

To the best of our knowledge, only the study conducted by Eray et al. [18], *no other investigations* have addressed the application of any evolving connectionist systems (ECoS) for modelling daily EP. Hence, in the present study, we investigated to demonstrate the usefulness and robustness of two different DENFIS methods for modelling daily EP from several dam reservoirs in Algeria country. The DENFIS methods were compared to the standard MLPNN and MLR. Four climatic variables T_{max}, T_{min}, RH, and W_S were used as inputs to the applied models in the current study.

2 Study Area and Data

In this study, we selected daily climatic data from two weather stations located at two dam reservoirs in the east of Algeria: (1) *El*-Agrem dam reservoir at *Jijel* Province and (2) Bou-Hamdane dam reservoir at *Guelma* Province. Consequently, in the remaining part of the present study, *we refer* to these *two stations as Guelma station* (latitude $43°19'29''$, longitude $122°59'55''$) *and Jijel station* (latitude $45°21'24''$, $122°41'02''$). The locations of the *Guelma* and *Jijel* stations are shown in Fig. 1. Guelma and Jijel are located in the semiarid and Mediterranean regions of Algeria, respectively. The data cover 13 years (Table 1, 2004–2016) for Guelma and

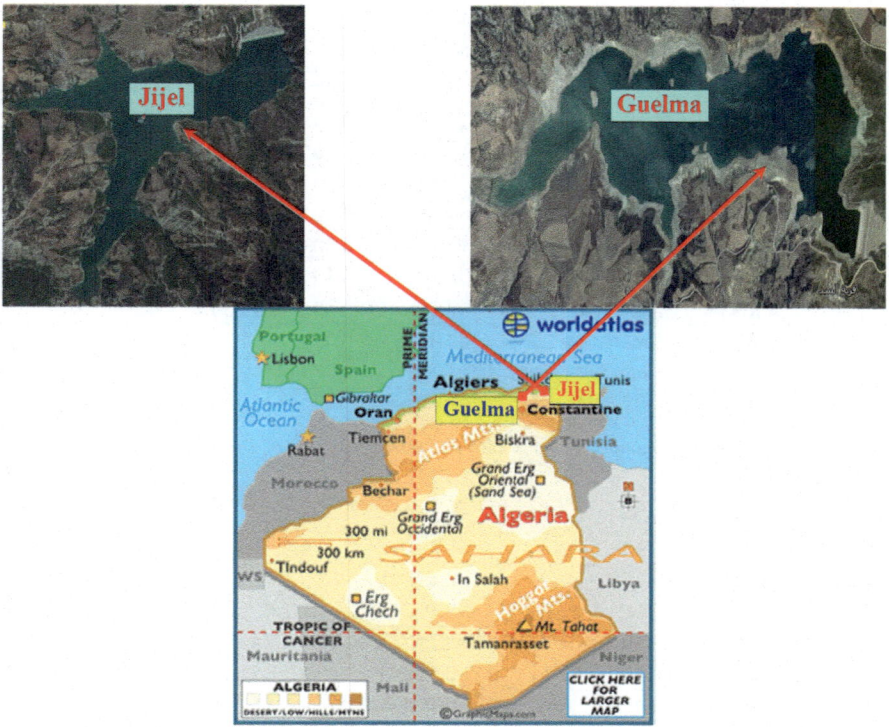

Fig. 1 Location of the *El*-Agrem dam reservoir (*Jijel*) and Bou-Hamdane dam reservoir (*Guelma*)

Table 1 Data set used for both stations

Station	Latitude	Longitude	Period of record	Total pattern	Incomplete pattern	Final pattern
Guelma	43°19′29″	122°59′55″	01/01/2004–31/12/2016	4,749	1,183	3,566
Jijel	45°21′24″	122°41′02″	01/01/2003–31/12/2015	4,748	4	4,474

(2003–2015) for Jijel stations. For each station, we selected daily records of maximum and minimum temperature (T_{max}, T_{min}), wind speed (W_S), relative humidity (RH), and the daily pan evaporation measured at Class A pan (EP). For each station, we randomly split the data into training (70%) and validation (30%). Hence, for Guelma station, we have total 2,497 patterns as a training subset and 1,069 as a validation subset. Similarly, for Jijel station we have total 3,132 patterns as a training subset and 1,342 as a validation subset. Daily statistical parameters of the climatic variables are given in Table 2. In the table, the X_{mean}, X_{max}, X_{min}, S_x, C_v, and R denote the mean, maximum, minimum, standard deviation, variation coefficient, and correlation coefficient with EP, respectively. T_{max} has the highest correlation with EP followed by the T_{min}, RH, and WS in both stations. WS has the highest variation and

Table 2 Statistical parameters of the used data sets for Guelma and Jijel stations

Station	Data set	Unit	X_{mean}	X_{max}	X_{min}	S_x	C_v	R
Guelma	T_{max}	°C	25.637	46.70	5.20	8.492	0.331	0.838
	T_{min}	°C	11.441	32.70	0.00	6.198	0.542	0.767
	RH	%	62.689	96.00	17.00	15.108	0.241	−0.639
	WS	m/s	1.990	12.667	0.00	1.528	0.768	0.044
	EP	mm	3.710	12.00	0.00	2.689	0.725	1.000
Jijel	T_{max}	°C	23.524	43.00	6.00	6.308	0.268	0.779
	T_{min}	°C	13.582	29.20	0.10	5.604	0.413	0.690
	RH	%	72.982	95.70	20.50	10.20	0.140	−0.409
	WS	m/s	2.535	14.10	0.00	1.726	0.681	−0.062
	EP	mm	4.047	18.30	0.070	2.752	0.680	1.000

Table 3 The input combinations of different models

Models				Input combinations
DENFIS_OF	DENFIS_ON	MLPNN	MLR	
DENFIS_OF1	DENFIS_ON1	MLPNN1	MLR1	T_{max}, T_{min}, RH, WS
DENFIS_OF2	DENFIS_ON2	MLPNN2	MLR2	T_{max}, T_{min}, RH,
DENFIS_OF3	DENFIS_ON3	MLPNN3	MLR3	T_{max}, T_{min}, WS
DENFIS_OF4	DENFIS_ON4	MLPNN4	MLR4	T_{max}, T_{min}
DENFIS_OF5	DENFIS_ON5	MLPNN5	MLR5	T_{max}, RH
DENFIS_OF6	DENFIS_ON6	MLPNN6	MLR6	T_{max}, WS

lowest correlation with EP. The climatic variables have a higher correlation with EP in Guelma compared to Jijel. In the present investigation, we evaluated several combinations of the input climatic variables, and in total six scenarios were compared (Table 3). Before applying the models, all the data were normalized *using the Z-score method* [19–21], calculated by the following formula:

$$x_{ni,k} = \frac{x_{i,k} - m_k}{Sd_K} \tag{1}$$

where $x_{ni, k}$ is the normalized value of the variable k (input or output) for each sample i. $x_{i,k}$ the original value of the variable k (input or output). m_k and S_{dk} are the mean value and standard deviation of the variable k (input or output).

3 Materials and Methods

In the present study, three kinds of models were developed and compared: multilayer perceptron neural network (MLPNN), dynamic evolving neural-fuzzy inference systems (DENFIS), and multiple linear regression (MLR). The flow charts for training and validation of the MLPNN, ANFIS, and MLR are shown in Fig. 2.

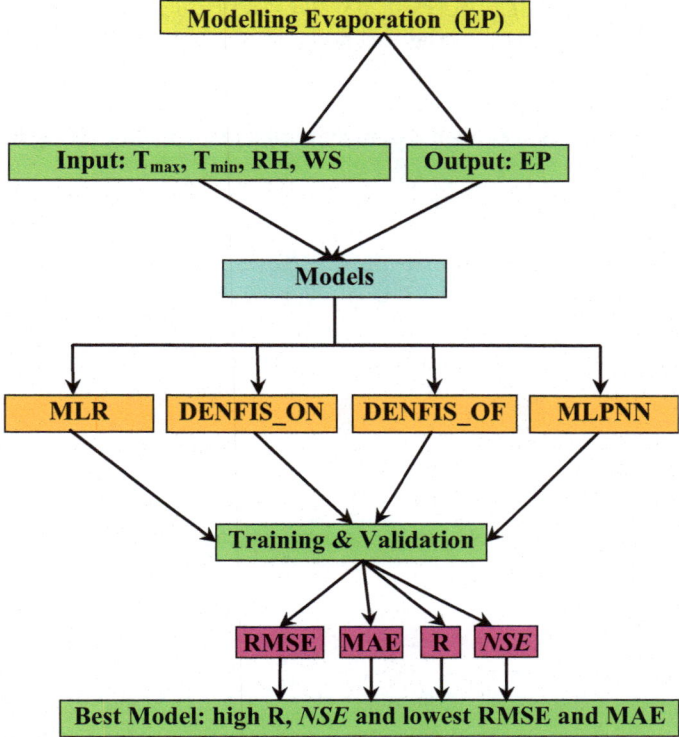

Fig. 2 Flow chart for the proposed DENFIS, MLPNN, and MLR models

3.1 Dynamic Evolving Neural-Fuzzy Inference Systems (DENFIS)

During the last decades, neuro-fuzzy models (NF) have been widely used for modelling several complex hydrological processes; this is no doubt because the NF models have high capabilities and efficiency. Since the development and the proposition of the first NF model, several modified versions have been introduced and published, and the training of any NF model is carried out in order to achieve the tasks: (1) identifying the linear (consequent) and nonlinear (premises) parameters of the fuzzy membership functions and (2) determining the structure of fuzzy rule base (FR). Since the complexity of the NF model depends directly on *the number* of *optimization parameters, all the improvements made were* centered on optimization of the FR. One of the most important NF models proposed during the last few years is certainly dynamic evolving neural-fuzzy inference systems (DENFIS) proposed by Kasabov and Song [22]. DENFIS is based on the so-called evolving clustering method (ECM). The ECM algorithm works with respect to two important conditions. The first *condition* is that, contrary to other NF models in which the FR is fixed at the beginning of the learning process, the number of rules is determined using

ECM during the evolving process, and new fuzzy rules are created during the learning process of DENFIS. The second condition is related to the structure of the rule nodes. Using the standard NF model, the rule nodes are connected to the input and output variables (x and y) at the beginning of the training. However, using the ECM, only a part of the rule nodes is connected to the x and y during the evolving process [23].

Similar to the well-known partitioned clustering methods such as fuzzy C-means (*FCM*), subtractive clustering, the gird partition methods, or the self-organizing maps, the ECM must be used *for partitioning the inputs space into several partitions based on the calculation of the E*uclidean distance between new data samples and an existing cluster [22]. Before the creation of the fuzzy rule of DENFIS model, the training data set must be clustered. The fuzzy rules base can be presented as follows [23]:

$$
\begin{cases}
\text{if } x_1 \text{ is } R_{11} \text{ and } x_2 \text{ is } R_{12} \text{ and} \ldots \text{and } x_q \text{ is } R_{1q}, \text{then } y \text{ is } f_1\left(x_1, x_2, \ldots, x_q\right) \\
\text{if } x_1 \text{ is } R_{21} \text{ and } x_2 \text{ is } R_{22} \text{ and} \ldots \text{and } x_q \text{ is } R_{2q}, \text{then } y \text{ is } f_2\left(x_1, x_2, \ldots, x_q\right) \\
\text{if } x_1 \text{ is } R_{m1} \text{ and } x_2 \text{ is } R_{m2} \text{ and} \ldots \text{and } x_q \text{ is } R_{mq}, \text{then } y \text{ is } f_m\left(x_1, x_2, \ldots, x_q\right)
\end{cases}
$$

$$(2)$$

where "x_j is R_{ij}", $i = 1,2 \ldots m; j = 1,2, \ldots q$ are $m \times q$ fuzzy propositions that form m antecedents for m fuzzy rules, respectively; $x_j, j = 1,2,\ldots,q$ are antecedent variables defined over universes of discourse $X_j, j = 1,2, \ldots, q$; and $R_{ij}, i = 1,2, \ldots m$; $j = 1,2, \ldots, q$ are fuzzy sets defined by their fuzzy membership functions $\mu_{Rij}: X_j \rightarrow [0,1], i = 1, 2, \ldots, m; j = 1,2, \ldots, q$ [23]. DENFIS has two versions: online called DENFIS_ON and offline called DENFIS_OF. T*wo versions* differ regarding the structure of the fuzzy rule base and the optimization of the consequent parameters, and the versions have the same triangular fuzzy membership functions. Our objective *is* to evaluate the differences between the *two versions*. More detail about DENFIS models can be found in [22, 23]. During the last few years, DENFIS has been applied successfully for solving various problems: Modelling daily pan evaporation [18], rainfall-runoff modelling and river routing forecasting [24], modelling hourly dissolved oxygen (DO) concentration [25], predicting DO in wastewater treatment plant [26], modelling coagulant dosage in water treatment plant [27], runoff forecasting [28], real data streams prediction [29], modelling and prediction of the Value-at-Risk [30], predicting banking efficiency [31], predicting interior spruce wood density [32], android malware classification using permission-based features [33], water level forecasting [34], solar radiation prediction [35], and river-channel confluence [36].

3.2   *Artificial Neural Network (ANN)*

Artificial neural networks are the most important categories of machine learning models used in engineering science. An ANN is a system composed of an ensemble of an element called neurons inspired by the function of the human brain. The ANN is structured in several parallel layers, and even though all layers are of equal importance from a mathematical point of view, only the input and the output layers are known, while the hidden layers can be of any number and with any activation function. ANNs with only one hidden layer are universal approximators [37, 38] and adopted in the present study. A simple three-layered network with an input layer, hidden layer, and output layer is shown in Fig. 3 and called as a multilayer perceptron neural network (*MLPNN*). In the first layer, we have four input variables (T_{max}, T_{min}, RH, and W$_S$). In the single hidden layer, we have several neurons determined by trial and error and, finally, only one neuron in the output layer and which correspond to the EP. The information flows from input to output layers and the ability of the MLPNN model for mapping the inputs to the output and the *construction* of the *robust* nonlinear *models is stored into the weight connection and biases, randomly initialized and optimized during the learning process. All neurons in the model except the neurons in input layer have two tasks that must be done:* (1) receiving the inputs from other neurons and m*ultiplying* each of these *inputs* by a *weight and adding a bias to the result and* (2) passing the results via an activation function to the neuron in the next layer. The activation function is generally the sigmoid. Determination of the best parameters (weights and biases) is achieved using a learning algorithm, generally the back propagation (BP) algorithm [39]. The BP algorithm belongs to the supervised training category and is based on the principle that weight and biases must be adjusted automatically during an *iterative process with the objective: minimizing the error function between* the *observed* and predicted value of daily EP. The most and broadly used error function is the mean square error (MSE). More detail about the theory of the ANN models can be found in [40].

Fig. 3 Multilayer perceptron neural network (*MLPNN*) structure for modelling daily EP

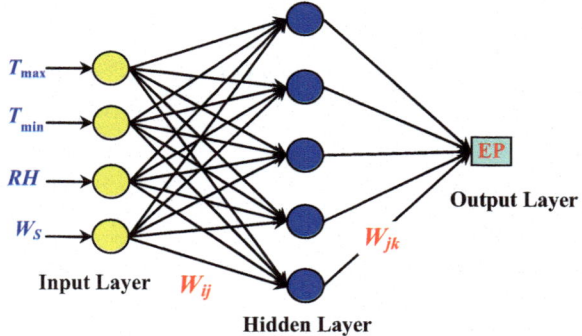

3.3   Multiple Linear Regression (MLR)

Multiple linear regression (MLR) is a widely used statistical modelling technique in which the dependent variable (EP) is related to a set of independent variables called predictors (T_{max}, T_{min}, RH, and W_S). The MLR model is written as:

$$EP = \beta_0 + \beta_1 \times T_{max} + \beta_2 \times T_{min} + \beta_3 \times RH + \beta_4 \times W_S \tag{3}$$

β_i are the parameters of the model that must be determined.

3.4   Performance Indices

In the present study, the following four performance indices were used, where N is the number of data points, O_i is the measured value, and P_i is the corresponding calculated value of EP. O_m and P_m are the average values of O_i and P_i, respectively: the coefficient of correlation (R), the Nash-Sutcliffe efficiency (NSE), the root mean squared error (RMSE), and the mean absolute error (MAE).

$$R = \left[\frac{\frac{1}{N}\sum(O_i - O_m)(P_i - P_m)}{\sqrt{\frac{1}{N}\sum_{i=1}^{n}(O_i - O_m)^2}\sqrt{\frac{1}{N}\sum_{i=1}^{n}(P_i - P_m)^2}} \right] \tag{4}$$

$$NSE = 1 - \frac{\sum_{i=1}^{N}[O_i - P_i]^2}{\sum_{i=1}^{N}[O_i - O_m]^2} \tag{5}$$

$$RMSE = \sqrt{\frac{1}{N}\sum_{i=1}^{N}(O_i - P_i)^2} \tag{6}$$

$$MAE = \frac{1}{N}\sum_{i=1}^{N}|O_i - P_i| \tag{7}$$

4 Results

4.1 Daily Pan Evaporation Estimation at Guelma Station

RMSE, MAE, R, and NSE values of the applied methods are shown in Table 4. For all the four models and all the six inputs combinations, the DENFIS_OF1 is the best model with the lowest RMSE and MAE and the highest R and NSE, while the MLR5 model is the worst in the training phase.

Moreover, better performances are obtained using the DENFIS_OF model, and, globally, the best accuracy was obtained using all the four climatic variables as inputs, even if some exceptions are represented by the MLPNN and DENFIS_ON models in which the best accuracy was obtained using only three input variables and when wind speed is excluded from the inputs. Between the two DENFIS models, the results obtained show that DENFIS_OF is more accurate than DENFIS_ON. For example, the best DENFIS_OF1 model had an R and NSE values equal to 0.891 and

Table 4 Performances of different models in modelling daily EP at Guelma station

	Training (70%)				Validation (30%)			
	RMSE	MAE	R	NSE	RMSE	MAE	R	NSE
Models	mm	mm	/	/	mm	mm	/	/
DENFIS_OF1	1.223	0.911	0.891	0.791	1.280	0.955	0.883	0.779
DENFIS_OF2	1.254	0.941	0.884	0.780	1.290	0.970	0.881	0.775
DENFIS_OF3	1.242	0.929	0.887	0.784	1.287	0.962	0.881	0.776
DENFIS_OF4	1.293	0.971	0.876	0.766	1.318	0.990	0.875	0.765
DENFIS_OF5	1.370	1.037	0.860	0.738	1.425	1.074	0.852	0.726
DENFIS_OF6	1.303	0.976	0.874	0.763	1.383	1.038	0.861	0.741
DENFIS_ON1	1.336	0.984	0.867	0.751	1.406	1.016	0.860	0.735
DENFIS_ON2	1.302	0.965	0.874	0.763	1.357	1.011	0.871	0.751
DENFIS_ON3	1.290	0.957	0.877	0.768	1.383	1.017	0.866	0.741
DENFIS_ON4	1.284	0.963	0.878	0.770	1.472	1.121	0.862	0.733
DENFIS_ON5	1.344	0.999	0.865	0.748	1.590	1.200	0.836	0.689
DENFIS_ON6	1.284	0.948	0.878	0.770	1.534	1.144	0.852	0.700
MLPNN1	1.237	0.922	0.887	0.786	1.283	0.964	0.882	0.777
MLPNN2	1.235	0.924	0.887	0.787	1.271	0.957	0.884	0.782
MLPNN3	1.227	0.913	0.889	0.790	1.283	0.954	0.882	0.777
MLPNN4	1.283	0.959	0.877	0.770	1.315	0.992	0.875	0.766
MLPNN5	1.354	1.013	0.863	0.744	1.418	1.056	0.853	0.728
MLPNN6	1.286	0.960	0.877	0.769	1.361	1.010	0.866	0.750
MLR1	1.342	1.033	0.865	0.748	1.398	1.088	0.858	0.733
MLR2	1.346	1.036	0.864	0.747	1.399	1.089	0.858	0.736
MLR3	1.348	1.041	0.864	0.746	1.403	1.093	0.857	0.734
MLR4	1.364	1.053	0.860	0.740	1.404	1.090	0.857	0.707
MLR5	1.449	1.129	0.841	0.707	1.515	1.181	0.831	0.658
MLR6	1.409	1.090	0.850	0.723	1.490	1.165	0.837	0.682

0.791, respectively, superior to the values obtained using the best DENFIS_ON2 ($R = 0.874$, NSE $= 0.763$). Overall, the DENFIS_OF performs better than the DENFIS_ON for the first three input combinations (combinations 1, 2 and 3, Table 3). However, using only two climatic variables as inputs (combinations 4, 5, and 6); DENFIS_ON models provide better accuracy compared to the DENFIS_OF models, and the best accuracy in terms of R, NSE, RMSE, and MAE was obtained using DENFIS_ON4 and DENFIS_ON6 with relatively similar performances and with comparable values. More in general, in the training phase, DENFIS_OF model is the best, followed by the MLPNN, the DENFIS_ON in the third place, and the MLR which has the poorest performance. Using only two input variables would generate *quite different* ranking of the models, the DENFIS_OF has the poorest performance compared to the DENFIS_ON and MLPNN models, and DENFIS_ON provided the best accuracy. According to Table 4, in the validation phase, and more in general, better performances are obtained using the MLPNN and DENFIS_OF models, and, globally, the best accuracy was obtained using all the four climatic variables as inputs. Therefore, to be more specific, the estimated daily EP obtained through DENFIS_OF is better than the results obtained using DENFIS_ON.

The R and the NS obtained by the best DENFIS_OF1 model ($R = 0.883$, NSE $= 0.779$) are higher than those obtained using the best DENFIS_ON2 model ($R = 0.871$, NSE $= 0.751$). Also the RMSE and MAE obtained using the DENFIS_OF1 (RMSE $= 1.280$ mm, MAE $= 0.955$ mm) are lower than those obtained using the DENFIS_ON2 model (RMSE $= 1.357$ mm, MAE $= 1.011$ mm). However, the results show a good estimate of daily EP in the validation phase using the MLPNN2 ($R = 0.884$, NSE $= 0.782$, RMSE $= 1.271$ mm), and it outperforms the DENFIS_ON and DENFIS_OF models. Moreover, using only two input variables (combinations 3, 4, and 5), the MLPNN model provides better accuracy compared to all other models. In addition to the all discussed results, the MLR model has the poorest performances. In all the four methods, 4th input combination (T_{max}, T_{min}) provides better accuracy than the 6th (T_{max}, WS) and 5th (T_{max}, RH) combinations. This implies that the daily EP can be predicted using only T_{max} and T_{min} inputs where other variables are not available in the studied station. Fig. 4 illustrates the comparison between the observed and estimated values of the daily EP using the four best models for Guelma station.

4.2 Daily Pan Evaporation Estimation at Jijel Station

For Jijel station, the performances of the four models (DENFIS_ON, DENFIS_OF, MLPNN, and MLR) were further evaluated mainly based on their predictability. The results of these models are reported in Table 5.

Similar to Guelma station, four measures were used to assess the predictions: RMSE, MAE, R, and NSE. From the results reported in Table 5, we can see that:

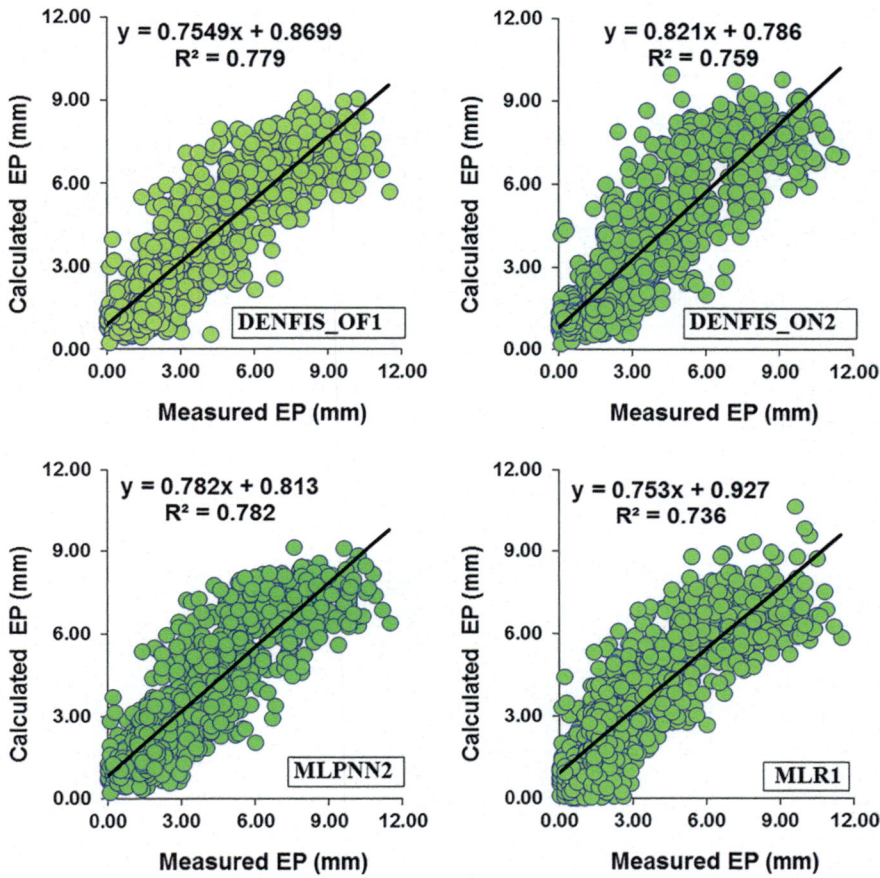

Fig. 4 Scatterplot of calculated and measured daily EP for the optimum developed models for the validation period at Guelma station

1. MLPNN and DENFIS_OF have relatively similar overall performances in the training and validation phases.
2. For different input combinations, their performances differ but perform similarly.
3. More in general and surprisingly, the most important point to note is that the MLR models perform consistently better compared to the DENFIS_ON models for all the input combinations in the validation phase.
4. The DENFIS_ON models are the poorest models compared to all others.

It is noteworthy that it was impossible to obtain better results using DENFIS_ON than those obtained using MLR, despite the increase of a number of epochs, a number of runs, and also by varying the important parameter of the DENFIS model: the distance threshold (*Dthr*). According to Table 5, in the training phase, it is clear that the MLPNN1 model is the best model and provides more accurate results than the DENFIS_OF1, DENFIS_ON1, and MLR1. Perhaps the only

Table 5 Performances of different models in modelling daily EP at Jijel station

Models	Training (70%)				Validation (30%)			
	RMSE (mm)	MAE (mm)	R/	NSE/	RMSE (mm)	MAE (mm)	R/	NSE/
DENFIS_OF1	1.662	1.266	0.797	0.635	1.636	1.244	0.804	0.647
DENFIS_OF2	1.672	1.282	0.794	0.631	1.643	1.253	0.802	0.643
DENFIS_OF3	1.669	1.273	0.795	0.632	1.649	1.257	0.801	0.641
DENFIS_OF4	1.683	1.289	0.791	0.626	1.653	1.264	0.800	0.639
DENFIS_OF5	1.691	1.299	0.789	0.622	1.657	1.266	0.799	0.637
DENFIS_OF6	1.688	1.293	0.790	0.623	1.658	1.267	0.799	0.637
DENFIS_ON1	1.815	1.359	0.757	0.565	1.763	1.330	0.770	0.589
DENFIS_ON2	1.755	1.337	0.772	0.593	1.738	1.332	0.777	0.601
DENFIS_ON3	1.748	1.313	0.774	0.596	1.714	1.305	0.784	0.612
DENFIS_ON4	1.710	1.282	0.784	0.614	1.732	1.329	0.782	0.604
DENFIS_ON5	1.729	1.314	0.779	0.605	1.708	1.299	0.785	0.615
DENFIS_ON6	1.721	1.301	0.781	0.609	1.709	1.305	0.784	0.614
MLPNN1	1.645	1.251	0.801	0.642	1.628	1.235	0.806	0.650
MLPNN2	1.666	1.270	0.796	0.633	1.643	1.251	0.802	0.644
MLPNN3	1.664	1.262	0.796	0.634	1.643	1.253	0.802	0.643
MLPNN4	1.671	1.274	0.794	0.631	1.668	1.269	0.795	0.633
MLPNN5	1.682	1.284	0.791	0.626	1.646	1.256	0.802	0.642
MLPNN6	1.693	1.292	0.788	0.622	1.660	1.266	0.797	0.636
MLR1	1.723	1.332	0.780	0.608	1.683	1.293	0.791	0.625
MLR2	1.723	1.333	0.780	0.608	1.684	1.294	0.791	0.624
MLR3	1.732	1.338	0.777	0.604	1.694	1.303	0.788	0.621
MLR4	1.736	1.342	0.776	0.602	1.698	1.305	0.787	0.619
MLR5	1.731	1.340	0.777	0.604	1.689	1.298	0.790	0.623
MLR6	1.732	1.339	0.777	0.603	1.694	1.303	0.788	0.621

disappointment is the poorest accuracy obtained using the DENFIS_ON models. In the validation phase as seen in Table 5, the MLPNN1 is the best model and gives an R the value of 0.806, which is consistent with those obtained by DENFIS_OF1 ($R = 0.804$), DENFIS_ON1 ($R = 0.770$), and MLR1 ($R = 0.791$) using the same input variables. Regarding the NSE, RMSE, and MAE, the MLPNN1 has 0.650, 1.628, and 1.235 mm, respectively. On the other hand, the MLPNN5 model based on the inclusion of only two input variables performs better than ($R = 0.804$) the other three models: DENFIS_OF4 ($R = 0.800$), DENFIS_ON5 ($R = 0.785$), and MLR5 ($R = 0.790$). In conclusion, results obtained at Jijel station demonstrate that MLPNN approach shows better results in modelling daily evaporation in both training and validation phases. Comparison of two stations' results reveals that the models are more successful in estimating EP of Guelma compared to Jijel. One reason for this may be the fact that the Guelma has higher correlations between climatic variables and EP than the Jijel. Fig. 5 illustrates the comparison between the observed and estimated values of the daily EP using the four best models for Jijel station.

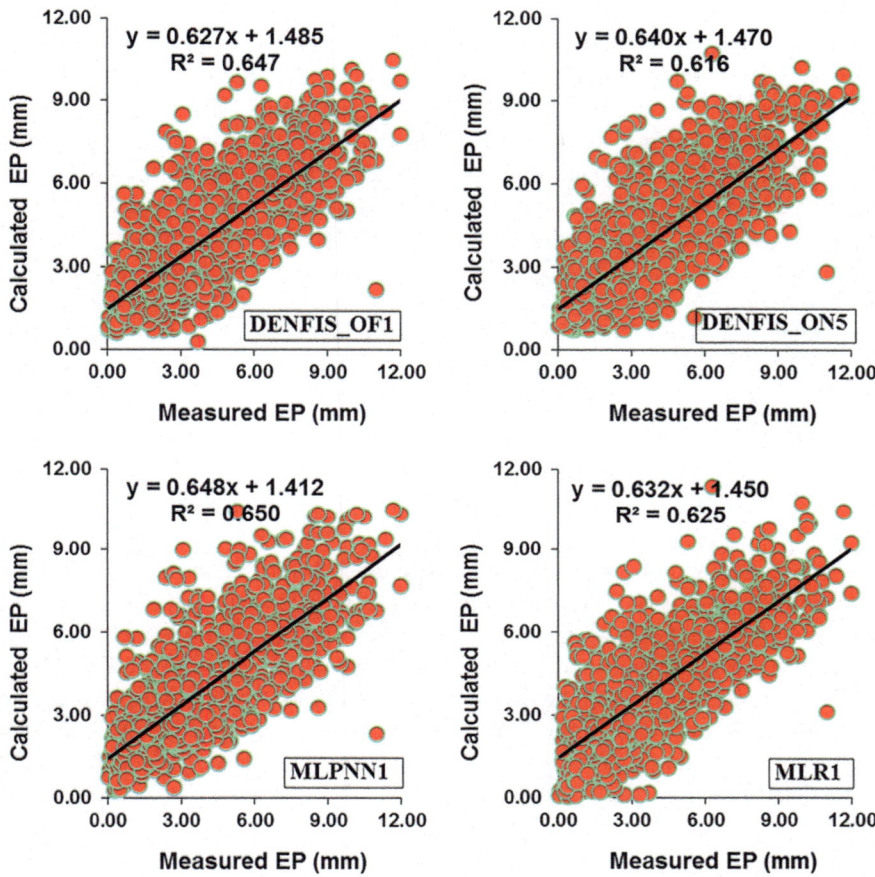

Fig. 5 Scatterplot of calculated and measured daily EP for the optimum developed models for the validation period at Jijel station

5 Discussion

In the present study, an evolving connectionist systems (DENFIS) was developed and applied for modelling daily evaporation (EP) from dam's reservoirs in Algeria country. A standard artificial neural network (ANN) and multiple linear regression models were also applied for farther analysis of the obtained results. Contrary to our expectations, DENFIS models did not significantly improve the accuracy of the fully available models in the literatures, and the obtained results reveal that the DENFIS models based on our data set are less accurate and have certain limitations. For comparison, [1, 3] reported that ANFIS model using solar radiation (SR) can predict daily EP very well with high precision accuracy: coefficient of determination (R^2) equal to 0.96 compared to an R^2 equal to 0.78 obtained using DENFIS models in our study. Although, SR is not available to be included in our analysis, that's not a good enough reason. Sanikhani et al. [4] reported that using SR combined with T_{mean}, RH, and WS, ANFIS model provided an R^2 equal to 0.99 in the validation phase, which has very high accuracy compared to our results. In similar studies, Rahimikhoob [2] and Kim et al. [5] stated that MLPNN model performed best compared to our DENFIS model and provide an R^2 equal to 0.92 and 0.896, respectively. Tabari et al. [7] applied another neuro-fuzzy model, the coactive neuro-fuzzy inference system (CANFIS), and reported very high accuracy with an R^2 equal to 0.865, largely superior to 0.78 obtained using our DENFIS model. Finally, for final discussion, Eray et al. [18] applied the DENFIS model for modelling monthly EP using the same inputs variables as in our study in addition to the SR. According to the obtained results, DENFIS model had high predicting accuracy with an R^2 equal to 0.925.

6 Conclusions

The proposition and development of robust models are the main objectives of any modelling strategies, and it is necessary to have a model with sufficient accuracy. Until now, several kinds of models have been proposed for modelling daily and monthly pan evaporations, and the existing models which have proved their effectiveness in practice must also be followed with other modelling approaches. On the other hand, to keep up with the continuously growing demand for models for the nonlinear and complex process, neuro-fuzzy approaches have proven their efficiency. Starting from this statement, a new kind of neuro-fuzzy models called DENFIS is proposed for modelling daily EP from dam reservoir of Algeria, using easily measured daily climatic variables such as T_{max}, T_{min}, RH, and W_S. The proposed DENFIS model is compared to the standard MLPNN and MLR. According to the obtained results, a number *of major* conclusions can be *drawn*. Firstly, compared to the standard MLPNN and MLR, we have obtained only a marginal improvement of the results for all stations, and overall MLPNN was found to be the best model. For example, at Guelma station, MLPNN was the best model compared to the DENFIS_ON and DENFIS_OF, and DENFIS_OF provided better accuracy only for the first combination when all the four input variables were included as inputs. Secondly, the best performance in validation phase was obtained using the MLPNN model with only three input variables: T_{max}, T_{min}, and RH. However, at Jijel station, the best performance was obtained using MLPNN with four input variables. Thirdly, DENFIS_OF model provided better accuracy for all input combination compared to DENFIQ_ON model. *Fourthly, and finally*, as the models were applied only for two stations, in the future, extending the investigation to other stations can help to draw more robust conclusions.

7   Recommendations

Results obtained in the present investigation highlighted a number of interesting points that may warrant further study. Firstly, the application of DENFIS models for modelling daily EP must be extended to other sites for in depth analysis. Secondly, selection of several others climatic variables as inputs can help to improve the accuracy of the models. Thirdly and finally, the models must be applied for modelling EP at monthly time steps.

References

1. Keskin ME, Terzi Ö, Taylan D (2009) Estimating daily pan evaporation using adaptive neural-based fuzzy inference system. Theor Appl Climatol 98:79–87
2. Rahimikhoob A (2009) Estimating daily pan evaporation using artificial neural network in a semi-arid environment. Theor Appl Climatol 98:101–105
3. Dogan E, Gumrukcuoglu M, Sandalci M, Opan M (2010) Modelling of evaporation from the reservoir of Yuvacik dam using adaptive neuro-fuzzy inference systems. Eng Appl Artif Intel 23:961–967
4. Sanikhani H, Kisi O, Nikpour MR, Dinpashoh Y (2012) Estimation of daily pan evaporation using two different adaptive neuro-fuzzy computing techniques. Water Resour Manag 26:4347–4365. https://doi.org/10.1007/s11269-012-0148-4
5. Kim S, Shiri J, Kisi O (2012) Pan evaporation modeling using neural computing approach for different climatic zones. Water Resour Manag 26(11):3231–3249
6. Kisi O, Pour-Ali Baba A, Shiri J (2012) Generalized neurofuzzy models for estimating daily pan evaporation values from weather data. ASCE J Irrig Drain Eng 138:349–362. https://doi.org/10.1061/(ASCE)IR.1943-4774.0000403
7. Tabari H, Talaee PH, Abghari H (2012) Utility of coactive neuro-fuzzy inference system for pan evaporation modeling in comparison with multilayer perceptron. Meteorol Atmos Phys 116:147–154. https://doi.org/10.1007/s00703-012-0184-x
8. Wu MC, Lin GF, Lin HY (2013) The effect of data quality on model performance with application to daily evaporation estimation. Stoch Environ Res Risk Assess 27:1661–1671. https://doi.org/10.1007/s00477-013-0703-4
9. Terzi Ö (2013) Daily pan evaporation estimation using gene expression programming and adaptive neural-based fuzzy inference system. Neural Comput Applic 23:1035–1044
10. Kim S, Singh VP, Seo Y (2014) Evaluation of pan evaporation modeling with two different neural networks and weather station data. Theor Appl Climatol 117:1–13
11. Kim S, Shiri J, Singh VP, Kisi O, Landerase G (2015) Predicting daily pan evaporation by soft computing models with limited climatic data. Hydrol Sci J 60(6):1120–1136. https://doi.org/10.1080/02626667.2014.945937
12. Malik A, Kumar A (2015) Pan evaporation simulation based on daily meteorological data using soft computing techniques and multiple linear regression. Water Resour Manag 29:1859–1872. https://doi.org/10.1007/s11269-015-0915-0
13. Kisi O, Genc O, Dinc S, Zounemat-Kermani M (2016) Daily pan evaporation modeling using chi-squared automatic interaction detector, neural networks, classification and regression tree. Comput Electron Agric 122:112–117
14. Allawi MF, El-Shafie A (2016) Utilizing RBF-NN and ANFIS methods for multi-lead ahead prediction model of evaporation from reservoir. Water Resour Manag 30:4773–4788. https://doi.org/10.1007/s11269-016-1452-1

15. Piri J, Mohammadi K, Shamshirband S, Akib S (2016) Assessing the suitability of hybridizing the Cuckoo optimization algorithm with ANN and ANFIS techniques to predict daily evaporation. Environ Earth Sci 75:246. https://doi.org/10.1007/s12665-015-5058-3
16. Tezel G, Buyukyildiz M (2016) Monthly evaporation forecasting using artificial neural networks and support vector machines. Theor Appl Climatol 124:69–80
17. Pammar L, Deka PC (2017) Daily pan evaporation modeling in climatically contrasting zones with hybridization of wavelet transforms and support vector machines. Paddy Water Environ 15:711–722. https://doi.org/10.1007/s10333-016-0571-x
18. Eray O, Mert C, Kisi O (2017) Comparison of multi-gene genetic programming and dynamic evolving neural-fuzzy inference system in modeling pan evaporation. Hydrol Res 49 (4):1221–1233. https://doi.org/10.2166/nh.2017.076
19. Olden JD, Jackson DA (2002) Illuminating the "black box": understanding variable contributions in artificial neural networks. Ecol Model 154:135–150
20. Houichi L, Dechemi N, Heddam S, Achour B (2013) An evaluation of ANN Methods for Estimating the Lengths of Hydraulic Jumps in U-shaped Channel. J Hydro Informatics 15 (1):147–154
21. Ladlani I, Houichi L, Djemili L, Heddam S, Belouz K (2014) Estimation of daily reference evapotranspiration (ET$_0$) in the North of Algeria using adaptive neuro-fuzzy inference system (ANFIS) and multiple linear regression (MLR) models: a comparative study. Arab J Sci Eng 39:5959–5969
22. Kasabov N, Song Q (2002) DENFIS: dynamic, evolving neural-fuzzy inference systems and its application for time-series prediction. IEEE Trans Fuzzy Syst 10:144–154
23. Kasabov N (2007) Evolving connectionist systems: the knowledge engineering approach. Springer, New York, p 465
24. Ashrafi M, Hock Chye Chua L, Quek C, Qin X (2016) A fully-online neuro-Fuzzy model for flow forecasting in basins with limited data. J Hydrol 545:424–435
25. Heddam S (2014) Modelling hourly dissolved oxygen concentration (DO) using dynamic evolving neural-fuzzy inference system (DENFIS) based approach: case study of Klamath River at Miller Island Boat Ramp, Oregon, USA. Environ Sci Pollut Res 21:9212–9227
26. Dovžan D, Škrjanc I (2011) Recursive fuzzy c-means clustering for recursive fuzzy identification of time-varying processes. ISA Trans 50:159–169
27. Heddam S, Dechemi N (2015) A new approach based on the dynamic evolving neural-fuzzy inference system (DENFIS) for modelling coagulant dosage: case study of water treatment plant of Algeria country. Desalin Water Treat 53-4:1045–1053. https://doi.org/10.1080/19443994.2013.878669
28. Talei A, Chua LH, Quek C, Jansson P (2013) Runoff forecasting using a Takagi-Sugeno neuro-fuzzy model with online learning. J Hydrol 488:17–32
29. Ge D, Zeng XJ (2018) Learning evolving T-S fuzzy systems with both local and global accuracy-A local online optimization approach. Appl Soft Comput 68:795–810. https://doi.org/10.1016/j.asoc.2017.05.046
30. Maciel L, Ballini R, Gomide F (2017) An evolving possibilistic fuzzy modeling approach for value-at-risk estimation. Appl Soft Comput 60:820–830
31. Wanke P, Azad AK, Emrouznejad A (2018) Efficiency in BRICS banking under data vagueness: a two-stage fuzzy approach. Glob Financ J 35:58–71
32. Demertzis K, Iliadis L, Avramidis S, El-Kassaby YA (2017) Machine learning use in predicting interior spruce wood density utilizing progeny test information. Neural Comput Applic 28:505–519. https://doi.org/10.1007/s00521-015-2075-9
33. Altaher A (2017) An improved Android malware detection scheme based on an evolving hybrid neuro-fuzzy classifier (EHNFC) and permission-based features. Neural Comput Applic 28:4147–4157. https://doi.org/10.1007/s00521-016-2708-7
34. Yu L, Tan SK, Chua LHC (2017) Online ensemble modeling for real time water level forecasts. Water Resour Manag 31:1105–1119

35. Kisi O, Heddam S, Yaseen ZM (2019) The implementation of univariable scheme-based air temperature for solar radiation prediction: New development of dynamic evolving neural-fuzzy inference system model. Appl Energy 241:184–195
36. Kisi O, Khosravinia P, Nikpour MR, Sanikhani H (2019) Hydrodynamics of river-channel confluence: toward modeling separation zone using GEP, MARS, M5 Tree and DENFIS techniques. Stoch Environ Res Risk A 33(4-6):1089–1107
37. Hornik K (1991) Approximation capabilities of multilayer feedforward networks. Neural Netw 4(2):251–257. https://doi.org/10.1016/0893-6080(91)90009-T
38. Hornik K, Stinchcombe M, White H (1989) Multilayer feedforward networks are universal approximators. Neural Netw 2:359–366. https://doi.org/10.1016/0893-6080(89)90020-8
39. Rumelhart DE, Hinton GE, Williams RJ (1986) Learning internal representations by error propagation. In: Rumelhart DE, PDP MC, Research Group (eds) Parallel distributed processing: explorations in the microstructure of cognition. Foundations, vol I. MIT Press, Cambridge, pp 318–362
40. Haykin S (1999) Neural networks a comprehensive foundation. Prentice Hall, Upper Saddle River

New Formulation for Predicting Daily Reference Evapotranspiration (ET$_0$) in the Mediterranean Region of Algeria Country: Optimally Pruned Extreme Learning Machine (OPELM) *Versus* Online Sequential Extreme Learning Machine (OSELM)

Salim Heddam ⓘ, Ozgur Kisi, Abderrazek Sebbar, Larbi Houichi, and Lakhdar Djemili

Contents

S. Heddam (✉)
Laboratory of Research in Biodiversity Interaction Ecosystem and Biotechnology, Hydraulics
Division, Agronomy Department, Faculty of Science, University 20 Août 1955, Skikda, Algeria
e-mail: heddamsalim@yahoo.fr

O. Kisi
School of Technology, Ilia State University, Tbilisi, Georgia
e-mail: ozgur.kisi@iliauni.edu.ge

A. Sebbar
Soil and Hydraulics Laboratory, Hydraulics Department, Faculty of Engineering Sciences,
University Badji-Mokhtar Annaba, Annaba, Algeria
e-mail: rsebbar@yahoo.fr

L. Houichi
Department of Hydraulic, University of Batna 2, Batna, Algeria
e-mail: houichilarbi@yahoo.fr

L. Djemili
Research Laboratory of Natural Resources and Adjusting, Hydraulics Department, Faculty of
Engineering Sciences, University Badji-Mokhtar Annaba, Annaba, Algeria
e-mail: lakhdardjemili@gmail.com

Abdelazim M. Negm, Abdelkader Bouderbala, Haroun Chenchouni, and
Damià Barceló (eds.), *Water Resources in Algeria - Part I: Assessment
of Surface and Groundwater Resources*, Hdb Env Chem (2020) 97: 181–200,
DOI 10.1007/698_2020_528, © Springer Nature Switzerland AG 2020,
Published online: 9 July 2020

Abstract This chapter aims to investigate the capabilities and usefulness of two new data-driven techniques: optimally pruned extreme learning machine (OPELM) and online sequential extreme learning machine (OSELM) newly applied and compared for predicting daily reference evapotranspiration (ET_0) in the Mediterranean region of Algeria. Using large data sets from east to west regions of Algeria, the models were developed using several well-known climatic variables as inputs: daily maximum and minimum air temperatures, wind speed, and relative humidity. The proposed models were compared using several well-known statistical indexes: root mean square error (RMSE), mean absolute error (MAE), and coefficient of correlation (R). The obtained results have shown that all the proposed models present high prediction accuracy and the OPELM models provide better overall performances compared to the OSELM models

Keywords Algeria, Climatic variables, ET_0, Extreme learning machine, Modelling, OPELM, OSELM

1 Introduction

Nowadays, reference evapotranspiration (ET_0) is one of the most important components of the hydrological cycle that have received great importance and has paid great attention by researcher's worldwide [1]. ET_0 is considered as an indicator of climate change and can be directly estimated from weather variables [2]. During the last decades, modelling ET_0 using data-driven models has been broadly discussed and widely reported in the literature, and *many models have been proposed. In-depth literature review demonstrates that generally the proposed models were based on the estimation of the ET_0 at several time steps using several climatic variables as inputs and the ET_0 is calculated* using the standards FAO56 Penman-Monteith model [3].

Keshtegar et al. [4] proposed a new scheme of the adaptive neuro-fuzzy inference systems (ANFIS) model for modelling daily ET_0 in Turkey. Using several climatic variables such as relative humidity (RH), solar radiation (SR), air temperature (T_{mean}), and wind speed (W), the authors compared the performances of the

ANFIS model with those of the M5 model tree (M5Tree), and the multilayer perceptron neural network (MLPNN) models, and demonstrated that the ANFIS is slightly better than the MLPNN and M5Tree. Karbasi [5] proposed forecasting ET_0 models using Gaussian process regression (GPR) and wavelet-GPR conjunction models at the first time. The proposed models were compared in forecasting daily ET_0 up to 30 days in advance using data from Zanjan (Iran). The authors demonstrated that wavelet decomposition improves the performances of the GPR model significantly, and overall, by increasing the forecasting period, the accuracy of the models was significantly decreased. For example, the root mean square error (RMSE) was increased from 0.068 mm/day (one day in advance) to 0.816 mm/day (30 days in advance). Mattar [6] applied gene expression programming (GEP) model for predicting monthly ET_0 in Egypt. Using large data sets from 32 weather stations, the authors demonstrated that the inclusion of the RH and W variables with the maximum and minimum temperatures (T_{max}, T_{min}) improve the performances of the GEP models significantly. Yin et al. [7] compared the *support vector machine* (SVM) and the extreme learning machine (ELM) for modelling monthly ET_0 in China. In the tropical region of Ghana, Landeras et al. [8] compared the accuracy of GEP and MLPNN with ancillary and external approaches in modelling ET_0. Sanikhani et al. [9] developed and compared six machines learning approaches for modelling monthly ET_0 using limited climatic variables from two stations in Turkey. The proposed models were generalized regression neural networks (GRNN), radial basis neural networks (RBNN), MLPNN, GEP, and ANFIS with grid partitioning (ANFIS-GP) and subtractive clustering (ANFIS-SC). *Results obtained* indicated that the accuracy of the proposed models varied from one station to another.

Shiri [10] applied GEP for modelling daily ET_0 in the dry region of Iran. Yin et al. [11] compared three data-driven methods for modelling daily ET_0 in a semi-arid region of China: (1) MLPNN, (2) SVM, and (3) genetic algorithm embedded support vector machine (GA-SVM) models. By comparing eight scenarios having several input combinations, the authors demonstrated that the GA-SVM model provided superior accuracy compared to SVM and MLPNN models. In another study, Feng et al. [12] compared ELM and GRNN for modelling daily ET_0 in China based on two scenarios. Firstly, two models were applied and compared for each station separately, and, secondly, two models were calibrated using pooled data and validated for each station separately. For the first scenario, ELM model provided the best results, while for the second scenario, the GRNN model was the best. Feng et al. [13] applied the random forests (RF) method for modelling daily ET_0 in the southwest China for the first time. Compared to the GRNN models using five climatic input variables, the RF provided slightly better accuracy. Mehdizadeh et al. [14] investigated the capabilities and robustness of four data-driven methods, namely, GEP, multivariate adaptive regression splines (MARS), SVM-polynomial (SVM_P), and SVM-radial basis function (SVM_R), for modelling monthly ET_0 in Iran. Results obtained demonstrated that the MARS and SVM_R were better than those obtained by GEP and SVM_P. Feng et al. [15] compared ELM, MLPNN optimized genetic algorithms (GMLP), and wavelet neural network (WNN), for modelling daily ET_0 in the humid region of China. According to the results obtained, the authors demonstrated that the ELM and GMLP were better than the WNN. Ladlani et al. [16] applied and compared two artificial neural network models, namely, GRNN and RBFNN, for modelling daily ET_0 in the Mediterranean region of Algeria country. According to the results obtained, the authors demonstrated that the GRNN model is better than the RBFNN model. Using the same climatic data, Ladlani et al. [17] demonstrated that ANFIS *embedded subtractive clustering* is a good and powerful tool for estimating daily ET_0 compared to the standard multiple linear regression (MLR). The potential of the MLPNN in modelling daily ET_0 in the vast Algerian Saharan region of Adrar was investigated by Laaboudi et al. [18]. The authors have demonstrated that the MLPNN is a good and robust method for modelling daily ET_0 with a coefficient of correlation equal to 0.97. Recently, Wu and Fan [19] proposed the extreme gradient boosting (XGBoost) model for modelling daily ET_0 in China. In another study, Huang et al. [20] introduced the CatBoost model as a new gradient boosting decision tree for modelling daily ET_0 in china.

To the best of our knowledge, no study in the literature has reported the application of the OPELM and OSELM for modelling daily ET0. Hence, the present study was aimed at evaluating and demonstrating the efficiency of OPELM and OSELM in modelling ET_0.

2   Materials and Methods

In the present study, two kinds of extreme learning machines were developed and compared: online sequential extreme learning machine (OSELM) and optimally pruned extreme learning machine (OPELM). The two models were developed and compared using the same data sets. The flow chart for training and validation of the OSELM and OPELM is shown in Fig. 1.

2.1   *Extreme Learning Machines*

Artificial neural network (ANN) models have gained much popularity over the last few decades, and several applications can be found in the literature. Among many types of ANN, a feedforward neural network (FFN) is well-known and became famous, in particular, due to its *capabilities for solving highly complex and nonlinear problems.* Since the Backpropagation algorithm (BP) has been proposed for training the ANN; the FFN with only one hidden layer called (SLFN) has become a universal approximator [21]. In the past two decades, several algorithms were proposed to improve the BP training algorithm. Recently an efficient algorithm called extreme learning machines (ELM) for training the SLFN was introduced by Huang et al. [22, 23]. SLFN model has two weight matrixes: (1) W_{ij} from the input layer to a single hidden layer and (2) β_{jk} from single hidden layer to the output layer. Contrary to the standard BP training algorithm in which all the weights and biases

Fig. 1 Flow chart for the proposed OSELM and OPELM models

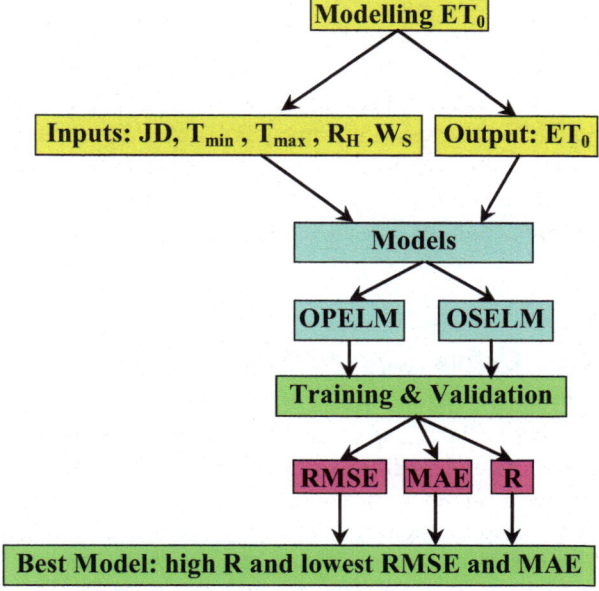

$(W_{ij}$ and $\beta_{jk})$ are optimized simultaneously, the ELM algorithm randomly chooses the input weight matrix W_{ij} and analytically determines the output weight matrix β_{jk} of the SLFN using least squares method [22, 23]. The standard SLFN with N hidden nodes with activation function $f(x)$ can be written as:

$$H\beta = T \tag{1}$$

where:

$$H = \begin{bmatrix} f_1(w_1 \cdot x_1 + b_1) & \cdots & f_N(w_N \cdot x_1 + b_N) \\ \vdots & \vdots & \vdots \\ f_1(w_1 \cdot x_N + b_1) & \cdots & f_N(w_N \cdot x_N + b_N) \end{bmatrix} \tag{2}$$

$$\beta = \begin{bmatrix} \beta_1^T \\ \vdots \\ \beta_N^T \end{bmatrix} \text{ and } T = \begin{bmatrix} t_1^T \\ \vdots \\ t_N^T \end{bmatrix} \tag{3}$$

where w is weight matrix from the input to the single hidden layer; x_i is the input variable; b_i is the bias for each hidden neuron; f is the activation function; T is the matrix of the target; or the output and β is the matrix of the output weights. ELM simply solves the function by:

$$\beta = H^+ T \tag{4}$$

where H^+ is the Moore-Penrose generalized inverse of matrix H [22, 23].

Since it has been proposed, the standard ELM has received several improvements in the last few years, and many types of ELM models have been published, among them: (1) the optimally pruned extreme learning machine (OPELM) [24–26] and (2) the online sequential extreme learning machine (OSELM) [27] were selected and considered in the present study.

2.2   Online Sequential Extreme Learning Machine (OSELM)

Introduced by Liang et al. [27], the online sequential extreme learning machine (OSELM) is a modified version of the standard ELM algorithm, proposed for SLFN. *The idea behind* the development and *the proposition of* the OSELM for training the SLFN is to overcome an important problem: the ELM model is trained only if all the patterns of the training subset are ready, and retraining the ELM model is necessary if a new data is available [27, 28]. The main idea of the OSELM is that through a *learning procedure* the available data are trained one-by-one or *chunk-by-chunk*, and it is necessary to remove each completed pattern. According to Liang et al. [27] and later Huang et al. [28], the OSELM algorithm is implemented in two stages: (1) the initialization phase and (2) the online sequential learning phase.

2.3   Optimally Pruned Extreme Learning Machine (OPELM)

Optimally pruned extreme learning machine (OPELM) is another kind of ELM proposed and presented as an improvement of the standard ELM [24–26]. Similar to the standards ELM, the OPELM can be trained with the same activation function such as Gaussian, sigmoid, and linear [26]. The advantage of the OP-ELM is the capability to cope with the presence of the irrelevant or and correlated variables in the training data set, by applying a pruning algorithm to decrease the complexity and the high dimension of the model. The OPELM is realized in three steps [26]:

1. A high dimensional ELM is presented having a high number of neurons,
2. Using the multi-response sparse regression (MRSR) [29] or least angle regression (LARS) [30] if the output is one-dimensional; the model undertaken a ranking of the hidden neurons taking into account the part of contribution of each neuron to the linear explanation of the ELM output.
3. Leave-one-out (LOO) validation is used to decide how many neurons to prune.

OPELM has been applied successfully for solving various problems. Daily streamflow forecasting [31]; forecasting dissolved oxygen in river [32]; nonlinear system identification [33]; prediction of time series using several world data set [34]; developing a full-noise model for shallow-sea application [35]; intelligent assistance positioning methodology [36]; modelling dissolved oxygen in river with and without water variables [37]; data streams classification [38] and financial time series prediction [39].

2.4 Performances Criteria

In the present study, the following four performance indices were used, where N is the number of data points, O_i is the measured value, and P_i is the corresponding calculated value of EP. O_m and P_m are the average values of O_i and P_i, respectively: the coefficient of correlation (R), the root mean squared error (RMSE), and the mean absolute error (MAE).

3 Study Area and Data Description

In the present investigation, we used data from four meteorological stations located in the Mediterranean region of Algeria. Figure 2 shows the location of the stations. *The used* meteorological *stations are* (1) Annaba (Latitude 36°50′33.55″, Longitude 7°49′34.71″), (2) Bejaia (Latitude 36°42′49.09″, Longitude 5°2′51.30″), (3) Mostaganem (Latitude 35°52′58.8″, Longitude 0°7′1.20″) and (4) Tizi Ouzou (Latitude 36°42′00.00″, Longitude 4°2′59.99″). For each station, we used a record period of about 13 years (2001–2013), and climatic data used in this study consisted of daily measured maximum and minimum temperatures (T_{max}, T_{min}), wind speed (W_S), and relative humidity (R_H), and the daily reference evapotranspiration ET$_0$ was calculated using the standard FAO56 Penman-Monteith model [1]. For each station, we split the data into 8 years for training (2001–2008) and 5 years for validation (2009–2013). Detailed description of the climatic stations with the period of records and splitting information are reported in Table 1.

Table 2 represents the daily statistical parameters of climatic variables for the four stations. In the table, the terms X_{mean}, X_{max}, X_{min}, S_x, C_v, and R denote the mean, maximum, minimum, standard deviation, coefficient of variation, and the coefficient of correlation between the variable and the ET$_0$, respectively. In all stations, T_{max} has the highest correlation with the ET$_0$ followed by T_{min}, R_H, and W_S. Variations of W_S data are higher for the Mostaganem and Tizi Ouzou compared to Annaba and Bejaia stations. This can be explained by the higher altitudes of the Mostaganem and Tizi Ouzou than the other stations. Climatic variables generally have the highest correlations with the ET$_0$ in Tizi Ouzou. In the present study, several combinations of the climatic variables were used, and in total ten scenarios were compared (Table 3). According to Table 3, the *Julian day (JD) number is included as input variables for*

Fig. 2 Location of the weather stations

eight models. To improve to the performances of the developed models, all variables in addition to the ET_0 were normalized using the Z-score method [40]:

$$x_{ni,k} = \frac{x_{i,k} - m_k}{Sd_K} \tag{5}$$

where $x_{ni,\,k}$ is the normalized value of the variable k (input or output) for each sample i. $x_{i,k}$ is the original value of the variable k (input or output). m_k and S_{dk} are the mean value and standard deviation of the variable k (input or output).

Table 1 Information about used data and stations

Description	Weather stations			
	Annaba	Bejaia	Mostaganem	Tizi Ouzou
Longitude	7°49'34.71"	5°2'51.30"	0°7'1.20"	4°2'59.99"
Latitude	36°50'33.55"	36°42'49.09"	35°52'58.8"	36°42'00.00"
Altitude, m	05	03	138	189
Begin date	01/01/2001	01/01/2001	01/01/2001	01/01/2001
End date	31/12/2013	31/12/2013	31/12/2013	31/12/2013
Training period	2001–2008	2001–2008	2001–2008	2001–2008
Validation period	2009–2013	2009–2013	2009–2013	2009–2013
Total pattern	4,743	4,501	4,536	4,197
Training	2,918	2,676	2,739	2,695
Validation	1,825	1,825	1,797	1,502

Table 2 Daily statistical parameters of data set for all stations

Station	Data set	Unit	X_{mean}	X_{max}	X_{min}	S_x	C_v	R
Annaba	T_{min}	°C	12.590	26.000	0.000	5.381	0.427	0.718
	T_{max}	°C	23.661	45.400	5.100	6.558	0.277	0.861
	R_H	%	75.673	96.100	27.000	9.334	0.123	-0.631
	W_S	m/s	3.720	11.900	0.000	1.228	0.330	0.085
	ET_0	mm	3.374	10.614	0.598	1.775	0.526	1.000
Bejaia	T_{min}	°C	13.909	27.600	0.000	5.627	0.405	0.716
	T_{max}	°C	23.355	44.400	2.000	5.981	0.256	0.839
	R_H	%	76.075	98.100	26.800	9.514	0.125	-0.558
	W_S	m/s	3.081	9.800	0.000	1.088	0.353	0.116
	ET_0	mm	3.192	11.351	0.477	1.516	0.475	1.000
Mostaganem	T_{min}	°C	12.827	32.700	0.000	5.919	0.461	0.682
	T_{max}	°C	23.641	45.600	6.400	6.694	0.283	0.836
	R_H	%	72.428	96.900	16.500	12.082	0.167	-0.672
	W_S	m/s	2.033	11.200	0.000	1.446	0.711	0.151
	ET_0	mm	3.230	11.124	0.534	1.775	0.549	1.000
Tizi Ouzou	T_{min}	°C	13.500	30.200	0.000	6.036	0.447	0.794
	T_{max}	°C	24.673	45.700	1.600	8.379	0.340	0.868
	R_H	%	69.737	99.300	20.500	14.749	0.211	-0.777
	W_S	m/s	1.432	11.500	0.000	1.419	0.991	0.401
	ET_0	mm	3.253	10.683	0.462	2.171	0.667	1.000

Abbreviations: X_{mean}, mean; X_{max}, maximum; X_{min}, minimum; S_x, standard deviation; C_v, coefficient of variation; R, coefficient of correlation with ET_0; T_{max}, maximum air temperature; T_{min}, minimum air temperature; W_S, wind speed; R_H, relative humidity; ET_0, reference evapotranspiration

Table 3 The input combinations for different models

Models		
OPELM	OSELM	Inputs combinations
OPELM1	OSELM1	JD, $T_{\min}$, $T_{\max}$, R_H, W_S
OPELM2	OSELM2	$T_{\max}$, $T_{\min}$, R_H, W_S
OPELM3	OSELM3	JD, $T_{\min}$, R_H, W_S
OPELM4	OSELM4	JD, $T_{\max}$, R_H, W_S
OPELM5	OSELM5	JD, $T_{\min}$, $T_{\max}$, R_H
OPELM6	OSELM6	JD, $T_{\min}$, $T_{\max}$, W_S
OPELM7	OSELM7	JD, $T_{\min}$, $T_{\max}$
OPELM8	OSELM8	JD, R_H, W_S
OPELM9	OSELM9	$T_{\min}$, $T_{\max}$
OPELM10	OSELM10	R_H, W_S

4 Results

In this section, we evaluate and compare the performances of the OPELM and OSELM models proposed in the present study for modelling daily ET_0. The results are shown in Tables 4 and 5 for each station and each model. Referring to Table 4, the proposed OPELM models provided higher accuracy in training and validation. According to the *results*, in the training phase, the best accuracy was obtained using OPELM4 for the Annaba, Mostaganem, and Tizi Ouzou; for the Bejaia station, however, the best results were obtained using OPELM5, in terms of R, RMSE, and MAE. OPELM10 provided poor predicting results for all the four stations. The results of the OPELM model indicate that the daily ET0 can be predicted with R values ranging from 0.686 to 0.992 for Annaba station, from 0.601 to 0.987 for Bejaia station, from 0.729 to 0.988 for Mostaganem station, and finally from 0.864 to 0.995 for Tizi Ouzou station. The overall *best accuracy* was *achieved by* OPELM4 at Tizi Ouzou station *with R* value equals to 0.995, which is 0.3% more than the result *obtained with* OPELM4 ($R = 0.992$) at Annaba station, and 0.8% more than the result *of* OPELM5 ($R = 0.987$) at Bejaia station, and 0.7% more than the result of OPELM4 ($R = 0.988$) at Mostaganem station. Regarding the RMSE and MAE values, the lowest values were obtained using OPELM4 at Tizi Ouzou station (RMSE = 0.231 mm/day and MAE = 170 mm/day). In general, the results in the training phase clearly show that the OPELM4 at Tizi Ouzou station has higher fitting accuracy than all the compared models. An important point to note is that the inclusion of the Julian day considerably improves the performances of the models. Moreover, in the training phase, OPELM1 decreased the RMSE and MAE of the OPELM2 by 52.42% and 56.751% at Annaba station, by 62.04% and 67.38% at Bejaia station, by 50.31% and 52.28% at Mostaganem station, and by 59.06% and 60.81% at Tizi Ouzou station, indicating that the OPELM model with Julian day have a potential use in daily ET_0 estimation.

In the validation phase as seen in Table 4, the ranges of the R, RMSE, and MAE are (0.668–0.990), (1.267–0.240), and (1.053–0.184) at Annaba station, respectively. Similarly, at Bejaia station, the ranges of the R, RMSE, and MAE are

Table 4 Performances of the OPELM models in training and validation phases for all stations

Stations	Models	Training			Validation		
		RMSE (mm)	MAE (mm)	R	RMSE (mm)	MAE (mm)	R
Annaba	OPELM1	0.324	0.237	0.984	0.341	0.254	0.980
	OPELM2	0.681	0.548	0.927	0.693	0.565	0.913
	OPELM3	0.360	0.264	0.980	0.366	0.272	0.977
	OPELM4	0.235	0.174	0.992	0.240	0.184	0.990
	OPELM5	0.260	0.182	0.990	0.264	0.180	0.988
	OPELM6	0.380	0.282	0.978	0.422	0.320	0.969
	OPELM7	0.462	0.339	0.967	0.472	0.362	0.961
	OPELM8	0.399	0.296	0.975	0.408	0.300	0.971
	OPELM9	0.846	0.672	0.884	0.864	0.704	0.861
	OPELM10	1.319	1.093	0.686	1.267	1.053	0.668
Bejaia	OPELM1	0.257	0.182	0.985	0.300	0.218	0.981
	OPELM2	0.677	0.558	0.894	0.701	0.580	0.887
	OPELM3	0.326	0.236	0.977	0.362	0.266	0.971
	OPELM4	0.239	0.176	0.987	0.300	0.221	0.981
	OPELM5	0.247	0.169	0.987	0.291	0.204	0.982
	OPELM6	0.352	0.249	0.973	0.398	0.293	0.965
	OPELM7	0.431	0.297	0.959	0.534	0.357	0.937
	OPELM8	0.446	0.328	0.956	0.466	0.338	0.952
	OPELM9	0.773	0.624	0.860	0.782	0.633	0.857
	OPELM10	1.209	1.028	0.601	1.238	1.053	0.578
Mostaganem	OPELM1	0.320	0.240	0.984	0.354	0.263	0.980
	OPELM2	0.644	0.503	0.932	0.722	0.564	0.914
	OPELM3	0.397	0.290	0.975	0.424	0.321	0.971
	OPELM4	0.269	0.204	0.988	0.289	0.219	0.987
	OPELM5	0.326	0.230	0.983	0.343	0.238	0.981
	OPELM6	0.350	0.244	0.980	0.387	0.282	0.977
	OPELM7	0.465	0.332	0.965	0.496	0.359	0.960
	OPELM8	0.397	0.287	0.975	0.448	0.324	0.968
	OPELM9	0.899	0.712	0.862	0.951	0.754	0.845
	OPELM10	1.214	0.989	0.729	1.191	0.961	0.743
Tizi Ouzou	OPELM1	0.287	0.212	0.992	0.399	0.305	0.981
	OPELM2	0.701	0.541	0.951	0.797	0.626	0.910
	OPELM3	0.393	0.284	0.985	0.464	0.338	0.975
	OPELM4	0.231	0.170	0.995	0.327	0.246	0.989
	OPELM5	0.501	0.326	0.975	0.463	0.337	0.972
	OPELM6	0.325	0.232	0.990	0.367	0.272	0.985
	OPELM7	0.543	0.372	0.971	0.501	0.358	0.966
	OPELM8	0.369	0.264	0.987	0.428	0.305	0.981
	OPELM9	0.322	0.236	0.990	0.444	0.336	0.975
	OPELM10	1.142	0.893	0.864	1.165	0.925	0.795

Table 5 Performances of the OSELM models in training and validation phases for all stations

Stations	Models	Training			Validation		
		RMSE (mm)	MAE (mm)	R	RMSE (mm)	MAE (mm)	R
Annaba	OSELM1	0.620	0.485	0.940	0.591	0.471	0.938
	OSELM2	0.731	0.597	0.915	0.729	0.608	0.904
	OSELM3	0.706	0.543	0.921	0.657	0.514	0.923
	OSELM4	0.627	0.497	0.938	0.619	0.495	0.932
	OSELM5	0.634	0.499	0.937	0.615	0.491	0.933
	OSELM6	0.629	0.487	0.938	0.629	0.499	0.929
	OSELM7	0.629	0.480	0.938	0.621	0.485	0.931
	OSELM8	1.010	0.772	0.831	0.961	0.746	0.825
	OSELM9	0.854	0.681	0.882	0.867	0.708	0.860
	OSELM10	1.326	1.107	0.682	1.270	1.060	0.666
Bejaia	OSELM1	0.610	0.466	0.915	0.629	0.482	0.910
	OSELM2	0.738	0.613	0.873	0.734	0.612	0.875
	OSELM3	0.754	0.597	0.867	0.721	0.583	0.881
	OSELM4	0.679	0.533	0.894	0.606	0.485	0.918
	OSELM5	0.625	0.477	0.911	0.569	0.449	0.928
	OSELM6	0.605	0.454	0.917	0.621	0.475	0.912
	OSELM7	0.667	0.488	0.897	0.664	0.477	0.899
	OSELM8	0.953	0.736	0.776	1.013	0.778	0.746
	OSELM9	0.790	0.639	0.853	0.784	0.636	0.856
	OSELM10	1.217	1.036	0.594	1.228	1.048	0.587
Mostaganem	OSELM1	0.604	0.450	0.940	0.645	0.492	0.932
	OSELM2	0.703	0.559	0.918	0.733	0.586	0.911
	OSELM3	0.757	0.592	0.904	0.749	0.584	0.907
	OSELM4	0.574	0.440	0.946	0.639	0.496	0.934
	OSELM5	0.577	0.430	0.946	0.610	0.458	0.939
	OSELM6	0.661	0.503	0.928	0.692	0.549	0.923
	OSELM7	0.728	0.552	0.912	0.731	0.562	0.912
	OSELM8	0.724	0.542	0.913	0.741	0.566	0.909
	OSELM9	0.913	0.725	0.857	0.907	0.725	0.860
	OSELM10	1.213	0.983	0.730	1.146	0.919	0.765
Tizi Ouzou	OSELM1	0.589	0.434	0.966	0.623	0.482	0.947
	OSELM2	0.819	0.662	0.933	0.867	0.703	0.892
	OSELM3	0.816	0.630	0.933	0.807	0.636	0.908
	OSELM4	0.524	0.390	0.973	0.604	0.456	0.950
	OSELM5	0.717	0.534	0.949	0.698	0.526	0.933
	OSELM6	0.537	0.404	0.972	0.523	0.410	0.965
	OSELM7	0.699	0.514	0.952	0.629	0.472	0.945
	OSELM8	1.103	0.887	0.874	1.102	0.886	0.819
	OSELM9	0.994	0.794	0.899	1.009	0.819	0.852
	OSELM10	1.164	0.916	0.859	1.159	0.924	0.797

(0.578–0.982), (1.238–0.291), and (1.053–0.204), respectively. The ranges are (0.743–0.987), (1.191–0.289), and (0.961–0.219) at Mostaganem station and (0.795–0.989), (1.165–0.327), and (0.925–0.246) at Tizi Ouzou station with respect to R, RMSE, and MAE, respectively. In the light of the results achieved in the training and validation phases, the inclusion of the Julian day considerably improves the performances of the models. The percent reduction in RMSE between OPELM1 and OPELM2 ranges from 49.94% to 57.20% with a mean of 52.23% in all the four stations. A wider range in percent reduction from 51.28% to 52.42% is observed for the MAE values with a mean of 55.53%. Improvement is also observed in R values and ranges from 6.6% to 9.4% with a mean of 7.45%. The high accuracy of the OPELM models obtained in the validation phase reflects the robustness of the models as good and powerful tools for modelling daily ET$_0$. OPELM4 model with JD, T_{max}, R$_H$, WS inputs generally has the best accuracy among all applied models in all stations. From the comparison of the OPELM4 and OPELM8 models, we can conclude that the T_{max} is a necessary variable for better mapping of daily ET$_0$ for all four stations. The scatterplots of observed and calculated daily ET$_0$ values for the best OPELM models at the four climatic stations are shown in Fig. 3.

Table 5 shows the performances of OSELM models in estimating daily ET$_0$. From the results of Table 5, we can conclude that the accuracy of the models is lower than the OPELM models. As it can be seen from Table 5, however, from one station to another, the proposed OSELM models achieve reasonable accuracy with high R values and low RMSE and MAE values. In the training phase, the best accuracy was obtained using OSELM1 at Annaba station, OSELM6 at Bejaia and Tizi Ouzou stations, and OSELM5 at Mostaganem station. This indicates that these models have smaller RMSE as well as MAE values and higher R values than the other models. At Annaba station, the OSELM1 provided RMSE of 0.620 mm/day, MAE of 0.485 mm/day, and R of 0.940. The OSELM6 provided RMSE of 0.605 and 0.537 mm/day, MAE of 0.454 and 0.404 mm/day, and R of 0.917 and 0.972 at Bejaia and Tizi Ouzou stations, respectively. Finally, the OSELM5 model provided RMSE of 0.577 mm/day, MAE of 0.430 mm/day, and R of 0.946 at Mostaganem station. In the validation phase as seen in Table 5, the range of the R, RMSE, and MAE are (0.666–0.938), (1.270–0.591) and (1.060–0.471) at Annaba station, respectively. Similarly, at Bejaia station, the ranges are (0.587–0.928), (1.228–0.569), and (1.048–0.449), respectively. The ranges are (0.765–0.939), (1.146–0.610), and (0.919–0.458) at Mostaganem station and (0.797–0.965), (1.159–0.523), and (0.924–0.410) at Tizi Ouzou station with respect to R, RMSE and MAE, respectively. Among all the ten OSELM models, the maximum RMSE *is less than* 1.270 mm/day, and the average MAE of prediction is *less than* 0.608 mm/day. The *OSELM models* showed a slightly better performance in the training phase compared to the *performances obtained* in the validation phase. From the results obtained during the validation phase, we can conclude that the OPELM models outperform the OSELM *models in* both training and validation phases. In the validation phase, by comparing the best models for each station, we can state that:

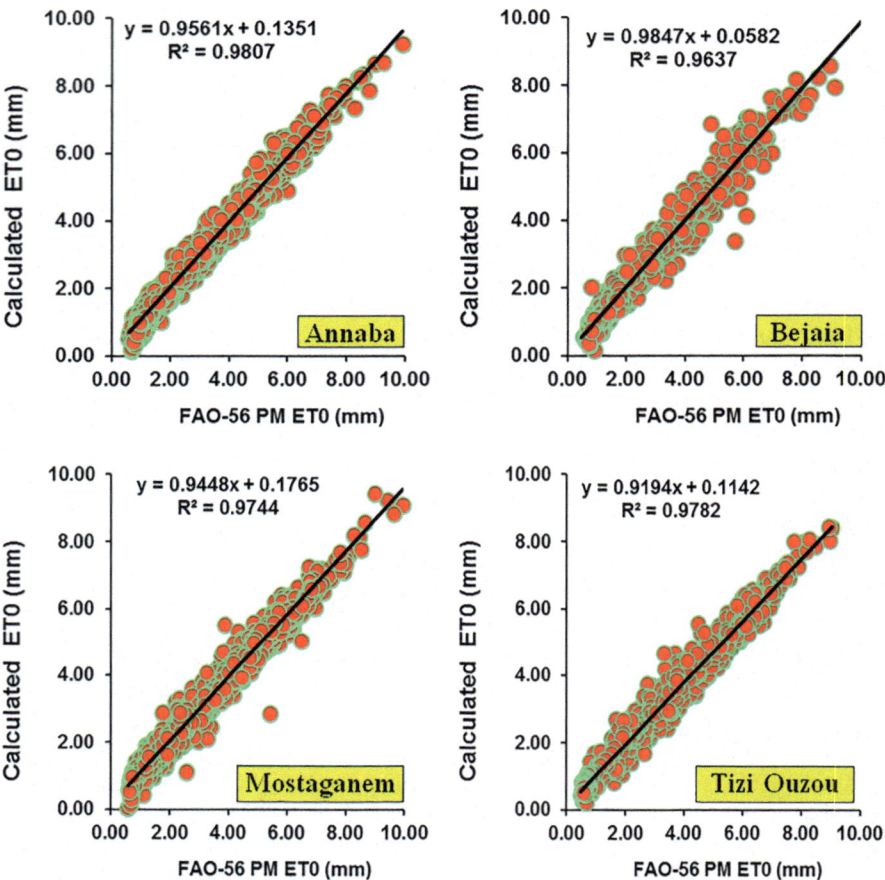

Fig. 3 Scatterplot of calculated and FAO-56 PM ET_0 for the optimum developed OPELM models for the validation period at all weather stations

1. OPELM decreased the RMSE and MAE of the OSELM by 59.40% and 60.93% at Annaba station.
2. By 48.86% and 54.57% at Bejaia station.
3. By 52.62% and 52.20% at Mostaganem station.
4. By 37.48% and 40.00% at Tizi Ouzou station, indicating that the OPELM models have potential use in daily ET_0 estimation compared to the OSELM models. The scatterplots of observed and calculated daily ET_0 values for the best OSELM models at the four climatic stations are shown in Fig. 4.

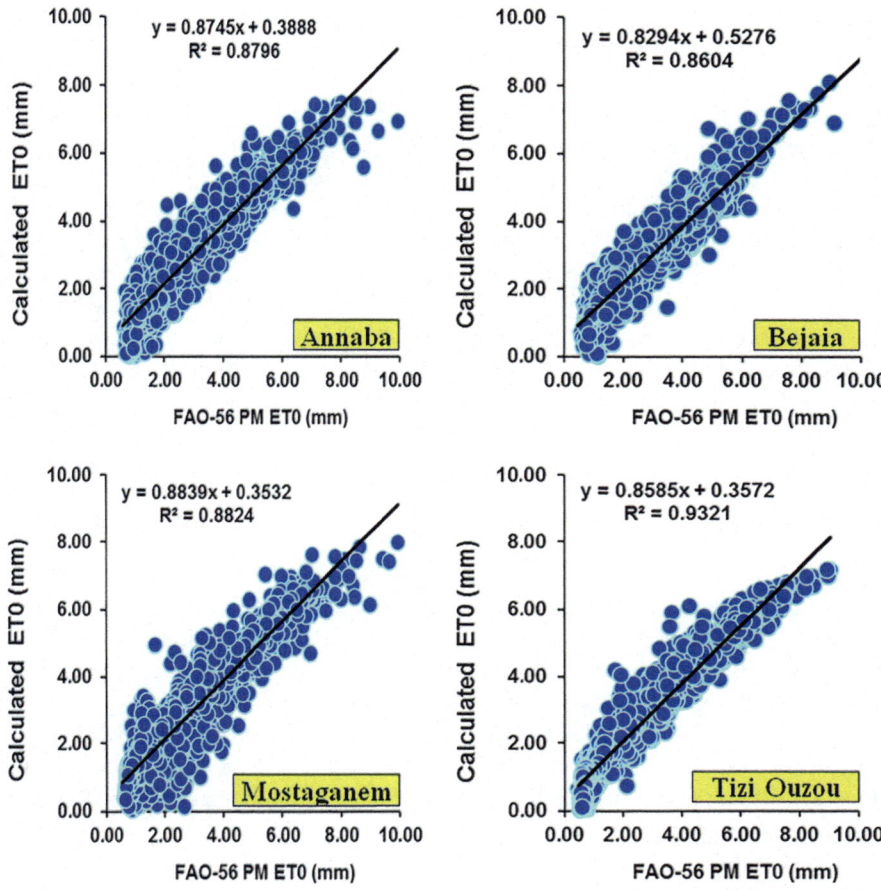

Fig. 4 Scatterplot of calculated and FAO-56 PM ET_0 for the optimum developed OSELM models for the validation period at all weather stations

5 Discussions

In the present study, two extreme learning machines (OPELM and OSELM) were developed and compared for modelling daily evaporation (ET_0) in Algeria country. Accuracy of the models was surprisingly very good at the four sites considering the height R^2 obtained. Results using the models described in our study compared to those obtained using several models reported in the literature revealed several important discussions that can be summarized as follows. Sanikhani et al. [9] applied and compared several models, namely, ANFIS, GRNN RBFNN, MLPNN, and GEP models, for modelling ET_0 and reported that the best model had an R^2 nearly equal to 0.93. Compared to our results, OPELM model using the Julian day as input variable achieve an R^2 nearly equal to 0.98, that is, significantly higher than 0.93. Keshtegar et al. [4] compared the same ANFIS approach to M5Tree and ANN model and reported very high accuracy with an $R^2 = 0.98$. Using the GEP model, Mattar [6] obtained an R^2 equal to 0.963 less than the value obtained using OPELM in our study. Using the original extreme learning machine (ELM) model, Yin et al. [7] reported that ELM model provided high accuracy with an $R^2 = 0.97$. Landeras et al. [8] applied the GEP programming for modelling ET_0 and obtained an RMSE $= 0.680$ mm compared to 0.240 mm obtained using the OPELM in our study. The GA-SVM model applied by Yin et al. [11] yielded a very high accuracy with R^2 nearly equal to 0.99 slightly superior to 0.98 obtained using our models. The random forest (RF) model applied by Feng et al. [13] provided an R^2 equal 0.97, while the MARS model proposed by Mehdizadeh et al. [14] had an R^2 equal 0.999 in the validation phase. In overall discussion, the OPELM and OSELM models applied in the present study have some potential application, and the provided accuracy was very encouraging compared to several models available in the literature.

6 Conclusion

The present study has proposed a new modelling approach of predicting daily ET_0 in the Mediterranean region of Algeria. The proposed model is based on the use of newly data-driven paradigm proposed during the last few years and called the extremes learning machines (ELM). Two modified version of the original ELM (OP-ELM and OS-ELM) were selected, applied, and compared for modelling daily ET_0 using well-known climatic variables and Julian day as inputs. One of the novelties of the present work is the inclusion of the Julian day as an input variable, and the performances of the models were improved a lot. Although the ET_0 can be estimated in different ways, through direct measurement *using* a *lysimeter or using empirical and semi-empirical formulas, the* use of data-driven models which use climatic variables as inputs is very welcome. Another important conclusion of the present study is that the OP-ELM provides more accurate estimates than the OS-ELM for all the four stations.

7 Recommendations

The applicability of the OPELM and OSELM models has been demonstrated in four sites with relatively similar climatic conditions. However, the applicability of the proposed models in other different regions needs to be investigated for robust conclusions. More in-depth investigation of the proposed approaches with several different inputs variables will be helpful for further analysis.

References

1. Allen RG, Pereira LS, Raes D, Smith M (1998) Crop evapotranspiration-guidelines for computing crop water requirements. FAO Irrigation and Drainage Paper 56FAO. Food and Agriculture Organization of the United Nations, Rome, Italy
2. Xing W, Wang W, Shao Q, Peng S, Yu Z, Yong B, Taylor J (2014) Changes of reference evapotranspiration in the Haihe River Basin: present observations and future projection from climatic variables through multi-model ensemble. Glob Planet Change 115:1–15. https://doi.org/10.1016/j.gloplacha.2014.01.004
3. Feng Y, Jia Y, Cui N, Zhao L, Li C, Gong D (2017) Calibration of Hargreaves model for reference evapotranspiration estimation in Sichuan basin of southwest China. Agric Water Manag 181:1–9. https://doi.org/10.1016/j.agwat.2016.11.010
4. Keshtegar B, Kisi O, Arab HG, Zounemat-Kermani M (2018) Subset modeling basis ANFIS for prediction of the reference evapotranspiration. Water Resour Manag 32:1101–1116
5. Karbasi M (2018) Forecasting of multi-step ahead reference evapotranspiration using wavelet-Gaussian process regression model. Water Resour Manag 32:1035–1052
6. Mattar MA (2018) Using gene expression programming in monthly reference evapotranspiration modeling: a case study in Egypt. Agric Water Manag 198:28–38. https://doi.org/10.1016/j.agwat.2017.12.017
7. Yin Z, Feng Q, Yang L, Deo RC, Wen X, Si J, Xiao S (2017) Future projection with an extreme-learning machine and support vector regression of reference evapotranspiration in a mountainous inland watershed in north-west China. Water 9(11):880. https://doi.org/10.3390/w9110880
8. Landeras G, Bekoe E, Ampofo J, Logah F, Diop M, Cisse M, Shiri J (2018) New alternatives for reference evapotranspiration estimation in West Africa using limited weather data and ancillary data supply strategies. Theor Appl Climatol 132:701–716. https://doi.org/10.1007/s00704-017-2120-y
9. Sanikhani H, Kisi O, Maroufpoor E, Yaseen ZM (2018) Temperature-based modeling of reference evapotranspiration using several artificial intelligence models: application of different modeling scenarios. Theor Appl Climatol 135:449–462
10. Shiri J (2017) Evaluation of FAO56-PM, empirical, semi-empirical and gene expression programming approaches for estimating daily reference evapotranspiration in hyper-arid regions of Iran. Agric Water Manag 188:101–114
11. Yin Z, Wen X, Feng Q, He Z, Zou S, Yang L (2017) Integrating genetic algorithm and support vector machine for modeling daily reference evapotranspiration in a semi-arid mountain area. Hydrol Res 48(5):1171–1191
12. Feng Y, Peng Y, Cui N, Gong D, Zhang K (2017) Modeling reference evapotranspiration using extreme learning machine and generalized regression neural network only with temperature data. Comput Electron Agric 136:71–78
13. Feng Y, Cui N, Gong D, Zhang Q, Zhao L (2017) Evaluation of random forests and generalized regression neural networks for daily reference evapotranspiration modelling. Agric Water Manag 193:163–173
14. Mehdizadeh S, Behmanesh J, Khalili K (2017) Using MARS, SVM, GEP and empirical equations for estimation of monthly mean reference evapotranspiration. Comput Electron Agric 139:103–114. https://doi.org/10.1016/j.compag.2017.05.002
15. Feng Y, Cui N, Zhao L, Hud X, Gong D (2016) Comparison of ELM, GANN, WNN and empirical models for estimating reference evapotranspiration in humid region of Southwest China. J Hydrol 536:376–383
16. Ladlani I, Houichi L, Djemili L, Heddam S, Belouz K (2012) Modeling daily reference evapotranspiration (ET$_0$) in the North of Algeria using generalized regression neural networks (GRNN) and radial basis function neural networks (RBFNN): a comparative study. Meteorol Atmos Phys 118:163–178. https://doi.org/10.1007/s00703-012-0205-9
17. Ladlani I, Houichi L, Djemili L, Heddam S, Belouz K (2014) Estimation of daily reference evapotranspiration (ET$_0$) in the North of Algeria using adaptive neuro-fuzzy inference system

(ANFIS) and multiple linear regression (MLR) models: a comparative study. Arab J Sci Eng 39:5959–5969. https://doi.org/10.1007/s13369-014-1151-2

18. Laaboudi A, Mouhouche B, Draoui B (2012) Neural network approach to reference evapotranspiration modeling from limited climatic data in arid regions. Int J Biometeorol 56:831–841. https://doi.org/10.1007/s00484-011-0485-7

19. Wu L, Fan J (2019) Comparison of neuron-based, kernel-based, tree-based and curve-based machine learning models for predicting daily reference evapotranspiration. PloS one 14(5): e0217520

20. Huang G, Wu L, Ma X, Zhang W, Fan J, Yu X, Zeng W, Zhou H (2019) Evaluation of CatBoost method for prediction of reference evapotranspiration in humid regions. J Hydrol 574:1029–1041

21. Hornik K, Stinchcombe M, White H (1989) Multilayer feedforward networks are universal approximators. Neural Netw 2:359–366. https://doi.org/10.1016/0893-6080(89)90020-8

22. Huang GB, Chen L, Siew CK (2006) Universal approximation using incremental constructive feedforward networks with random hidden nodes. IEEE Trans Neural Netw 17(4):879–892. https://doi.org/10.1109/TNN.2006.875977

23. Huang GB, Zhu QY, Siew CK (2006) Extreme learning machine: theory and applications. Neurocomputing 70(1–3):489–501

24. Miche Y, Sorjamaa A, Lendasse A (2008) OP-ELM: theory, experiments and a toolbox. In: Proceedings of the international conference on artificial neural networks. Lecture notes in computer science, vol 5163. Springer, Prague, pp 145–154

25. Miche Y, Bas P, Jutten C, Simula O, Lendasse A (2008) A methodology for building regression models using extreme learning machine: OP-ELM. In: ESANN 2008, European symposium on artificial neural networks, Apr 23–25, Bruges, Belgium

26. Miche Y, Sorjamaa A, Bas P, Simula O, Jutten C, Lendasse A (2010) OP-ELM: optimally pruned extreme learning machine. IEEE Trans Neural Netw 21(1):158–162

27. Liang NY, Huang GB, Saratchandran P, Sundararajan N (2006) A fast and accurate online sequential learning algorithm for feedforward networks. IEEE Trans. Neural Netw 17:1411–1423. https://doi.org/10.1109/TNN.2006.880583

28. Huang GB, Wang DH, Lan Y (2011) Extreme learning machines: a survey. Int J Mach Learn Cybern 2:107–122. https://doi.org/10.1007/s13042-011-0019-y

29. Similä T, Tikka J (2005) Multiresponse sparse regression with application to multidimensional scaling. In: Artificial neural networks: formal models and their applications-ICANN, vol 3697, Springer, Berlin, pp 97–102

30. Efron B, Hastie T, Johnstone I, Tibshirani R (2004) Least angle regression. Ann Stat 32:407–499. https://doi.org/10.1214/009053604000000067

31. Rezaie-Balf M, Kisi O (2018) New formulation for forecasting streamflow: evolutionary polynomial regression vs. extreme learning machine. Hydrol Res 49(3):939–953. https://doi.org/10.2166/nh.2017.283

32. Heddam S (2016) Use of optimally pruned extreme learning machine (OP-ELM) in forecasting dissolved oxygen concentration (DO) several hours in advance: a case study from the Klamath River, Oregon, USA. Environ Process 3:909–937

33. Shihabudheen KV, Mahesh M, Pillai GN (2018) Particle swarm optimization based extreme learning neuro-fuzzy system for regression and classification. Expert Syst Appl 92:474–484. https://doi.org/10.1016/j.eswa.2017.09.037

34. Alencar AS, Neto ARR, Gomes JPP (2016) A new pruning method for extreme learning machines via genetic algorithms. Appl Soft Comput 44:101–107

35. Guo J, He B, Sha Q (2018) Shallow-sea application of an intelligent fusion module for low-cost sensors in AUV. Ocean Eng 148:386–400

36. Guo J, He B, Duan H (2018) Intelligent assistance positioning methodology based on modified iSAM for AUV using low-cost sensors. Ocean Eng 152:36–46

37. Heddam S, Kisi O (2017) Extreme learning machines: a new approach for modeling dissolved oxygen (DO) concentration with and without water quality variables as predictors. Environ Sci Pollut Res 24(20):16702–16724. https://doi.org/10.1007/s11356-017-9283-z

38. Xu S, Wang J (2016) A fast incremental extreme learning machine algorithm for data streams classification. Expert Syst Appl 65:332–344

39. Xue J, Zhou S, Liu Q, Liu X, Yin J (2018) Financial time series prediction using $\ell_{2,1}$ RF-ELM. Neurocomputing 277:176–186. https://doi.org/10.1016/j.neucom.2017.04.076

40. Houichi L, Dechemi N, Heddam S, Achour B (2013) An evaluation of ANN methods for estimating the lengths of hydraulic jumps in U-shaped channel. J Hydro Informatics 15 (1):147–154

Part IV
Aquifer Characterization and Assessment of Groundwater Resources

Water Resources in Coastal Aquifers of Algeria Face Climate Variability: Case of Alluvial Aquifer of Mitidja in Algeria

Abdelkadar Bouderbala and Nacéra Hadj Mohamed

Contents

Abstract Algeria is considered as a vulnerable country in the world regarding its water resource availability, especially in front of the changing climate conditions. Because the availability of water resources contributes strongly to the socioeconomic development of the country, the water supply is the main task challenge of the public institutions under the severe natural conditions of climate variability represented by the decrease of rainfall with the increase of evaporation and also the different anthropogenic pollutions.

Our study area concerned the alluvial aquifer of Mitidja plain. In the later decades, the trend of rainfall was decreased at about 20%, with important annual irregularity in time, which had a negative impact on the groundwater resources of the alluvial aquifer.

A. Bouderbala (✉) and N. Hadj Mohamed
Department of Earth Sciences, University of Khemis Miliana, Khemis Miliana, Algeria
e-mail: bouderbala.aek@gmail.com

Abdelazim M. Negm, Abdelkader Bouderbala, Haroun Chenchouni, and
Damià Barceló (eds.), *Water Resources in Algeria - Part I: Assessment
of Surface and Groundwater Resources*, Hdb Env Chem (2020) 97: 203–224,
DOI 10.1007/698_2020_529, © Springer Nature Switzerland AG 2020,
Published online: 12 July 2020

The analysis of piezometric map shows a drawdown level of groundwater from 1974 to 2010 with a decrease in water levels more than 10 m in average, and in the coastal sector, the wells have high salinity due to seawater intrusion after an overexploitation of groundwater in the catchment field, which induced a reverse flow of groundwater from the sea toward the aquifer.

The analysis of physicochemical parameters of groundwater shows high concentrations of nitrate for the major part of the plain; they are moderately higher than the standard value (50 mg/l). It is due to the anthropic activities in Mitidja plain such as the agricultural origin, primarily linked to the intensive and abusive uses of fertilizers. We can also report urban pollution in this plain; it comes from the discharge sewerage networks without treatment.

Keywords Anthropogenic pollution, Climate variability, Coastal aquifers of Algeria, Water Quality Index, Water resources, Seawater intrusion

1 Introduction

Algeria is Africa's second most water scarce country, after Libya; it is in the category of the poorest countries in terms of hydric potentiality. It is in the theoretical limit of drinking water recommended fixed by WHO of 400 m^3 per inhabitant and per year, but the water supply still has continued to decrease since the 1980s, which may decrease lower than the international standards [1].

The Government recognizes that the right to water is inseparable from the right to life, health, and sustainable development. According to some experts, in 2025, Algeria will know a reduction in rainfall between 5 and 10% and an increase in temperatures between 0.5 and 1.1°C, and it will increase from 2 to 4°C over the next 100 years, which has a direct effect on the hydrologic cycle by increasing evaporation and indirect impact on groundwater.

Algeria will suffer recurrent droughts and floods, as well it will become in front of water supply, which will be double under population growth and agricultural development [2]. The country is already experiencing the increase of drought, which had as consequences the aggravation of desertification phenomena, high evapotranspiration, soil salinization, soil erosion, pollution of surface water and groundwater, and such other consequences of climatic variations and climate variability. Similarly, the floods continue to rage in the north as in the south, and it would be more important in terms of frequency, especially during the spring and autumn [3].

There has been a lot of concern about the impact of climate variability on the socioeconomic activities in Algeria since the 2000s [4, 5].

The potential impacts of climate variability on water resources and food security are receiving growing attention from some researcher, especially in arid and semiarid regions that face of high water demands for agricultural, domestic, and environmental uses [6].

The relationship between the climate variability and water resources in coastal aquifers is more complicated. The high variability of precipitation for a longer period has an impact on the decrease of the groundwater piezometric levels which arise marine intrusion risk in coastal aquifers [7]. Understanding climate variability is vital to the ecosystems and our environment, particularly with regard to the changes affecting the sustainability and availability of groundwater resources [4].

This work aims at assessing the impact of climate variability on water resources within the alluvial aquifer of Mitidja.

2 Impact of Climate Variability on the Hydrological Parameters in Algeria

Algeria is among African countries vulnerable to climate variability. The variation of the two fundamental parameters, temperature, and precipitation, during the last century, is a good indicator of the evolutionary aspect of the climate in Algeria. Changes in the frequency and severity level of extreme events have been significant consequences for natural and human systems. There is a trend of increase of temperature in Algeria [8].

The annual temperatures range from 21.5°C (1916) to 23.5°C (1985). The coldest years are 1904, 1907, 1916, 1917, 1925, 1974, and 1976, respectively, and the hottest years are 1945, 1954, 1985, 1988, and 1997 (Fig. 1).

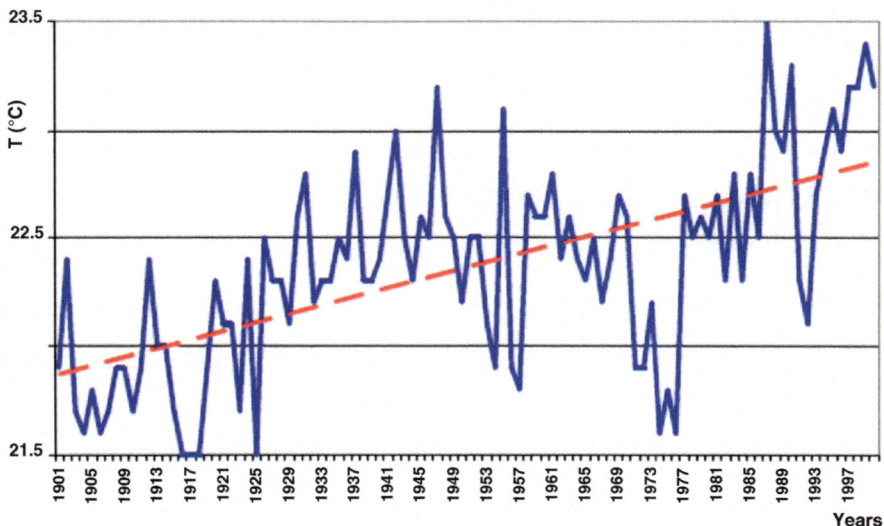

Fig. 1 Evolution of average annual temperatures in Algeria (1901–2000) [9]

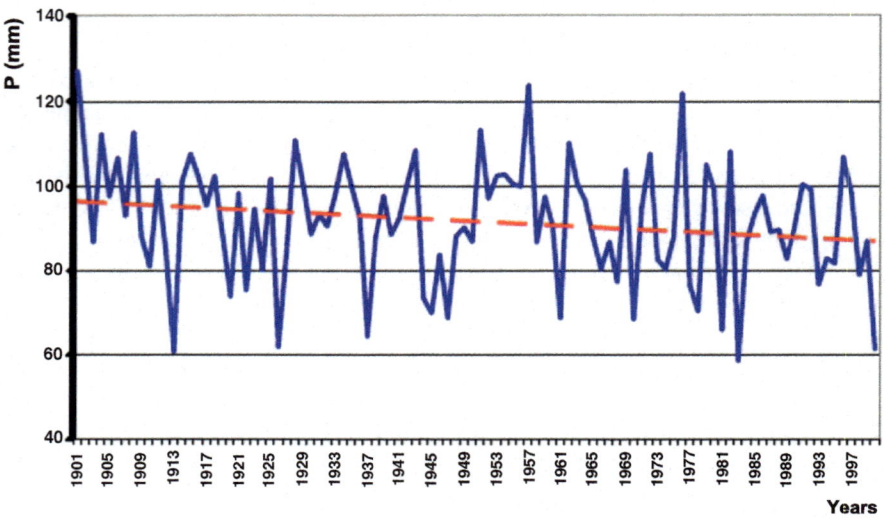

Fig. 2 Evolution of average annual precipitations in Algeria (1901–2000) [9]

The evolution of temperatures over the last 100 years shows that average annual temperatures have increased by an average of 1.0°C, with a maximum during the summer seasons.

The general trend of precipitation is decreasing in Algeria, where the annual rainfall varies between 59 mm (1984) and 128 mm (1901). The years with minimum rainfall amounts are 1914, 1926, 1938, and 1984. The years with high values of precipitation are 1901, 1952, 1957, and 1976 (Fig. 2). Algeria has experienced in recent decades remarkable rainfall deficit recorded throughout the country.

These trends of rising of temperatures and declining of rainfalls have inevitably an effect on the components of our environment (natural resources as water, soil, forests, vegetation, etc.), and they contribute to the re-emergence of certain diseases.

3   The Vulnerability of the Agricultural Sector to Climate Variability

The notable impact of climate variability in agricultural systems is the tool acceleration of soil degradation through water and wind erosion and the salinization due to the higher evaporation and the lower rainfall. These two factors also contribute to the limitation of water resources, the loss of fertile lands, the degradation of forests, and the degradation of natural and pastoral areas. Land degradation is manifested by a deterioration of the vegetation cover and the destruction of their biological potential (depletion of the genetic diversity of fauna and flora) or their ability to support the populations living there. However, the increase of temperature stimulated the decomposition of humus and organic matter and reduced the fertility of certain soils. The process of climate variability was resulting in a northward shift of Mediterranean bioclimatic stages leading to a rise in arid and desert areas [6].

The increase in temperatures and the decrease in rainfall have reduced the growth of vegetation in the dry periods.

Indeed, the temperature directly influences the plants by acting on their metabolism: the crops need a certain quantity of temperature, variable according to the species, to reach the different stages of growth such as flowering and maturity by the photosynthesis phenomena [10].

This amount of temperature expressed in degree-days is a good measure of the growth rate achieved so that high temperature is usually accompanied by a reduction in the growth period. Agronomic studies often use this measure to simulate the variation of the growth period in the case when the temperature increase due to various causes [11].

The risks of climate variability have penalized the agricultural resources in the coastal areas through sea level rise effects and the seawater intrusion in the aquifers, resulting in the disappearance of coastal farming areas or the intensification of extreme events (droughts, floods). In fact, the agricultural sector occupies a crucial place in the daily survival of millions of people, and its sensitivity to climate fluctuations has created a great deal of concern about the ability of global agriculture to the food needs of the entire population. This is why the issue of agriculture received the most attention until today [6].

Thus, fundamental factors such as the length of the growing season, the timing of frost, the accumulation of temperature, the precipitation, the evapotranspiration, the hours of sunshine, and the available moisture, as well as the concentration of carbon dioxide, directly affect the yield of a crop. In addition, there are indirect factors such as a potential increase in insect infestations and pathogens and a change in soil characteristics. Climate variances will not only affect the average surface temperature of the planet but will also affect the seasonal temperature, extreme weather events, and water resources gap. These changes will have impacts on the quantity and quality of agricultural production and on the environment (soils, water, biodiversity, etc.) and will amplify the areas of action of certain pests.

4 Impact of Climate Variability on Water Resources in Coastal Aquifers

The most important impacts of climate variability on the groundwater are the decrease of their levels and quantity and the degradation of their quality.

The aquifers are mainly recharged by precipitation or through interaction between surface and groundwater, and the reduction of precipitation resulted from the climate variability that affects ultimately the groundwater systems.

It is important to consider the impact of climate variability on groundwater as part of the hydrologic cycle because it will be touched by changes in recharge and changes in the different uses of water supplies.

Groundwater is the major resources, particularly in rural regions in arid and semiarid areas. Aquifers are generally recharged by effective rainfall, rivers, or lakes. The precipitation may reach the aquifer quickly, through fissures, or more slowly by infiltrating through soils. The reduction of the effective rainfall will change the amount of recharge and in the duration of the recharge season. The quantification of recharge is complicated by the characteristics of the aquifers themselves. It is dependent on the characteristics of the aquifer media and the properties of the overlying soils. Determining the potential impact of climate variability on groundwater resources is difficult due to the complexity of the recharge process and the variation of recharge in different climatic zones [12].

For water resources in coastal aquifers, they are considered as important fresh-water sources. However, marine intrusion is a major problem in these areas. It is the movement of saline water into freshwater, which leads to a reduction of available fresh groundwater resources. Salinization by the saline intrusion into coastal aqui-fers, in general, depends on the overexploitation of groundwater which reduces of fresh groundwater resources. A reduction of rainfall coupled with sea level rise would not only cause a diminution of the harvestable volume of water; it also would reduce the size of the narrow freshwater lens. A link between the rising sea level and changes in the water balance is suggested by a general description of the hydraulics of groundwater discharge at the coast. Fresh groundwater rides up over denser, salt water in the aquifer on its way to the sea, and groundwater discharge is focused into a narrow zone that overlaps with the intertidal zone. The width of the zone of groundwater discharge measured perpendicular to the coast is directly proportional to the discharge rate [12].

5   Study Area of Mitidja Plain

5.1   Geographic and Climatic Situation

The alluvial Mitidja plain is the largest sub-coastal plain of Algeria, used exclusively for agriculture, with an average altitude of 100 m a.s.l. It is an elongated depression which extends from the east to west for a length of 100 km and a width between 3 and 18 km. It covers an area of 1,450 km². It is limited in the north by the Mediterranean Sea and the Sahel of Algiers region (260 m a.s.l), in the south by the Blidian Atlas Mountains (1,630 m a.s.l.), and in the west by the mountains ranges of Chenoua and Hadjout (1,560 m a.s.l). In the east the plain borders on a series of hills between Boudouaou and Boumerdes. It lies between latitudes 36° 24′ N and 36° 49′ N and between longitudes 2° 33′ E and 3° 19′ E (Fig. 3).

The plain of Mitidja hosts a vivid agricultural economy, which is strengthened by the existing water resources and topography features, endowed with vast fertile and

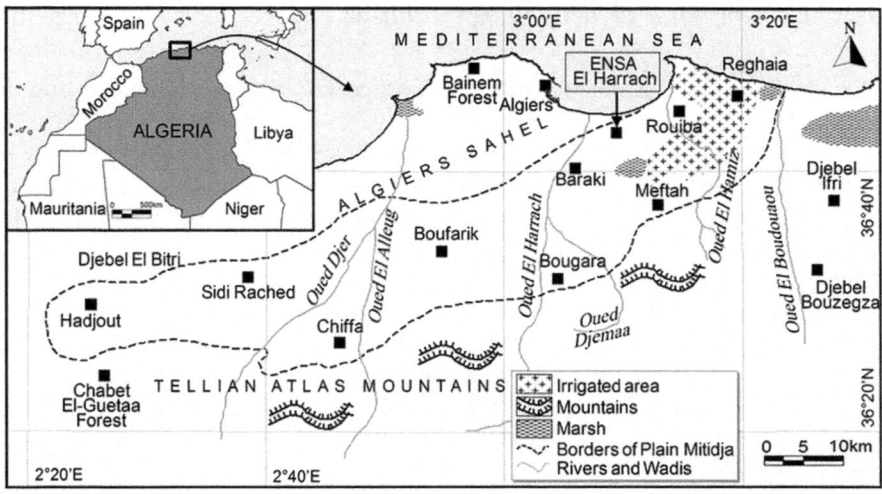

Fig. 3 Limit of Mitidja plain (in [13])

gently sloping lands. It is occupied by cereals, vegetables, fruit trees, and other crops [14]. The climate of the study area corresponds to the Mediterranean type, according to Emberger classification, it is characterized by hot and dry summers extending over 4 months per year and mean annual temperatures exceeding 15°C, and by a rainy and wet winters. The coldest month is January, and the warmest one is August. The average rainfall in the plain is about 600 mm, the average air temperature is 18.3°C, and the annual rate of evapotranspiration is 1,200 mm. The hydrology of the basin is characterized by a set of watercourses that drain alluvium outcroppings; they are drained mainly by the great rivers that flow into the plain to reach the Mediterranean Sea like wadi Nador, wadi Mazafran, wadi El Harrach, and wadi Hamiz. These rivers constitute the main sources of groundwater recharge in the region.

In Algeria, water resources are vulnerable to climate variability. The water management is already causing problems regardless. A great interest must be granted to the water resources because any development policy is focused on the availability of water. Semiarid regions are reported to be the most affected due to the increase of droughts. A reduction in the availability of water for the irrigated agriculture needs the mobilization of unconventional water resources (desalination of seawater, treatment of wastewater for irrigation, etc.).

Other hydrological characteristics of water resources were also impacted by climate variability. The evapotranspiration related to the temperature was increased; it is conditioned by the quantity of water existing in the soil and plants. Temperature and precipitation directly affect soil moisture, but the strongest influence is normally due to precipitation, which has an influence on crop growth and the need for irrigation. Runoff is affected by the hydrological variables, in particular, by precipitation [15].

5.2 Geology and Hydrogeology Context

The lithological succession in Mitidja plain appears from the bottom to top as follows (Fig. 4):

The Pliocene: it is divided into the upper and lower Pliocene. Lower Pliocene is formed of gray or blue-gray marls, including a high layer of blue marls, sometimes sandy, attributed to the Piacenzian. The upper Pliocene is composed of yellow marls, sandy, limestone and sandstone limestone, as well as the molasses attributed to the Astian [14].

The quaternary: it is formed by consolidated sedimentary formation, fluvial siliceous gravel, and sandstone gravel with the red clay of Cretaceous origin, alluvial deposits with silty lens alternating with pebbles; gravel and sand.

The geophysical study conducted in 1973 revealed the existence of two superimposed aquifers in the Mitidja plain, which are formed by subsidence and sedimentation (Fig. 5):

1. The Pliocene aquifer (upper Pliocene) is a confined aquifer formed by the sandstone and sandy limestone. Its substratum is composed of the blue marls (Plaisancian marls of lower Pliocene), and its top is composed of semipermeable yellow marls named Villafranchian marls of El Harrach. This aquifer is very deep, generally located between 250 and 300 m, in the major part of the plain.
2. The quaternary alluvial aquifer is overlain almost the entire basin. It is mainly composed of sand, gravel, and pebbles alternating with silts and clays. Apart from the zone of Mazafran, this aquifer is entirely unconfined and based on the marls of El Harrach which constitute the substratum of the alluvial aquifer. Its thickness varies from 100 to 150 m. Its eastern and western limit is ensured by the rise of the blue marls of the Pliocene. The depth of the water table ranges between 4 and 30 m.

The recharge of the aquifers is mainly being done by effective rainfall, by the infiltration of river water in some places, and also by the excess of irrigation water, and by the underground supply from the Blidean Atlas. The different pumping tests conducted in the alluvial quaternary aquifer indicate values of transmissivities ranging from 1.0 to 2×10^{-2} m^2/s, and the hydraulic conductivities are varying between 10^{-4} and 10^{-2} m/s. For the industrial, irrigation, and drinking water supplies, the aquifer provides near than 300 million m^3 per year, with more than 3,000 wells existing in the plain.

5.3 Methodology

This study is based on available data of precipitations, piezometry, and hydrochemistry. This data could be used to explain the impact of climate variability on groundwater mainly in the alluvial coastal aquifer of Mitidja. This work exploits

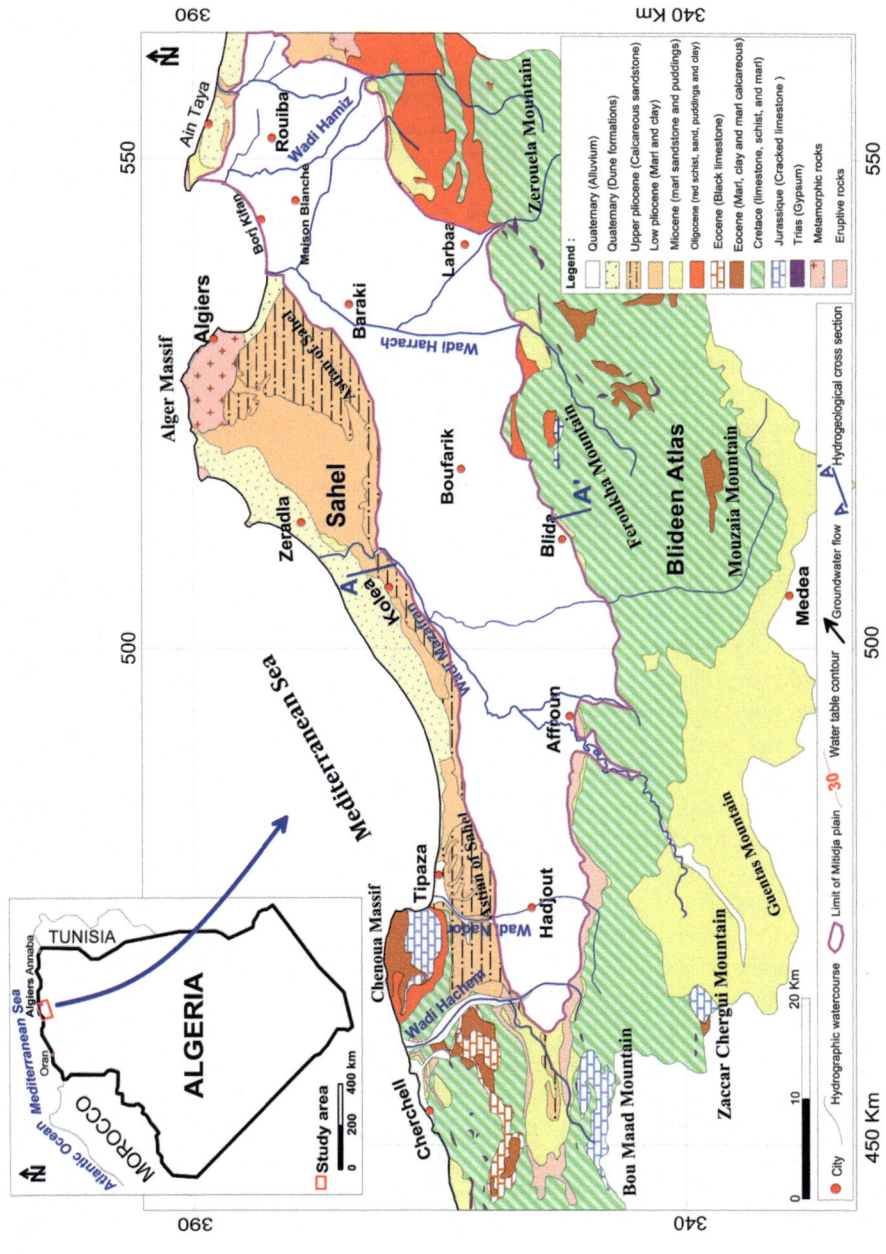

Fig. 4 Geological map of the study area (in [4])

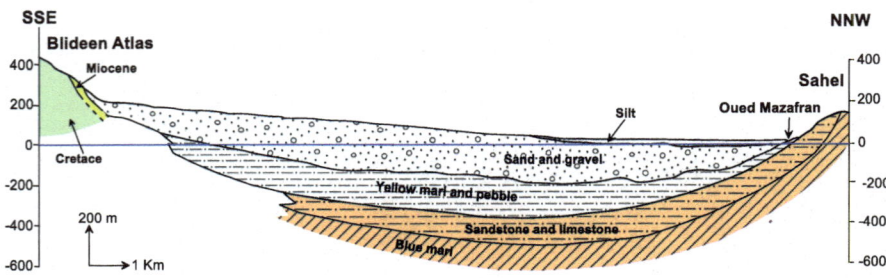

Fig. 5 Hydrogeological cross section A-A′ in the Mitidja plain (in [4])

the analytical results of 36 water samples, which were collected in Juin 2010 in different parts of the plain, simultaneously with piezometric measurements. The major ionic parameters analyzed were Ca^{2+}, Mg^{2+}, Na^+, K^+, Cl^-, SO_4^{2-}, HCO_3^-, and NO_3^-, and the physical parameters were pH and electrical conductivity. Water samples were collected from wells after a minimum of several minutes of pumping prior to sampling. Samples were collected into 1, 5 polyethylene bottles. All the samples were stored in an ice chest at a temperature lower than 4°C and later transferred to the laboratory. Immediately after sampling, physical parameters such as the electric conductivity (EC), water temperature (T), and pH were measured in the field using a multiparameter WTW universal conductivity meter. Chemical analysis was carried out in the laboratory of the National Agency of Hydraulic Resources (ANRH-Blida). All the results are compared with standard limits recommended by the World Health Organization [16]. Nitrate was analyzed by colorimetry with a UV-visible spectrophotometer. Calcium, magnesium, chloride, and bicarbonate were analyzed by volumetric titrations. Concentrations of sodium and potassium were measured using a flame photometer and that of sulfate by the turbidimetric method. The accuracy of the chemical analysis was checked by calculation of the ionic balance, which was generally less than 5%. All concentration values were expressed in milligram per liter (mg/l) unless otherwise indicated.

For the assessment of groundwater quality, the Water Quality Index is used. This index is used by several users [17, 18].

The calculations of WQI were based on the standards suggested for uses, where 11 water quality parameters were chosen (pH, EC, TDS, Ca^{2+}, Mg^{2+}, Na^+, K^+, HCO_3^-, Cl^-, SO_4^{2-}, and NO_3^-).

In the first step, the relative weights (WI) were assigned to each parameter based on their relative importance in the overall water quality of drinking purposes (Table 1). The maximum weight of 5 has been assigned for parameters like TDS, Na^+, Cl^-, NO_3^-, and SO_4^{2-} due to their importance in water quality assessment, and the minimum weight of 1 has been given to HCO_3^- because it plays a comparatively less significant role in water quality assessment [18, 19].

In the second step, the relative weights (R_{wi}) were calculated by using the following equation:

Table 1 The weight (wi) and relative weight (Wi) of each chemical parameter [18]

Parameters	WHO (2006)	Weight (wi)	Relative weight (Wi)
EC (μS/cm)	1,500	5	0.119
TDS (mg/l)	1,000	5	0.119
Cl^- (mg/l)	250	5	0.119
SO_4^{2-} (mg/l)	200	5	0.119
Na^+ (mg/l)	150	5	0.119
NO_3^- (mg/l)	50	5	0.119
Mg^{2+} (mg/l)	75	4	0.095
Ca^{2+} (mg/l)	100	2	0.048
HCO_3^- (mg/l)	300	2	0.048
K^+ (mg/l)	12	1	0.024
pH	8.5	3	0.071
		$\sum wi = 42$	$\sum Wi = 1$

$$R_{wi} = wi \Big/ \sum_{i=1}^{n} wi$$

where R_{wi} is the relative weight, wi is the weight of each parameter, and n is the number of parameters.

In the third step, the quality rating scale (qi) for each parameter was attributed by dividing its concentration in each water sample by its respective standard according to the WHO guidelines (WHO 2006):

$$qi = \left(\frac{Ci}{Si}\right) \times 100$$

where qi is the quality rating; Ci is the concentration of each chemical parameter in each water sample in mg/l; and Si is the concentration permissible, for each chemical parameter according to the guidelines in mg/l

For computing the WQI, the following equation was used:

$$\text{WQI} = \sum R_{wi} \times qi$$

Water quality types were determined based on WQI (Table 2). The spatial distribution maps for values of WQI were prepared using inverse distance weighting (IDW) interpolation technique.

Table 2 Classes proposed for drinking water quality based on the Water Quality Index (WQI) (Chatterjee and Raziudd in 2002)

Class	The range of WQI for drinking purposes	Type of water quality
1	<25	Excellent water quality
2	25.1–50	Good water quality
3	50.1–75	Permissible water quality
4	75.1–100	Doubtful quality
5	>100	Water unsuitable for drinking uses

5.4   Results and Discussion

5.4.1   Precipitation Analysis

The analysis of the annual rainfall data of Hamiz dam station located in this area for the period from 1905 to 2006 shows a marked decrease in annual precipitation, as shown by the trend line (Fig. 6).

This reduction is estimated at about more than 20% in this rainfall station. It shows an important annual irregularity in time, with an alternation of drought and wet years. These drought years reduce the recharge of the alluvial aquifer and consequently overexploitation of groundwater, which deteriorates the groundwater quality and advancement of the seawater intrusion in the coastal area of Mitidja plain that caused a hydrodynamic disequilibrium. While the wet and rainy years contributed to the recharge and the dilution of the aquifer, and when there are very heavy rainfalls, they cause flooding (e.g., Flood of Algiers in 2001). We may say that the climate variability observed during the last years is characterized by a rainfall return but with greater and rapid intensity.

Annual rainfall (mm)

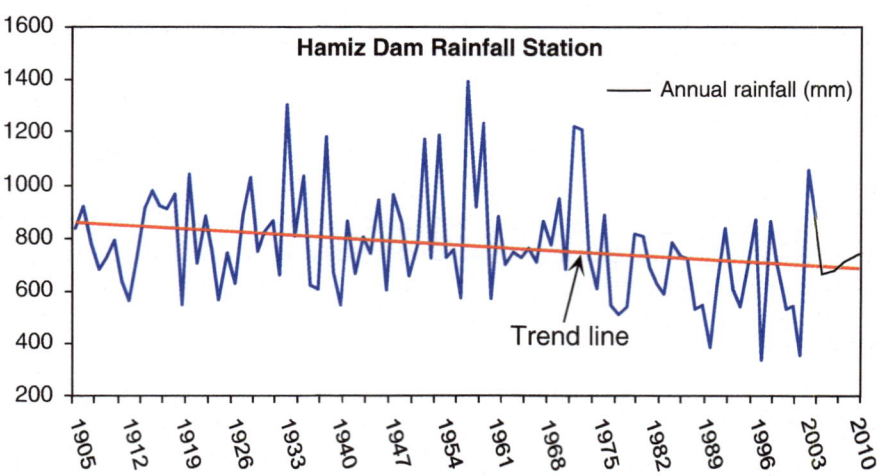

Fig. 6 Annual rainfall variability for Hamiz dam rainfall station (1905–2006)

The hydrological balance in the basin of the Mitidja shows that the rate of infiltrating water and the runoff in this area is 6% and 8.5% of precipitation respectively. This is due to the geomorphological characteristics of the watershed and the predominance of impermeable layers in the surface, which thus promotes runoff. The interaction river aquifer has a high influence on water recharge; however, in the last years, the hydraulic construction of the river has reduced the effective recharge via rivers.

5.4.2 Piezometric Perturbation

The potentiometric map of Mitidja plain, for the dry period 2010, shows groundwater flow from south to north (Fig. 7). A line of groundwater divide is located between the cities of Boufarik and Baraki. This map shows two depressions in this plain; it is due to overexploitation of groundwater in the catchment fields (on the Mazafran catchment fields, and the other in the catchment fields of Baraki and of Maison Blanche).

The high hydraulic gradients in the southern parts of the plain are related to the high slope of substratum, and the low thickness of the alluvial aquifer added to that this part of the plain is considered as a recharge area. However, the high hydraulic gradient in the west part of the plain (near Affroun city) is explained by the low permeability of the aquifer. The high hydraulic gradient in the center of the plain is due to the high thickness and permeability of the aquifer. The direction of the flow reverses and takes the direction from the sea to the continent; it is the seawater intrusion that has been observed for three decades.

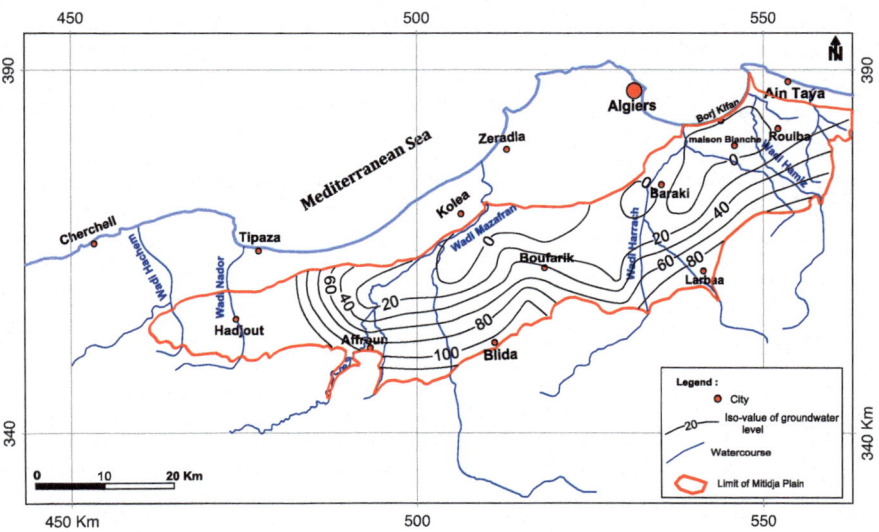

Fig. 7 The potentiometric surface map in the Mitidja Plain of the dry period 2010

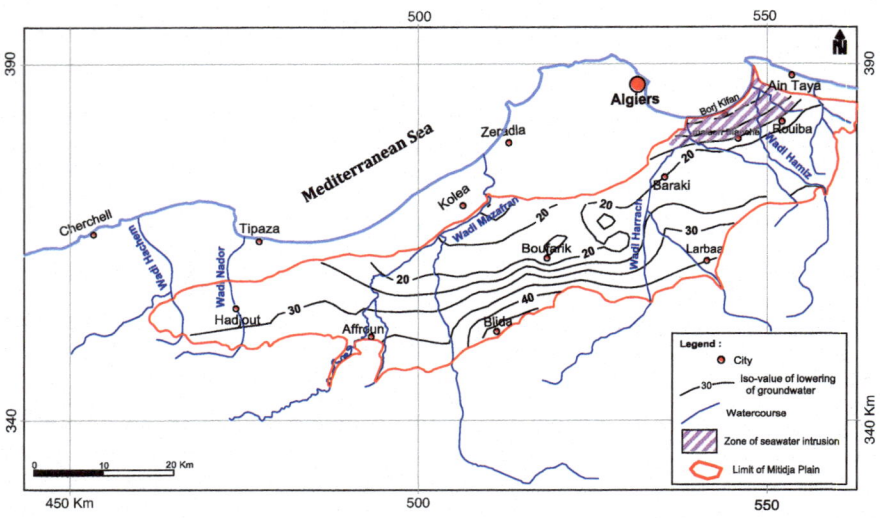

Fig. 8 Reducing the groundwater level in the alluvial plain of Mitidja from 1974 to 2010 [18]

The map of groundwater piezometric drawdown level of the alluvial plain of Mitidja from 1974 to 2010 shows a decrease in water levels in this aquifer (Fig. 8). In the coastal sector, near the cities of "Maison Blanche" and "Rouiba," the wells in this area have high salinity due to seawater intrusion after overexploitation of groundwater in this catchment field, which induced a reverse flow of groundwater from the sea toward the aquifer.

The consequence of the lowering of the piezometric level in this aquifer is due mainly to the long period of drought (since the 1980s), the significance of wells that extracted groundwater (more than 5,000 wells, according to ANRH), the proliferation of illicit boreholes that are never inventoried, as well as the continuous and intensive pumping into the well fields in this plain.

According to the Hydrographic Basin Agency of Algiers [20], the groundwater resources in the aquifer of Mitidja is near 330 Hm^3 per year; the exploitation of these resources is essentially represented by pumping for drinking water supply for about 100 Hm^3/year, for irrigation for about 220 Hm^3/year, and for the industry for about 6 Hm^3/year.

5.4.3 Impact on Groundwater Geochemistry

The geochemistry allows us to understand the chemical process that forms in the aquifer and the interaction between water and rocks in this aquifer. The groundwater quality and the origin of different chemical elements depends mainly on the lithological composition of saturated and non-saturated zones, on the residence time in the aquifer, and on the flow direction of groundwater. The quality of groundwater is influenced also by climatic conditions and by the anthropogenic activities. These interactions influence the concentrations of the major elements (Na^+, Mg^{2+}, Ca^{2+}, K^+, Cl^-, HCO_3^-, etc.).

The analysis of the annual average of physicochemical parameters of groundwater in the Mitidja plain over a period of 6 years (2005–2010) shows concentrations in the limit of the Algerian standard of drinking water, except the nitrate concentrations, they are moderately higher than the standard value of nitrate (50 mg/l) (Table 3). The probable origin of this ion is due to the anthropic activities in Mitidja plain such as the agricultural origin because this plain has an agricultural vocation, where the rate of fertilization has exceeded the norms and it reaches the value 400 kg of nitrogen/hectare. The discharge of sewage coming from domestic sewerage networks and different industrial units without treatment can also pollute the groundwater in this plain, and it can increase the concentrations of nitrate.

The physicochemical analyses of groundwater of the Mitidja aquifer were carried out in the ANRH laboratory. They covered 36 samples for the wet period 2010 (the period with a massive number of sampling compared to the dry water period 2010 (Fig. 9), and the samples are more distributed over the plain).

Table 4 gives the descriptive statistical analysis of the physicochemical parameters collected from the Mitidja Plain in the wet period 2010

The pH parameter determines the acidity or alkalinity of water, which corresponds to the activity of the H+ ions that exist in water. The pH values of the wet period 2010 (Fig. 10) show that all samples are in the range of drinking water (between 6.8 and 8.8), indicating groundwater is slightly alkaline in this alluvial aquifer. It can be explained by the influence of anthropogenic pollution on groundwater.

The results show that the electrical conductivity is strongly dependent on the chemical composition of water and to its temperature. On the Mitidja plain, the majority of samples has EC > 1,500 µS/cm (71% of samples), but based on the Algerian norm of EC, there are only 20% of samples exceed the norm (2,800 µS/cm). The highest values of EC (EC > 5,000 µS/cm) are observed mainly in the coastal area; near Rouiba-Maison Blanche and Bordj Kiffan. They are due to the problem of seawater intrusion, after the overexploitation of groundwater (Fig. 11). However, the high values of EC in the upstream and in the center of the plain are due to the anthropological pollution of groundwater or due to the water-rock interaction (the geology of the aquifer).

The sodium and chloride concentrations show that the high values (more than 27% for Na^+, and 18% for Cl^- exceeds the norm of drinking water) are observed near the coastal area in the northeast of the plain, which is due to the phenomenon of seawater intrusion in this coastal sector of the aquifer (Table 4).

The concentrations of calcium in this area indicate that more than 35% of the wells have contents higher than the limit of drinking water quality. The highest values are observed near the coastal region of the study area, which can be explained by the flow direction and the dissolution of the carbonate formations. However, the rest of the wells located in the center and upstream of the plain have values lower than 200 mg/l for the majority of wells. The presence of bicarbonates in groundwater is due to the dissolution of the carbonated formations which border the plain.

Table 3 The annual average of chemical parameters in Mitidja Plain from 2005 to 2010 [21–23]

Year	EC (µS/cm)	pH	Na$^+$ (mg/l)	Mg^{2+} (mg/l)	Ca^{2+} (mg/l)	K$^+$ (mg/l)	Cl$^-$ (mg/l)	HCO$_3^-$ (mg/l)	SO$_4^{2-}$ (mg/l)	NO$_3^-$ (mg/l)
Wet period										
2005	1,799	7.3	116	48	163	1	297	277	211	75
2006	1,753	7.3	100	44	187	2	185	339	232	86
2007	1,622	7.4	75	50	171	0	162	356	226	63
2008	1,382	7.4	66	44	145	1	151	327	183	34
2009	1,850	6.7	99	43	131	2	189	291	201	53
2010	1,702	7.8	114	41	147	4	166	426	178	37
Dry period										
2005	1,345	7.4	81	40	153	1	178	281	165	48
2006	1,517	7.4	94	36	174	1	183	312	192	64
2007	1,397	7.6	53	46	167	1	154	343	199	47
2008	1,374	7.4	103	39	126	1	148	336	164	37
2009	1,464	7.4	107	47	140	3	176	351	211	45
2010	1,404	7.9	74	45	132	2	133	340	164	35
Algerian norm 2011	*2,800*	*6.5–9.5*	*200*	*150*	*200*	*12*	*500*	*–*	*400*	*50*

The italic numbers correspond to "the drinking water standards of Algeria of 2011"

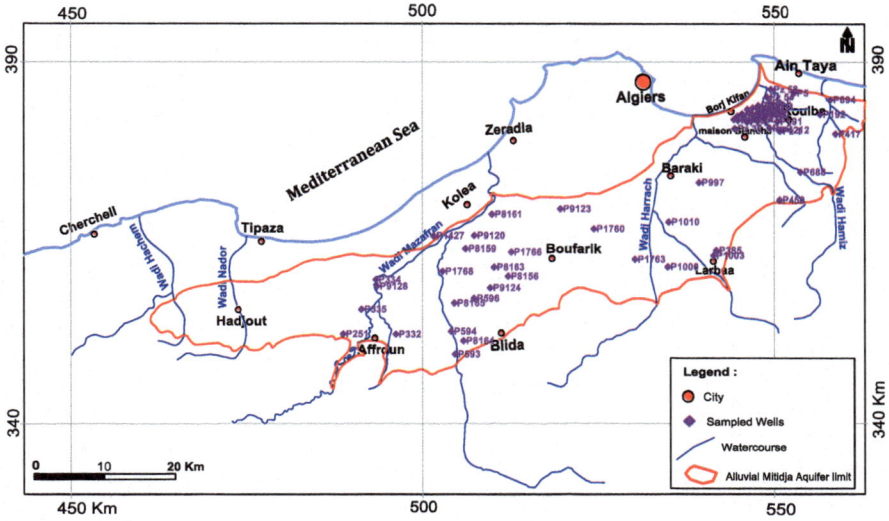

Fig. 9 Location map of sampled wells

The analysis results reveal that the potassium values exceed 12 mg/l in some wells in this plain. The potassium results generally from the alteration of potassium clays and the dissolution of chemical fertilizers (NPK) which were used massively by farmers. The presence of this ion may also be related to the discharge of domestic wastewater into the plain. Local contamination by the septic tanks and systems sewerages could be also responsible.

The highest values of sulfates concentrations of groundwater are located in the northeast of Reghaïa city and in the coastal area, which can be explained by effect of the pollution from the Lake of Reghaïa (nature reserve located in the Mitidja plain), the industries exist in the southwest of the study area, the discharge of wastewater into the different wadi in the plain without prior treatment, the use of fertilizers in agriculture practices, the evaporation, and also from the seawater intrusion in the coastal area.

The nitrate concentrations in the study area indicate more than 45% are between 50 and 231 mg/l, exceeding the limit of 50 mg/l established by the World Health Organization standards [16]. They were due to the excess of using fertilizers in agricultural activities (the arboricultural and vegetable cultures occupy the most part of the Mitidja plain). The geology of Mitidja in the shape of basin associated with an aquifer made up with permeable formations (gravels and sands), supported the immigration of the nitrate toward the saturated zone and the contamination of the groundwater.

We can say that spatial distribution of sodium, chlorides, calcium, magnesium, and nitrates show that the zones with higher concentrations are observed in northeast of the plain near the coastal area, which can have been attributed to anthropogenic pollution after overexploitation of groundwater in this aquifer process (seawater

Table 4 Physicochemical parameters in the Mitidja plain (wet period 2010)

Parameter	EC (µS/cm)	pH	Na^+ (mg/l)	Mg^{2+} (mg/l)	Ca^{2+} (mg/l)	K^+ (mg/l)	Cl^- (mg/l)	HCO_3^- (mg/l)	SO_4^{2-} (mg/l)	NO_3^- (mg/l)
Wet period										
Min	370	6.8	21	6	13	0	30	1	14	0
Max	7,800	8.8	1,800	148	521	40	2,754	669	696	231
Average	2,202	7.52	218	50.1	183	5.5	386.4	260.2	223.2	57.7
SD	1,627.5	0.44	322.5	29.2	106.9	6.8	579.3	227.7	154	54
Algerian norm 2011	2,800	6.5–9.0	200	150	200	12	500	–	400	50
> Exceed norm Algerian	20.3	0	27.1	0	35.6	8.5	18.6	–	10.2	45.8

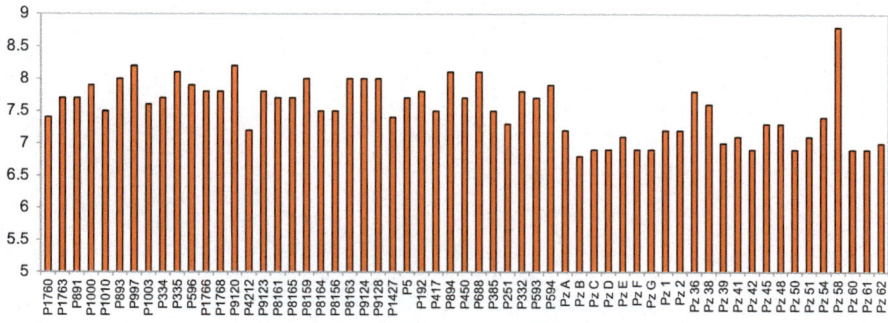

Fig. 10 Variation of groundwater pH in the Mitidja alluvial aquifer (wet period 2010)

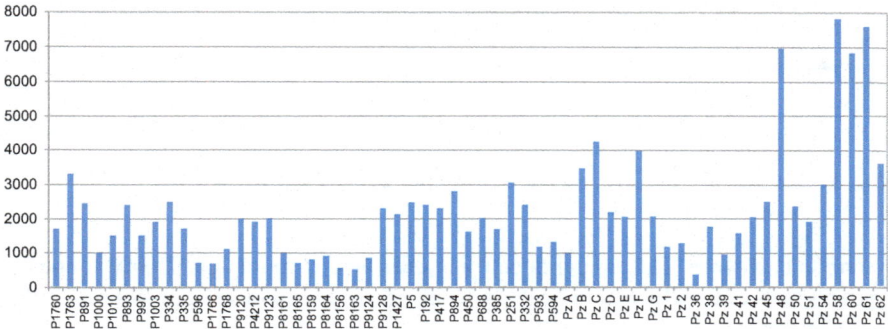

Fig. 11 Variation of groundwater EC in the Mitidja alluvial aquifer (wet period 2010)

intrusion). The consequences of seawater intrusion are principally the increase contents of chloride and sodium ions.

The spatial distribution map of the WQI shows that the western and eastern of the study area have a poor groundwater quality. In general, WQI decreases from the center to the periphery of the Mitidja aquifer, under the effects of the direction of groundwater flow, especially toward the coastal area in Bordj Kiffan and Rouiba (Fig. 12). These results indicate that the salinity by seawater intrusion is the principal factor which degrades the chemical quality of the aquifer of Mitidja in the east part of the plain. However, the anthropogenic pollution and the natural processes are responsible for the deterioration of the groundwater quality in the center and west parts of the plain.

Our investigation of socioeconomic activities in the plain helped us to explain the origin of nitrate pollution in this area. In fact, the effect of industrial activities with more than 300,000 m^3 of industrial wastewater discharged daily in wadi El Harrach and wadi Hamiz or in their tributaries, added to this the urban discharges, agricultural effluents, chemical fertilizers, intensive livestock farming and the several wild dumps concentrated in this area are responsible for the nitrogen pollution. Wastewater is discharged as such into the receiving environment without a serious control and treatment. The quantity of nitrate resulting from these discharges continues to

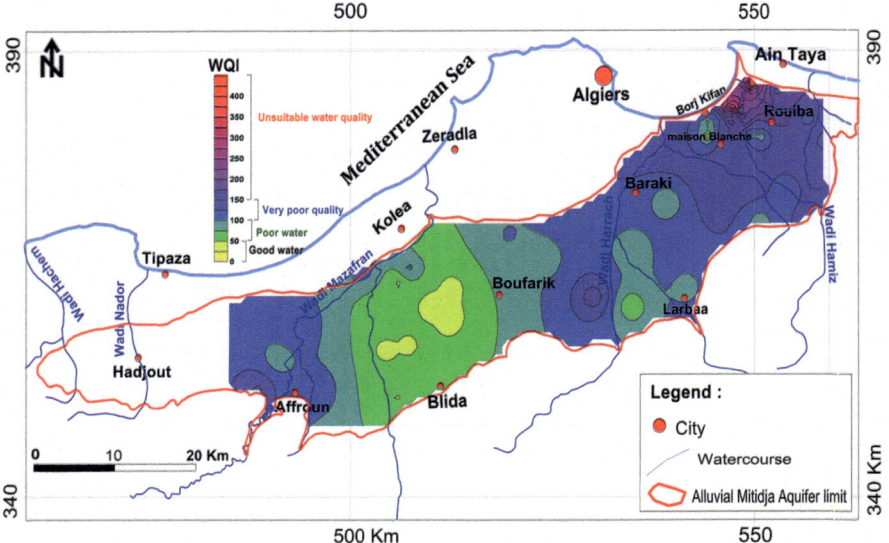

Fig. 12 The spatial distribution map of the WQI (wet period 2010)

increase each year. The annual quantity is estimated at more than 75,000 tonnes of nitrates per year, which is a real problem of groundwater contamination in this alluvial aquifer.

6 Conclusion

Water resources in Algeria, particularly in the Mitidja plain, are limited, vulnerable, and unequally distributed spatially in this area. This sensitive situation inevitably requires new actions to exploit these resources in a rational way.

The purpose of this work is to study the spatial variability of some chemical properties of groundwater in the Mitidja plain by a parametric approach of the quality index in order to evaluate the quality of water for drinking and irrigation purposes, in a relationship with climate variability, and to analyze the impact of climate change on groundwater quantity.

Analysis of the chemical data showed the predominant of groundwater facies: sodium chloride, calcium bicarbonate, and mixed facies. In addition, the salinity of the water in this plain varies from average to very high salinity. It can be said that the groundwater in this aquifer is facing a huge risk of nitrate pollution and seawater intrusion and becomes unfit for domestic and agricultural use.

It appears that the northeastern part of Mitidja is the most vulnerable to the salinity because the contents of some elements greatly exceed the international standards set by WHO.

The consequence of the lowering of the piezometric level in this aquifer is due mainly to the long period of drought (since the 1980s), the significance of wells that extracted groundwater, the proliferation of illicit wells that are never inventoried, as well as the continuous and intensive pumping into the well fields in this plain.

Thus, this work will be a statement of place, and it can be a tool for decision-making regarding the exploitation of this resource from the groundwater quality and the risk of contamination, whether for human health.

7 Recommendations

In the light of the above results, we can present some recommendations for the managers in the goal to protect of the alluvial aquifer of Mitidja:

- Implementation of a new monitoring network wells distributed in this plain.
- Respect the perimeter of protection map when installing new drilling wells, in order to reduce the quantitative and qualitative degradation of groundwater resources.
- Implement future polluting anthropogenic activities in low vulnerability areas to limit their negative impact on water resources in the groundwater.
- It is necessary to control the industrial discharges and to sensitize the farmers to use fertilizers in a rational way.
- Review the pricing of water, especially industrialists to be economical and protective toward the water.
- A hydrogeological modeling study of the alluvial aquifer and the transfer of nitrates in this aquifer are very important in order to properly manage the groundwater in this aquifer.

References

1. Dudgeon D, Arthington AH, Gessner MO, Kawabata ZI, Knowler DJ, Lévêque C et al (2006) Freshwater biodiversity: importance, threats, status and conservation challenges. Biol Rev 81 (2):163–182
2. Pavelic P, Giordano M, Keraita B, Ramesh V, Rao T (2012) Groundwater availability and use in Sub-Saharan Africa: a review of 15 countries. International Water Management Institute (IWMI), Colombo, p 274
3. Verner D (ed) (2012) Adaptation to a changing climate in the Arab countries: a case for adaptation governance and leadership in building climate resilience. World Bank Publications, Washington
4. Bouderbala A (2018) Effects of climate variability on groundwater resources in coastal aquifers (case of Mitidja Plain in the North Algeria). Groundwater and global change in the Western Mediterranean Area. Springer, Cham, pp 43–51
5. Meddi MM, Assani AA, Meddi H (2010) Temporal variability of annual rainfall in the Macta and Tafna catchments, Northwestern Algeria. Water Resour Manag 24(14):3817–3833

6. Bhattarai U (2017) Impacts of climate variability on biodiversity and ecosystem services: direction for future research. Hydro Nepal J Water Energy Environ 20:41–48
7. Kumar CP (2013) Recent studies on impact of climate change on groundwater resources. Int J Phys Soc Sci 3(11):189
8. Bessaklia H, Ghenim AN, Megnounif A, Martin-Vide J (2018) Spatial variability of concentration and aggressiveness of precipitation in North-East of Algeria. J Water Land Dev 36 (1):3–15
9. Abderrahmani B (2015) Les risques climatiques et leurs impacts sur l''environnement. Thèse de doctorat université USTO-MB, Algérie, p 170
10. Tardieu F (2013) Plant response to environmental conditions: assessing potential production, water demand, and negative effects of water deficit. Front Physiol 4:17
11. Barlow KM, Christy BP, O'leary GJ, Riffkin PA, Nuttall JG (2015) Simulating the impact of extreme heat and frost events on wheat crop production: a review. Field Crops Res 171:109–119
12. Kumar CP (2012) Climate change and its impact on groundwater resources. Int J Eng Sci 1 (5):43–60
13. Bendjoudi D, Voisin JF, Doumandji S, Merabet A, Benyounes N, Chenchouni H (2015) Rapid increase in numbers and change of land-use in two expanding Columbidae species (Columba palumbus and Streptopelia decaocto) in Algeria. Avian Res 6(1):18
14. Khouli MR, Djabri L (2011) Impact of use of agricultural inputs on the quality of groundwater case of Mitidja plain (Algeria). Geographia Technica 11(2):35–44
15. Da Cunha LV, De Oliveira RP, Nascimento J, Ribeiro L (2007) Impacts of climate change on water resources: a case-study for Portugal, vol 310. IAHS Publication, Wallingford, p 37
16. WHO World Health Organization (2008) Guidelines for drinking water quality.2nd edn. WHO, Geneva
17. Bouderbala A (2015) Groundwater salinization in semi-arid zones: an example from Nador plain (Tipaza, Algeria). Environ Earth Sci 73(9):5479–5496
18. Bouderbala A (2017) Assessment of water quality index for the groundwater in the upper Cheliff plain, Algeria. J Geol Soc India 90(3):347–356
19. Khosravi R, Eslami H, Almodaresi SA, Heidari M, Fallahzadeh RA, Taghavi M et al (2017) Use of geographic information system and water quality index to assess groundwater quality for drinking purpose in Birjand City, Iran. Desalin Water Treat 67:74–83
20. ABH (2000) Groundwater resources in the aquifer of Mitidja, rapport, p 20
21. ARNH (2010) Rapport sur la lutte contre l'intrusion marine dans la baie d'Alger. Rapport de mission, p 66
22. Djoudar D (2014) Approche methodologique de la vulnerabilite de la ressource en eau souterraine en milieu fortement urbanise: exemple en Algerie des plaines littorales (Mitidja). Thèse de doctorat USTHB, p 178
23. Zamiche S, Hamaidi-Chergui F, Demiai A, Belaidi M (2018) Identification of factors controlling the quality of groundwater in mitidja plain using indexing method and statistical analysis. J Fundam Appl Sci 10(1):248–267

Assessment of Groundwater Resources in the Jurassic Horst (Western Algeria)

Fouzia Bensaoula, Bernard Collignon, and Mohammed Adjim

Contents

F. Bensaoula (✉) and M. Adjim
Hydraulic Department, Faculty of Technology, Tlemcen University, Tlemcen, Algeria
e-mail: fbensaoula@gmail.com; moh.adjim@gmail.com

B. Collignon
Hydroconseil, Chateauneuf de Gadagne, France
e-mail: collignon@hydroconseil.com

Abdelazim M. Negm, Abdelkader Bouderbala, Haroun Chenchouni, and
Damià Barceló (eds.), *Water Resources in Algeria - Part I: Assessment
of Surface and Groundwater Resources*, Hdb Env Chem (2020) 97: 225–266,
DOI 10.1007/698_2019_406, © Springer Nature Switzerland AG 2019,
Published online: 28 September 2019

Abstract The karst aquifers of the Tlemcen Mountains are the region's main groundwater resource. More than 270 boreholes have been drilled in this region and have a total production capacity of 40 million m^3/year.

Surface karstic forms are not very developed. However, the numerous boreholes drilled in the region showed that the carbonate reservoirs are well karstified (and this to a depth of more than 500 m).

The increasing demand for water, combined with insufficient rainfall over the last few decades, has led to groundwater mining and a significant drop in the piezometric level. This problem was solved by using desalinated seawater as an alternative water source to reduce groundwater abstraction.

A more successful Integrated Water Resources Management (IWRM) would aim to abstract less water than the average annual recharge, by modulating abstraction according to seasonal recharge. In addition, it would be advisable to study the feasibility of artificial recharge of these aquifers during periods of heavy rainfall.

Groundwater quality is generally quite good, but these aquifers are vulnerable. Mapping intrinsic vulnerability to pollution is therefore necessary to improve water resource protection. Karst vulnerability mapping methods (such as the RISK method) have proven useful in documenting decisions related to drilling location and wastewater collection and treatment.

Résumé Les aquifères karstiques des Monts de Tlemcen constituent la principale ressource en eau souterraine de la région. Plus de 270 forages ont été réalisés dans cette région et ont une capacité de production totale de 40 millions de m^3/an.

Les formes karstiques de surface ne sont pas très développées. Cependant, les nombreux forages réalisés ont démontré que les réservoirs carbonatés sont bien karstifiés (et cela jusqu'à plus de 500 m de profondeur).

La demande en eau croissante, conjuguée à l'insuffisance des précipitations au cours des dernières décennies, a entraîné la surexploitation de ces ressources et une baisse significative du niveau piézométrique. Ce problème a été résolu en utilisant l'eau de mer dessalée comme source d'eau alternative afin de réduire les prélèvements d'eau souterraine.

Une gestion intégrée des ressources en eau (GIRE) plus aboutie viserait à prélever moins d'eau que la recharge interannuelle moyenne, en modulant les prélèvements en fonction de la recharge annuelle. De plus, il serait souhaitable d'étudier la faisabilité de la recharge artificielle de ces aquifères pendant les périodes de pluies abondantes.

La qualité de l'eau souterraines est généralement assez bonne, mais ces aquifères sont vulnérables. Une cartographie fine de la vulnérabilité à la pollution intrinsèque est donc indispensable. Les méthodes de cartographie de la vulnérabilité du karst (telles que la méthode R.I.S.K.), se sont révélées utiles pour documenter les décisions relatives à l'implantation des forages et à la collecte et au traitement des eaux usées.

Keywords COP and RISK method, Karst water resources, Managed aquifer recharge (MAR), Mounts of Tlemcen, Strategic reserves, Vulnerability

Mots clés Monts de Tlemcen, Aquifères karstiques, Réserves stratégiques, Vulnérabilité, Méthode COP et RISK, Recharge artificielle des nappes

1 Introduction

Karstic terrains cover approximately 12% of the Earth's continental surface, and 25% of the world's population is supplied partially or entirely by karst water resources. In the Mediterranean and Southeast Asia, karst aquifers are the primary water resources [1].

In Algeria, karst aquifers play a very important role in supplying the country's largest springs with water. For thousands of years, they have fed many cities and countless villages during the summer season, when most wadis dry up.

In western Algeria, surface water resources are very limited, as rainfall is scarce. It is therefore tempting to try to mobilise groundwater resources. This is not possible everywhere, as the geological context is often unfavourable. The most extensive outcrops consist in Cretaceous and Cenozoic clay, marl and marly limestone. Such types of rock are not suitable for groundwater abstraction. Hydrogeological prospecting focuses on the reliefs where the Jurassic geological formations, and in particular the limestones and dolostones of the Malm, are exposed.

The main geological structure with abundant Jurassic limestone is the Oranese Meseta. This is a horst structure, which extends from the Moroccan border to Ksar Chellala and where carbonate rocky outcrops cover an area of about 10,000 km^2, making it Algeria's most extensive karst area.

This article focuses on the karst hydrogeology, water reserves, vulnerability and constraints for groundwater resource management in the western section of Oranese Meseta: the Mounts of Tlemcen which extend over 4,000 km^2, 50% of which are limestone and dolomite outcrops.

These aquifers were used to bring about significant improvements in urban water supply during the 1980s and the 1990s. However, increasing population growth and urbanisation are now jeopardising the sustainable management of the aquifers, and there is also a need to ensure water resource protection. Sustainable water resource management means making the best use of the various water resources available (dams, wells, wastewater reuse and desalination) and considering the specific value of the karstic aquifers in the regional context (good quality water, low turbidity, huge storage volume, limited investment needs).

2 Geographical and Geological Context

This section presents the geographical, hydro-climatic and geological context of the Mounts of Tlemcen, constituted by a horst of Jurassic rocks, which dominates the Algerian West and acts as a water tower for most watersheds of the region (Fig. 1).

2.1 Mounts of Tlemcen Geographical Context

See Fig. 1.

2.2 Scarce and Irregular Rainfall

The Mounts of Tlemcen are one of the wettest regions in the west of Algeria, which means they act as a natural source of water supply for the majority of towns in the region [2].

Several Algerian isohyet contour maps have been published with data covering long time periods: 1913–1938 [3], 1913–1963 [4] and 1922–1989 [5]. Laborde studied 120 rainfall stations in northern Algeria. According to his findings, annual rainfall patterns are characterised by four phases: two periods of surplus rainfall from 1922 to 1938 and 1947 to 1972 that are interspersed with periods where there was a lack of rain. In 1973, a long period of drought set in (Laborde 1993 in [6]).

The rainfall distribution pattern is influenced by three factors. These are altitude, distance from the sea and longitude. Figure 2 partially shows this influence. In fact, the wettest stations are those located at the highest altitude (the Hafir and Meffrouch stations), and they receive over 600 mm of rain annually in years when the weather is humid. Rainfall decreases rapidly when moving further south.

2.3 An Inherited and Well-Integrated Hydrographic Network

The source of the Tafna River lies in the Ain Taga, and it rises at the foot of the Djebel Guern. After heavy storms, it overflows into the Ghar Boumaza. It opens out into a valley set within Jurassic terrain and continues on its journey as the Khemis Oued (river) before reaching Sidi Medjahed, after which it flows through the Neogene formations.

The central part of the Mounts of Tlemcen is drained by the Sikkak Oued, which begins on the Terni plateau with the Ennachef Oued. It continues to the end of the breakdown zone of Tlemcen, picks up the few tributaries that drain the neighbouring land reliefs and flows with the main Tafna River, as well as with the Zitoun Oued and

Fig. 1 Mounts of Tlemcen location

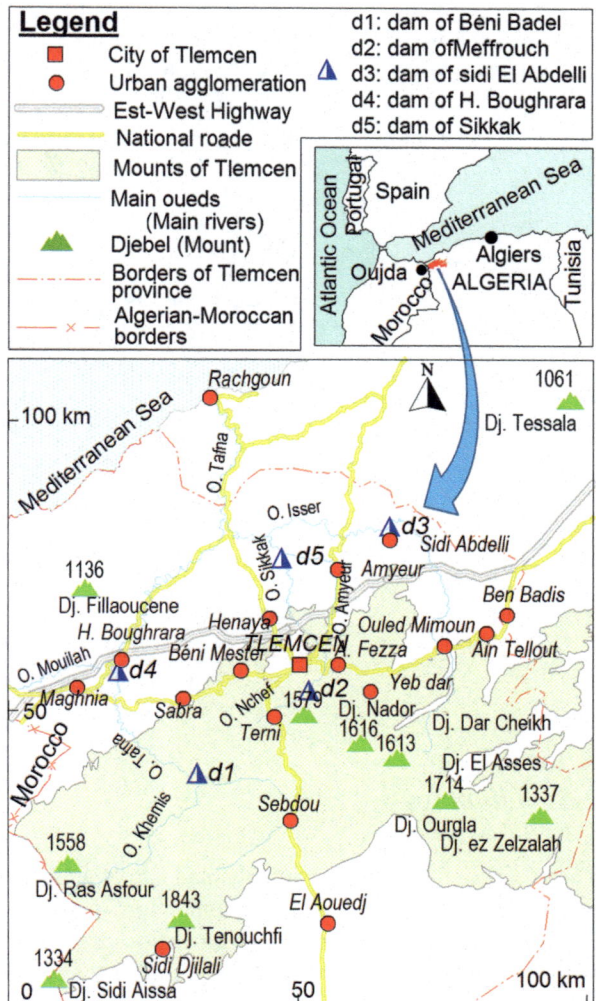

the el Atchane Oued, which are left-bank tributaries. The eastern part of the Mounts of Tlemcen is initially drained by the Lakhdar Oued (formerly the Chouly Oued), the source of which lies within the southern and eastern massif that borders the Terni plateau. It then flows into the left bank of the Isser Oued, cutting through the Abdellys' plateau (Fig. 1).

Surface water tends to run off only occasionally, predominantly after short rainy periods.

Fig. 2 Map of average
yearly rainfall in western
Algeria [5, modified]

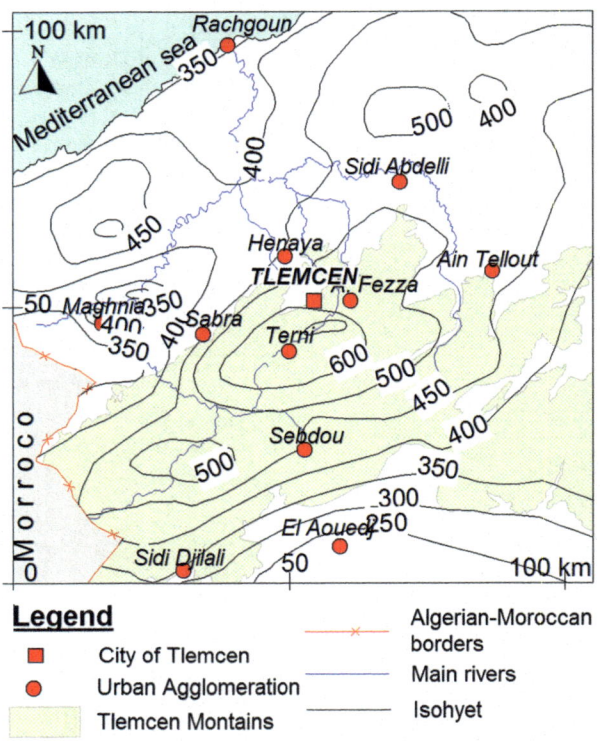

Legend

■ City of Tlemcen

● Urban Agglomeration

   Tlemcen Montains

—×— Algerian-Moroccan borders

——— Main rivers

——— Isohyet

2.4 Lithology

The Mounts of Tlemcen are mainly made up of upper Jurassic and lower Cretaceous
formations. Kimmeridgian and Tithonian limestone and dolostones account for over
80% of the plateau summits. These formations are masked by thick tertiary sediment,
with Eocene fluviatile deposits on the south side and Miocene marls and continental
Plio-Quaternary deposits to the north (Fig. 3).

The lithostratigraphic log in Fig. 4 shows the series of the Mounts of Tlemcen
formations. It is particularly interesting to note that these formations have facies with
lateral changes, varying thicknesses and highly irregular dolomitisation (according
to research by several authors [2, 7–9].

2.4.1 The Palaeozoic

Palaeozoic rocks outcrop in the horst of Ghar Roubane. These Palaeozoic formations
were the focus of work conducted by Lucas [11, 12]. In general, these are formations
that are considered to be an impermeable substratum.

Fig. 3 Regional geological map (after the geologic map of Algeria, scale 1/500,000)

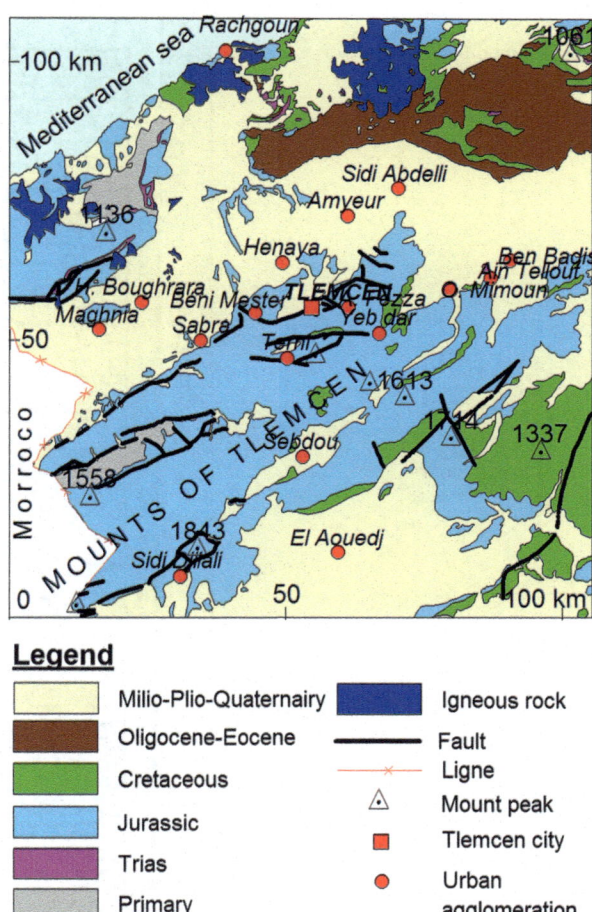

Legend

▢	Milio-Plio-Quaternairy	▣		Igneous rock
▢	Oligocene-Eocene	———		Fault
▢	Cretaceous	—×—		Ligne
▢	Jurassic	⚠		Mount peak
▢	Trias	▪		Tlemcen city
▢	Primary	●		Urban agglomeration

2.4.2   The Triassic

This is a Keuper facies (red and gypsiferous soft clays). The Triassic partially outcrops and tends to be diapiric in structure (A. Tellout, Beni Bahdel).

2.4.3   The Lias and Dogger

The Lias and Dogger formations appear only in two horsts: Ghar Roubane and Dj Tenouchfi. These formations are characterised by significant variations in lateral facies. The work conducted by Lucas [11, 12] makes the most important reference to these.

The lower and middle Lias are massive limestones over 200 m thick. These huge karstified limestones are known for their blende and galena seams, which used to be

Fig. 4 Synthetic geological
log [10, modified]

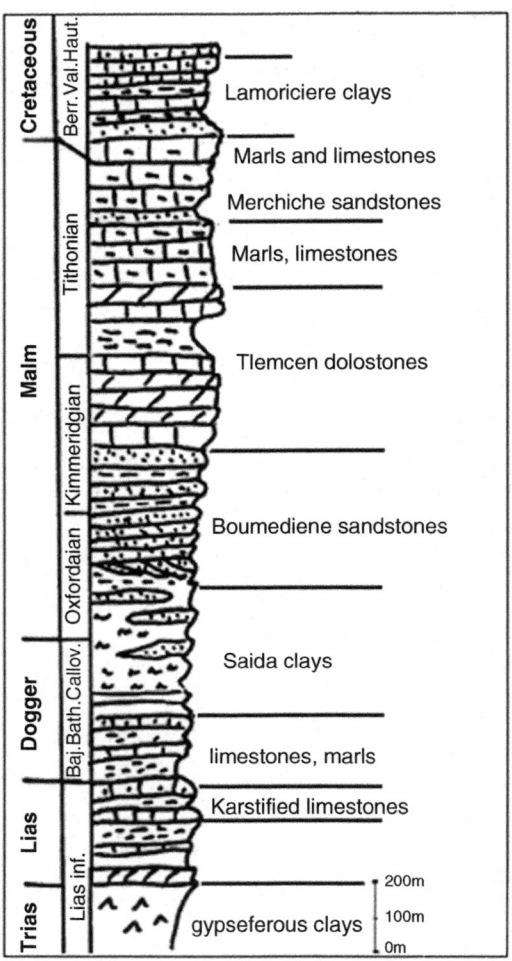

mined at the southwest foot of Koudiat Er Ressass. Through the middle Lias,
initially, formations of calcareous flint were first deposited, interspersed with marl
and followed by a series of several ammonite levels from the middle Toarcian to the
Dogger.

Variations in lateral facies have also been observed in the lower Dogger of the
Djebel Tenouchfi region and southwest of this massif towards Sidi Aissa [13].

2.4.4 The Callovo-Oxfordian (Saida Clays)

These form the Jurassic base and can be found in the Sabra region and closer to the
Beni Bahdel dam. They are deposits of clay, marl and sometimes schistose, with

sandstone layers. This formation can be up to 300–500 m thick and is part of the Callovo-Oxfordian.

2.4.5   Upper Jurassic Sandstone (Boumédiène Sandstone)

This is white sandstone that can sometimes also be brown with a ferruginous surface. It is hard with multicoloured marly bed intercalations. This series is thick, and there is a blue limestone bed made up of Zarifet limestones at the top. This formation was identified in the Lusitanian by Lucas [11, 12] and in the Oxfordian by Auclair and Biehler [7] and was replaced in the upper Oxfordian and Kimmeridgian by Benest [8, 14].

2.4.6   Upper Jurassic Limestones and Dolostones

This formation is commonly called Tlemcen Dolomies. It has the most far-reaching outcrops (more than 1,000 km^2) and also benefits from the best transmissivity properties [2].

Auclair and Biehler [7] identify three main parts to this formation:

- A lower calcareo-dolomitic section (250–350 m)
- A middle marl-limestone section (70–120 m)
- An upper calcareo-dolomitic section (100–200 m)

It is particularly worth noting that these three layers match the Tlemcen dolostones, Raourai marl-limestones and Terni Dolostones, respectively, as described by Benest [8].

The Tlemcen Dolomies characterise the Mounts of Tlemcen, giving them a very individual morphological style. In places, erosion has created highly picturesque ruiniform reliefs with chimneys. The lower limit of the formation is hardly ever synchronised in the Mounts of Tlemcen region. There are two reasons for this:

- The first is related to tectonic dolomitisation, which affects the underlying formation of Zarifet limestones, either in full or in part.
- The second relates to the lateral facies variations that were predisposed to dolomitisation.

In fact, boreholes reveal that dolomitic facies mainly dominate limestone facies.

2.4.7   Cretaceous Basal Alternations

These formations were defined by Auclair and Biehler [7]. They are a series of 200–300 m alternate layers of marl and marl-limestone with a layer of sandstone at their base (Merchiche sandstone), which is the only level containing groundwater.

This formation covers large areas in the eastern Mounts of Tlemcen, where it masks massive carbonate formations and impedes infiltration [2].

Benest identified this marly calcareous formation as the calcareous formation of Ouled Mimoun.

2.5 Geological Structure

The primary resistant moles in Ghar Roubane to the west and Tiffrit to the east have had a major influence on the region's structural development. This is an area where rock competence leads to brittle tectonics. However, this does not prevent soft deformations such as synclines and anticlines.

Above all, the Mounts of Tlemcen are affected by:

- A brittle distensive tectonic with a SW-NE to WSW-ENE longitudinal fault system (in the Tellian direction) and the formation of grabens
- A softer compressive tectonic with folded structures of varying complexity

The Tlemcen and Ghar Roubane Mountains form the western part of the northern edge of the high Oran plains [15]. They are divided by three main transversal fault systems. To the north, horizontal movements and collapsed areas mark where the Jurassic land begins, with three raised spurs on the edge of the Maghnia-Ben Badis depression. The western area is still oriented towards the North [15].

- The Tafna transversal, with the Dj Tefatisset spur in the west of Tlemcen that extends from Dj Tenouchfi (south-southwest) to Beni Mester
- The Chouly Oued transversal to the west, from where the Dj Ramlya massif begins. It runs from El Arbi Dj (to the south-southwest) to the plain of Sidi Abdelli in the north-northeast.
- The Ain Tellout transversal that separates the Mounts of Tlemcen and the Mounts of Daia. The western section is significantly offset to the north and is highly prominent between Ain Tellout and Dj Ez Ziait. It continues to the south to level with Dj. Ouargla.

As a consequence, the Meseta Oranaise is dissected by numerous faults that have a vertical displacement of several hundred metres. These faults border a series of smaller compartments, some of which are collapsed (such as the Sebdou graben where Jurassic terrain is covered by Cenozoic continental deposits), whereas others are elevated (such as the Ghar Roubane horst, where Palaeozoic outcrops can be seen).

This structure has a major impact on groundwater flow as the fragmentation of the Mounts of Tlemcen into several dozen blocks has led to the creation of the same number of distinct aquifers, none of which are linked to each other.

3   Karst Landforms and Underground Rivers

After the Mounts of Saida, the Mounts of Tlemcen are the most extensive karstified carbonate massif in northwestern Algeria. The first indication of their karstification came to light in 1880 during initial earth-moving work for the Meffrouch Dam (located 8 km south of Tlemcen) when the first September flood gushed and disappeared into the open excavations of the dam [16]. The project was abandoned and could not be resumed until 1946. Similar observations were made during the construction of the Beni Bahdel dam, located 30 km southwest of Tlemcen [17]. Numerous drilling campaigns in the carbonated massif of the Mounts of Tlemcen have highlighted this karstification and enabled quantitative analysis to be carried out [18]. Several inventories have been made. These authors have only focused on large cavities and underground rivers. Since 2014, the Speleological Club of Algiers (SCA) has been logging inventories in conjunction with the University of Tlemcen and the Tlemcen National Park (TNP). This work complements all previous studies (Fig. 5).

3.1   Lapiaz and Other Karst Forms

The surface karstic forms are surprisingly underdeveloped although most of the outcrops are limestone and dolostones. The extension of limestone pavement is limited, and there are few sinkholes (Figs. 6 and 7). This is likely to be a consequence of the very advanced, ancient dolomitisation process, which has resulted in large dolomite crystal forming (often described as *sugar-like* texture dolostone), which produce coarse dolomitic sand when eroded and this sand obliterates surface karstic landforms.

3.2   Sink Holes and Swallow Holes

There has so far been no report or record of any dramatic or concentrated swallow hole. Temporary watercourse beds will certainly have disappeared from time to time, but only one has been reported (by villagers during field surveys near Tal Terni village), and one swallow hole/resurgence on the southern flank of the Djebel Nador has also been recorded.

This can be compared to dolomitic sand that obstructs karst forms (where there is a predominance of dispersed rather than concentrated infiltration).

In contrast, the resistance of dolomitised beds is not in question; these can support large vaults (and very large cavities) such as the Yebdar cave, the Seghendouna cave or the underground Ghar Boumaza network.

Fig. 5 Karst landforms

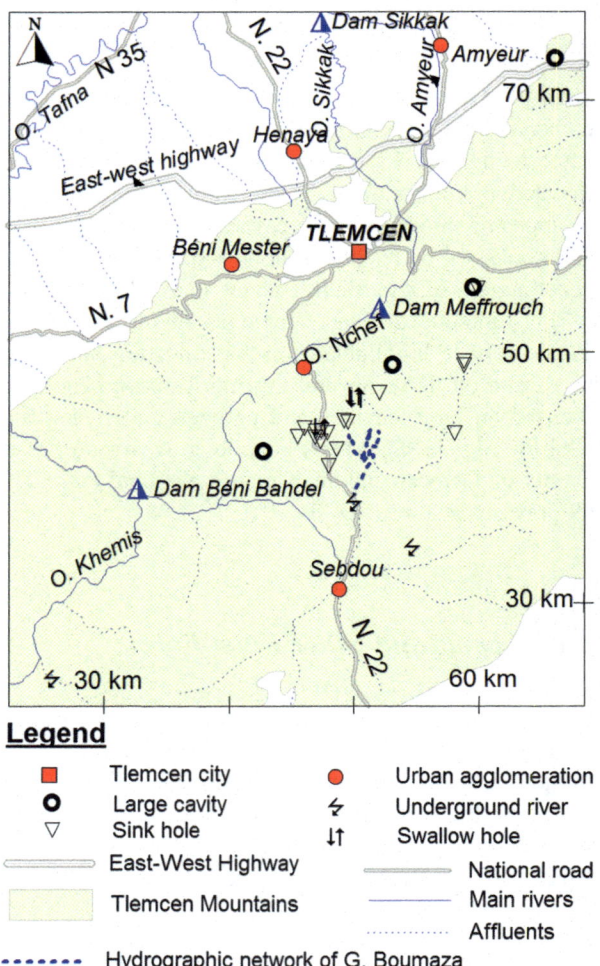

Legend

■	Tlemcen city	●	Urban agglomeration
O	Large cavity	⚡	Underground river
▽	Sink hole	⬍	Swallow hole
	East-West Highway		National road
	Tlemcen Mountains		Main rivers
			Affluents
- - - - -	Hydrographic network of G. Boumaza		

Many sinkholes that are part of the karstic system have been explored in the Ghar Boumaza area. These are tens of metres – and sometimes up to about a hundred metres – deep (the Oraf Ain chasm, Notenboom) [19]. They can be seen on the Lalla Setti plateau, Terni plateau, Dj Oum El Allou and Dj Fernane Achour and further to the east in the Ouled Mimoun region [9, 20]. Two of these sink holes are pictured in Figs. 8 and 9.

3.3   Underground Rivers

Although the karst surface may be unimpressive, the Mounts of Tlemcen have extensive underground rivers.

Fig. 6 Lapiaz field under snow in the southwest of Terni. (Cliché Bensaoula)

Fig. 7 Deep lapiaz evolving in a sink hole (Dar Lahnech) in Tal Terni. (Cliché Bensaoula)

Fig. 8 Ghar Lehmam, Tagma. (Photo S.C.A., 2016), 20 m deep

Fig. 9 Ghar Dar Etaous (Terni) (Photo S.C.A., 2017) 7 m deep

According to the work of Birebent [21], Notenboom [19] and Collignon [22], several underground rivers have been explored in the Mounts of Tlemcen. These include Ghar Boumaza, one of the longest (18.4 km) caves in Africa to have been explored to date and one of the longest underground rivers in the world (17 km) [23] (Fig. 10), Ghar el Kahal (2,210 m), Hassi Dermam (150 m) and Ain Bir Tessaa (1,805 m).

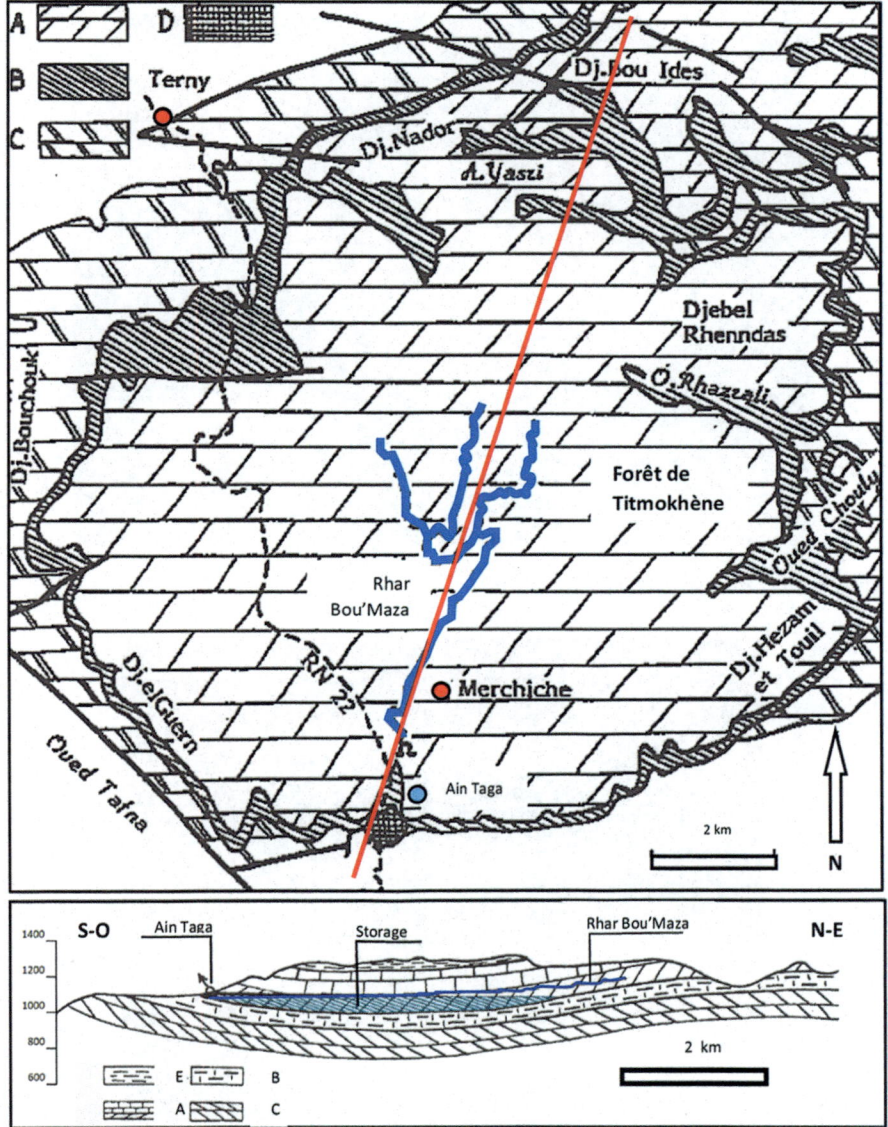

Fig. 10 Ghar Boumaza underground river and its geological context ($X = 1.312°$W; $Y = 34.7°$N) [23]. *A* Terni limestones and dolostones, *B* Hariga shales, *C* Tlemcen dolostones, *D* Travertines, *E* Zegla B marls, cross-section along the red line

3.4 Large Cavities

The country's famous Yebdar cave (385 m) is the only cave that is publicly accessible and has spectacular concretions. However, other equally interesting

Fig. 11 Beni Add cave (Photo Bensaoula)

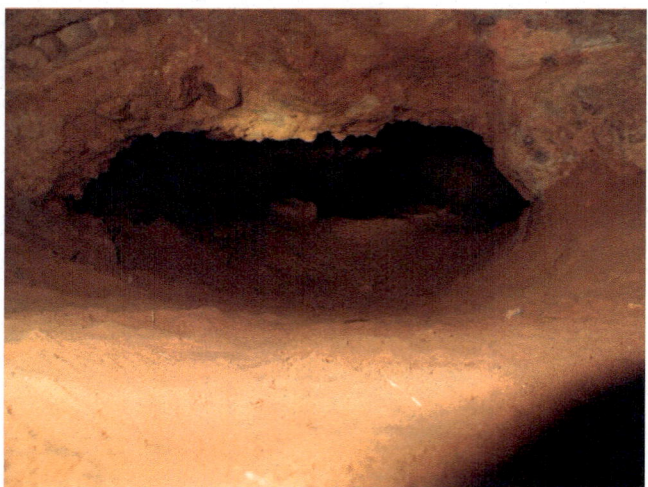

Fig. 12 Ghar Benchikhi cave (Meffrouch, SCA Avril 2015), an extension of about 40 m

cavities have been documented within the Mounts of Tlemcen. These include Ghar
Eddaghra (Tagma), Ghar Lehmam, etc. (see the location map), which have beautiful
concretions but are unfortunately hard for the public to access (Figs. 11 and 12).

Fig. 13 Travertine massifs at Lalla Setti plateau [9]

3.5 Deep and Inaccessible Karstification

Drilling has further confirmed the widespread development of the underground karst and has intersected numerous underground cavities.

In fact, drilling is an excellent karst observation method, especially in flooded areas [24]. Using the rotation system, mud loss can be recorded through drilling, and this is an excellent indicator of fissured or even karstified levels. Studying mud loss has shown that limestone tends to promote cavity formation [24], and these karstic cavities can be found even at great depths (619 m at SAL drilling in Zouia) [25].

3.6 Abundant Travertines

The large travertine massif that has developed along the escarpments is one of the specific features of the Mounts of Tlemcen landscape. This massif is made up of significant mass deposits (50 Mm^3 for Lalla Setti, 15 Mm^3 for Ouchba) that stand like tall sheer cliffs (60 m thick according to Doumergue [26]). They can be extremely vacuolar or even cavernous (Figs. 13 and 14).

These massifs are sometimes tiered and extend over significant areas (1 km^2 in the case of the Ouled Mimoun massif). This produces a vertical outlet (Ain Bentsoltane to Ouled Mimoun), which currently emerges at the foot of these travertines.

These are easily identifiable because they are rarely covered with vegetation and cause slope rupture. They highlight the location of current or former springs associated with Jurassic massifs. Some punctuate the Tlemcen Dolostone contact with Boumediene sandstones.

Fig. 14 Ghar Boumaza, the source of the Tafna

More than 30 travertine outcrops have been successfully recorded across the Mounts of Tlemcen (Fig. 15). These can be considerably large, and, interestingly, they are more abundant in the western part of the Mounts of Tlemcen than in the eastern section [9].

This travertine massif shows that large water sources once existed at this level before they dried up. The fact that watercourses no longer exist on the surface, and water no longer reaches the surface, could mean that the basic water circulation level runs deeper [27]. Indeed, drilling carried out all over the Mounts of Tlemcen has measured quite low piezometric levels.

Tectonic significance and hydrogeological consequences:

In the Mounts of Tlemcen region, travertine deposits are highly correlated with slope rupture (El Ourit, Lalla Setti, Ain Ouchba, Ain Fezza, Beni Bahdel, etc.). This can be explained by the fact that, in the areas where these deposits form, water is extremely turbulent (with waterfalls) and loses its CO_2, causing a pH increase and Ca CO_3 precipitation.

4   Groundwater Resources

In this section, we provide details of the hydrogeological features of the carbonate formations found in the study area. It can be seen that this region is well-watered and experiences high water infiltration feeding many springs. The physicochemical quality of the groundwater is also discussed.

Fig. 15 Map of the main travertine uplands and associated springs

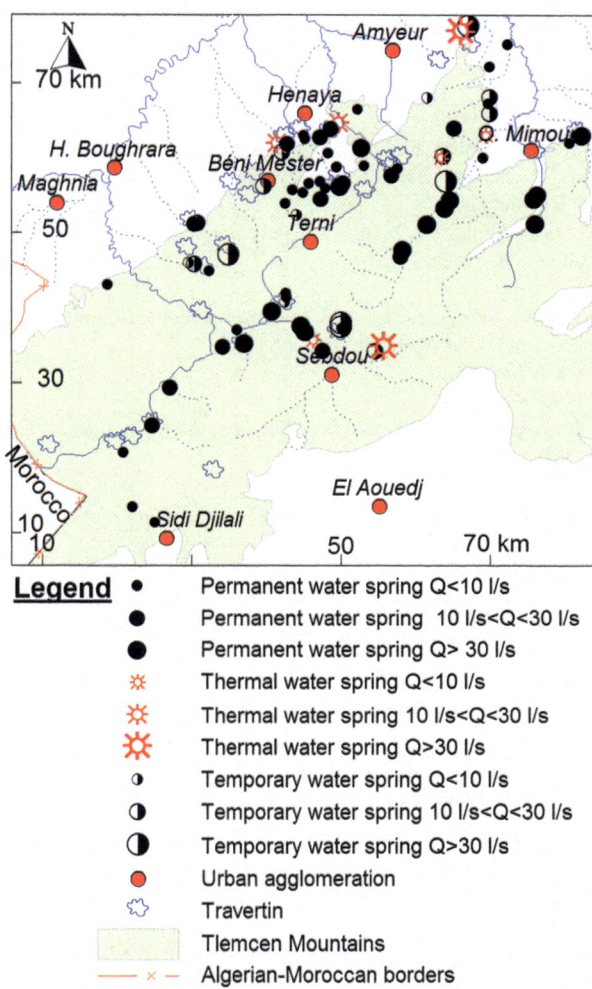

Legend

•	Permanent water spring Q<10 l/s
•	Permanent water spring 10 l/s<Q<30 l/s
●	Permanent water spring Q> 30 l/s
☼	Thermal water spring Q<10 l/s
☼	Thermal water spring 10 l/s<Q<30 l/s
☀	Thermal water spring Q>30 l/s
◑	Temporary water spring Q<10 l/s
◑	Temporary water spring 10 l/s<Q<30 l/s
◑	Temporary water spring Q>30 l/s
●	Urban agglomeration
✿	Travertin
	Tlemcen Mountains
— ×—	Algerian-Moroccan borders

4.1   Main Productive Lithological Horizons

The Mounts of Tlemcen are composed of sedimentary rock, and, for this reason, most wells produce water. However, productivity varies greatly from one geological formation to another, and the only geological formations suitable for supplying water to large settlements are the Tlemcen dolostones, the Terni dolostones and the Bajocian and Bathonian limestones and dolostones. The Boumediene sandstone and Helvetian sandy marl areas have very low productivity, and any boreholes sunk in these formations are suitable for supplying individual households or small settlements only (Table 1).

Table 1 Lithology and aquifers in the Mounts of Tlemcen

Geological layer	Lithology	Specific yield (m³/ h per m)ᵃ	Borehole productivity
Miocene's marls north of the Meseta Oranaise (Helvetian)	Marls with thin sandstone layers	Poor	Many boreholes are productive, but with a very limited yield
Ouled Mimoun marly lime-stones (Berriasian)	Marly limestone	Very poor	No recorded bore-hole with a signifi-cant yield
Lato limestones, Terni dolostones and Hariga marly limestones (Tithonian)	*Limestones and dolostones*	*Excellent*	*Majority of high-yield wells*
Raourai marly limestones (Tithonian)	Marls and marly limestones	Very poor	No recorded bore-hole with a signifi-cant yield
Tlemcen dolostones and Zarifet limestones (Kimmeridgian)	*Large banks of limestones, and especially dolostones, often with a sugar-like texture*	*Excellent*	*Majority of high-yield wells*
Boumediene sandstones (Oxfordian-Kimmeridgian)	Quarzitic sandstone, fine-grained, highly cemented	Low	Many boreholes are productive but with a very limited yield
Bajocian Bathonian dolostones and limestones	*Dolostones and limestones*	*Excellent*	*Many boreholes are productive with high yield*

Data source: [25, 28]
ᵃAverage specific yield expressed in m³/h per metre of drawdown

4.2   Structural Position of Karstic Aquifers

The Mounts of Tlemcen do not constitute a single karstified unit drained by a limited number of large springs (unlike the Mounts of Saida and the Mounts of Chellala). The numerous faults that divide them enclose compartments that comprise just as many independent aquifers. Figure 16 below shows a north to the south cross-section, from Hennaya to El Aouedj, which illustrates this complex hydrogeological structure. Viewing the cross-section from north to south, there are around a dozen situations that are more or less conducive to the formation of the large aquifers that are described below (Table 2).

4.3   Main Aquifers

The Mounts of Tlemcen are not a single aquifer. On the contrary, they are divided into numerous small blocks separated from each other either by erosion or by

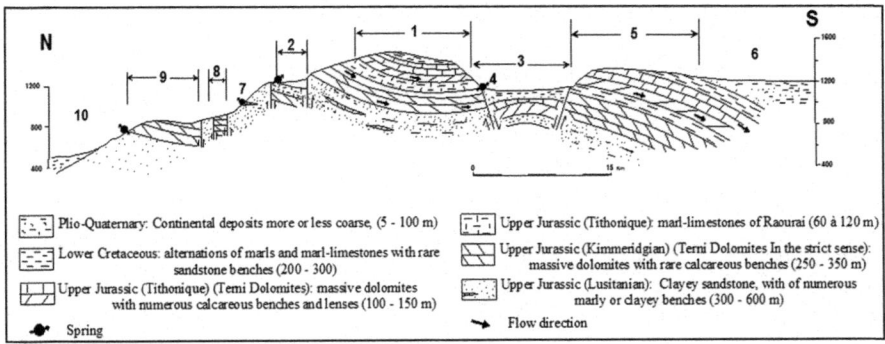

Fig. 16 N-S geological cross-section [2]

Table 2 The various structural situation impact on water resources

1	Merchiche plateau	Extensive karst with good infiltration and large dynamic storage but limited static storage. Most of this karst is drained by a underground river: Rhar Bou Maza
2	Terni plateau	The central part of this small karst supplies the Meffrouch dam. Although it is covered by Miocenous marls, dam construction has been difficult because of leakage through karstified limestones
3	Sebdou graben	Deeply buried karst, where productive wells have been sunk. As a consequence of aquifer depth, there are some thermal springs with typical karstic facies. Existence of hypogene karst is suspected
4	Cliff, south of Merchiche plateau	A large spring (Ain Taga) drains most of Merchiche plateau. Travertine formations are still growing downstream of this spring
5	Djebel Maiter monocline	No springs recorded. For this reason, deep infiltration towards the High Plains is suspected. Exploration boreholes have revealed the existence of a deep karstic aquifer
6	High Plains	Cenozoic continental sediments recover Mesozoic carbonates. Existence of deep karstic aquifers is suspected that could contribute to Dayet El Ferd water balance
7	Cliff, north of Terni plateau	Presently, there are only small springs in Boumediene sandstones. Extremely thick and extensive travertine formations testify to the past existence of high-flow karstic springs (possibly carbo-gas springs)
8	Bou Hallou graben	A very small structural unit with prodigious karstification (high yield boreholes: $>1{,}000$ m^3/h)
9	Sidi Abdelli monocline	No springs recorded. For this reason, deep infiltration towards Hennaya plain is suspected
10	Hennaya plain	Jurassic carbonates are deeply buried under Miocenous marls. Deep boreholes are very productive

tectonic activity (graben and horsts separated by faults with strong vertical displacement).

Table 3 Renewable reserves of the karstic aquifers according to their structural position (Mm^3/year)

	Karstic area (km^2)	Spring discharge (Mm^3/year)	Borehole number	Renewable reserves (Mm^3/year)
Northern cliffs and slopes	335	30	76	70
Plateau aquifer	590	75.3	18	76
Valley aquifer	295	13.5	0	35.5
Sebdou graben	5	0	5	0.5
Southern slopes	505	0	10	29
Total	1,730	119	109	211

It is required to delineate some 30 independent aquifers [2], with very different sizes (from 4 to 300 km^2). These aquifers can be grouped into five categories, depending on their hydraulic behaviour (next section and Table 3).

4.4   Aquifer Recharge and Water Balance

The water balance and the recharge through rainfall for the main karst aquifers in the Mounts of Tlemcen have been assessed by reviewing the following data [2]:

(a) Calculation of average precipitation on each limestone outcrop based on regional rainfall maps
(b) Estimated infiltration based on control basins
(c) Estimated average annual recharge, basin per basin
(d) Outflow measure at all known outlets (springs and boreholes)
(e) Water balance and underground flow calculation, by calculating the difference between c and d

This assessment was carried out for each of the hydrogeological compartments, and the results were grouped into three main areas (Fig. 17 and Table 3).

- *Northern blocks* correspond to best watered areas (>600 mm/year). They benefit from large renewable water reserves (70 Mm^3/year), and they are drained both by the springs located at the foot of the escarpments and by lateral contact with the Miocene sandstones. They are the most used aquifers in the region because they are located near high water demand areas (the most populated regions) and host large static water resources (permanent reserves).
- *High-altitude karsts* (such as the one drained by Ghar Boumaza) are very extensive, but less watered than the northern slopes (300–600 mm/year). They have very large dynamic water resources (renewable reserves) (80 Mm^3/year). These resources are harnessed by spring catchments (such as those that feed small irrigation networks in the Oued Chouly valley) and also by dams that store river flow a few km downstream (e.g. Beni Bahdel dam). These aquifers have small

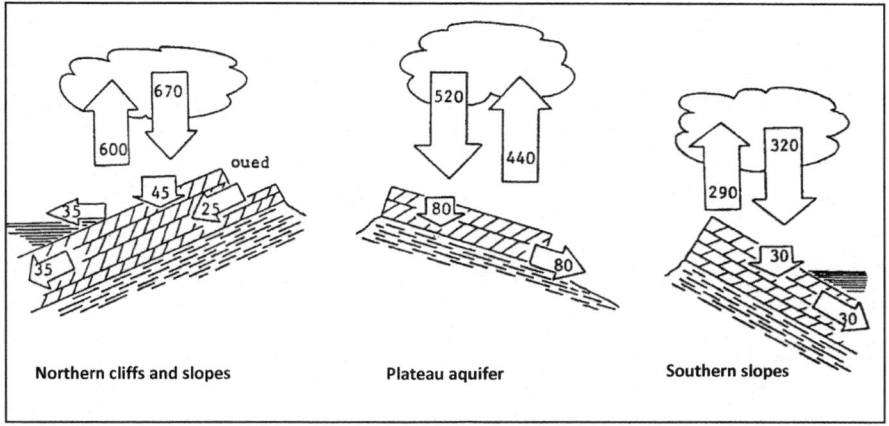

Fig. 17 Hydrologic balance of the karstic aquifers according to their structural position (Mm³/year)

static water resources (permanent reserves) because of the structural position of the reservoirs. Some boreholes are dry, even when they intersect karstified rock as the water table is too deep.

- The *southern blocks* are the least watered areas (< 300 mm/year), and, because there are no springs to estimate a good water balance, their renewable reserves are much more difficult to assess than for the other two groups. According to the infiltration model, these resources have been estimated at 30 Mm³/year. The main outlet of these aquifers is underground transfer towards the high plains, where the water ultimately evaporates (in Dayet El Ferd and similar areas). Very little water from these aquifers has been abstracted thus far due to the low local demand for water in the Chott ech Chargui area and high abstraction costs (deep boreholes, far from the areas to be supplied, such as the Sebdou region). Static water resources (permanent reserves) are probably very large (several thousand Mm³), which more than justifies the development of high-cost projects, such as the 60 new deep wells drilled around Chott ech Chargui to withdraw water from deep aquifers.

4.5 Storage Capacity (Static Reserves)

Static reserves are much more difficult to assess than renewable reserves due to the lack of boreholes and long-duration pumping tests in many aquifers. Moreover, there is currently no complete information available on the extension of karst aquifers to the north and south of the Mounts of Tlemcen.

We have produced a rough estimate of these reserves using the following assumptions (Table 4):

Table 4 Static reserves of the main karstic aquifers

	Aquifer extension (km^2)	Water thickness (m)	Storativity (%)	Static reserves (Mm^3)
Northern cliffs and slopes	500	150	3	2,200
Plateau aquifer	590	20	3	300
Valley aquifer	295	50	3	400
Sebdou graben	100	150	3	400
Southern slopes	1,000	100	3	3,000
Total	2,485			6,300

5 Sustainable Management of Groundwater Resources

In this section, we will discuss the various water management options that have been implemented since 1950 to cope with the rapidly increasing demand for water:

- The use of deep boreholes to rapidly harness karst water resources
- Groundwater mining to offset drinking water shortages in the region
- The substitution of groundwater with water from desalination plants since 2011
- Production costs for these two options (groundwater and seawater desalination)

5.1 1950–1980: Conventional Water Resources Management – Spring Intakes and Large Dams

Since ancient times, people have obtained drinking water in the Tlemcen region by harnessing water from the main springs. These include the upper and lower Ain Fouara, Ain Bendou, Ain Tellout, Ain Sabra, etc. Numerous efforts have been made to tap into groundwater and surface water to better support the region's socio-economic development and to offset the water shortages recorded after several periods of drought (Table 5).

5.2 1980–2000: Deep Boreholes to Mobilise Rapidly Additional Groundwater Resources

5.2.1 Tremendous Increase in the Number and Depth of Boreholes in the Mounts of Tlemcen Since 1980

During the 1960s and 1970s, most drilling works were carried out north of the Mounts of Tlemcen and in the foothills and the Maghnia and Remchi plains. However, the outcomes of this work have been relatively disappointing as these

Table 5 Main characteristics of the water supply sources of the population of the Tlemcen region (D.R.E.[a] Tlemcen 2017)

Type of facility	Name of dam	Location	Capacity	Year commissioned	Water use
Dam	Beni Bahdel	Beni Bahdel	56 Hm^3	1952	Water supply and irrigation
	Meffrouch	Terni Beni Hdiel	15 Hm^3	1963	Water supply and irrigation
	Sidi Abdelli	Sidi Abdelli	110 Hm^3	1987	Water supply and irrigation
	Boughrara	Hammam Boughrara	177 Hm^3	1998	Water supply and irrigation
	Sikkak	Ain Youcef	27 Hm^3	2005	Water supply and irrigation
Well field (groundwater)	Zouia I and II (19 boreholes)	Zouia (border area)	40,000 m^3/day	From 2003 to 2006	Water supply

[a]Direction des Ressources en Eau (Water resources Direction of Tlemcen)

areas predominately consist of sandy Miocene marls, along with some thin sandstone layers, which are not conducive to very productive aquifers. Seventy-five boreholes were drilled before 1980, totalling 8,800 m (average depth of 120 m – [29]), none of which revealed a highly productive aquifer.

As part of their search for additional resources to supplement water supply within the Tlemcen region, the Wilaya Water Department (DHW) and the National Water Resources Institute (INRH) launched large-scale prospecting and drilling campaigns in the 1980s and 1990s that directly targeted the Jurassic carbonate aquifers.

These new hydrogeological target areas had previously been discounted as they were considered both unpromising (elevated karsts with few permanent reserves) and too uncertain (productivity varies greatly from one well to another and is difficult to predict, which complicates investment planning).

The outcomes of these drilling campaigns have exceeded all expectations. 203 boreholes (Fig. 18) were drilled between 1980 and 2010 for a total length of 57,000 m (average depth of 280 m – [29]). These deep boreholes provided robust evidence regarding:

- The very good average productivity of the Kimmeridgian carbonate formations (limestone and dolostones of Tlemcen and Terni): 1.2 l/s and per metre of drawdown
- The good chemical quality of the water in these aquifers (despite the intrinsic vulnerability of karst aquifers): calcium and magnesium bicarbonate waters, high hardness but low in nitrates
- The very large permanent reserves (several hundred million m^3 – see d), which enable large-scale production schemes to be set up

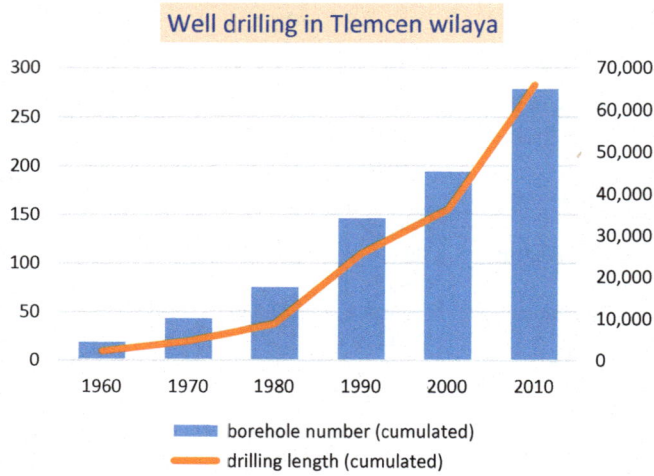

Fig. 18 The impressive development of drilling works in the region over the past 30 years [29]

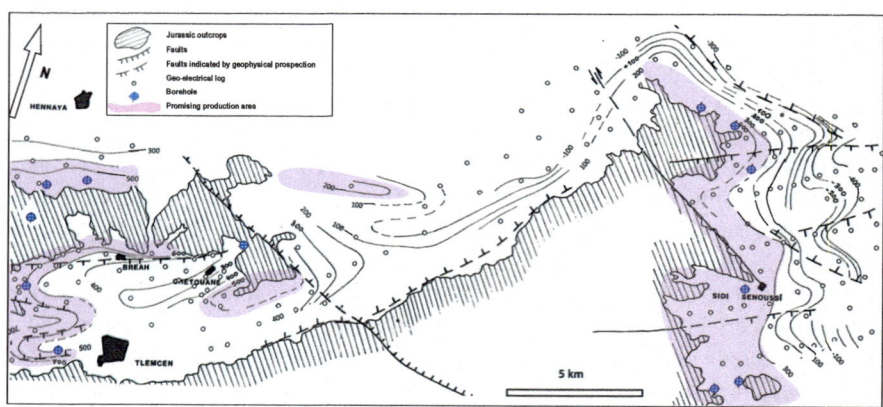

Fig. 19 Geophysical prospecting on Mounts of Tlemcen northern slopes based on ENAGEO report [2]

5.2.2 Most Suitable Drilling Areas

The northern foothills of the Mounts of Tlemcen are the area where demand for high-yield boreholes is high because these are the most populated areas, where demand for good quality water is rising very quickly.

To delimit the exploitable zones on the northern foothills of the Mounts of Tlemcen, the INRH has contracted ENAGEO to carry out an extensive geophysical survey. The prospecting was carried out by resistivity measurements (Schlumberger device, with AB lines 1,000–2,500 m long, i.e. a penetration depth of 500 m). This measurement campaign demonstrated (Fig. 19):

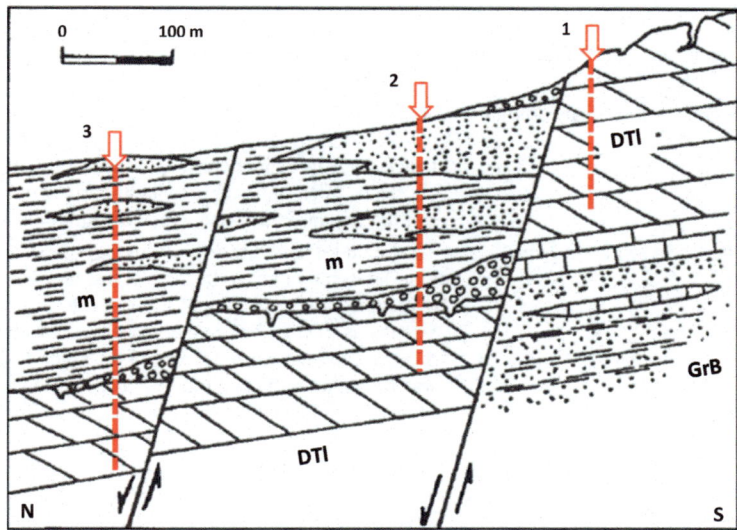

Legend GrB: Boumediene sandstones, DTl: Tlemcen dolostones, m: Miocene shales

Fig. 20 Borehole siting on Northern slopes [2, modified]. *GrB* Boumediene sandstones, *DTl* Tlemcen dolostones, *m* Miocene shales

- The existence of a very resistive layer (>200 Ω m) at depth, under the Miocene marls, which was assimilated to the Jurassic limestones and dolomites
- The strong vertical displacement along the faults bordering the Mounts of Tlemcen (200–500 m) and which accentuate their character of a horst dominating the Hennaya plain where prevails a thick cover of Miocene sandy marls
- The existence of numerous faults (N60 at 80°E and N130 at 150°E), which divide this horst into sub-compartments that constitute as many small aquifers
- The existence of zones where limestones and dolomites are accessible with limited investments (300–500 m deep boreholes, through the Miocene marls)

The model built from the geophysical prospecting results was validated by drilling works. More than 20 wells were drilled through the Miocene marls in the presumed favourable zones (in pink in Fig. 19), and they reached the Jurassic limestones and dolomites, at a depth of less than 200 m. All these boreholes have good productivity.

Borehole Siting

Within this promising area, three drilling site options are possible (Fig. 20):

1. Drilling directly into Tlemcen dolostones, at the edge of outcrops; these are medium-depth boreholes (100–200 m deep) difficult to drill (total mud losses); these areas are very vulnerable to pollution.

Table 6 Obstacles and solutions for drilling work

Specific constraints of drilling works in highly karstified aquifers	Technical solution for drilling works
Extreme toughness of many dolomitic layers	Select hard rock tools, including tungsten carbide pellet bit- and PCD (polycrystalline diamond)-type tools. Down-the-hole hammer drilling is effective in hard rock, but this is often impossible in highly productive karstic aquifers because the blowing power required is too high
Total mud losses while drilling	Total mud losses while drilling are the bane of drillers in the Tlemcen dolomites. In some cases, they can be overcome with conventional sealants (fibres, highly viscous mud, etc.), but, in the majority of cases, drilling must be continued in conditions of total mud losses with water circulation (which implies good site logistics). Down-the-hole hammer drilling often results in accidents due to jamming of the drill rods and bits. The use of reverse circulation could be an interesting alternative
Large depth of static level (>100 m or even >150 m)	These large depths involve installing the pumping chamber at great depth and thus large diameter drilling ($17''1/2$) up to large depths (>200 m). This involves the use of rotary drills with high torque (or, if necessary, machines equipped for reverse circulation)
Hole deviations when cavities occur	Drill with several drill collars, interspersed with stabilising tools

2. Drilling a few hundred metres from the escarpments, through the Miocene sandstone and clayey sediments, to reach the Tlemcen dolostones between 100 and 200 m deep; these are deeper boreholes (200–300 m), but quite easy to drill.
3. Drilling further away from the escarpments through the mainly clayey Miocene sediments to reach the Tlemcen dolostones between 200 and 400 m deep; these are deeper boreholes (300–500 m), providing a high level of protection against pollution.

5.2.3 Well Drilling Constraints

There are a number of difficulties involved in drilling in mid-depth karst aquifers, and these must be taken into account when planning the work, selecting drilling equipment and estimating the cost of the work (Table 6).

These difficulties can be overcome by conducting rigorous drilling operations and by using experienced drillers and sufficiently powerful drilling machines [2, 28].

5.3   2000–2010: Groundwater Mining to Compensate Drinking Water Shortage

5.3.1   Karst Aquifers: An Easily Useable Resource

The very high productivity of the boreholes drilled in well-karstified aquifers makes it relatively easy to extract large volumes of water (one to two million m^3 per year per well). This makes it possible to harness significant water resources within a short time and thus to cope with water supply crisis situations.

This was the option chosen by DHW and INRH to address the 1985 water crisis in Oran and Tlemcen, when the largest dams in the region (Beni Bahdel and Meffrouch) almost ran dry. Twenty deep boreholes and accompanying water supply systems were installed in less than 2 years, a period three times shorter than that required to study and build a large dam or seawater desalination plant. Water from these wells (30 million m^3 per year) was then injected into the Beni Bahdel water main.

5.3.2   Limited Renewable Reserves of Karst Aquifers

As it is relatively easy to develop well fields within karst aquifers, the risk of over-exploitation is high. Once a new well has been sunk, it is tempting to exploit it to obtain the maximum possible yield.[1] However, this maximum flow is completely independent of the aquifer's natural recharge. Sizing pumps on the basis of borehole features, without taking into account the rate of aquifer recharge, can thus lead to exceeding this recharge rate. This is known as "mining water resources", i.e. water abstraction that can lead to resource exhaustion.

5.3.3   Mining Karst Aquifers in the Mounts of Tlemcen

The successful drilling campaigns in the Mounts of Tlemcen (between 1983 and 2010) encouraged operators (first and foremost the ADE – Algérienne des Eaux) to increase water abstraction in order to cover growing water demand, despite the growing scarcity of water resources caused by a rainfall deficit during the period 1987–2002 [30].

This increase in water abstraction has inevitably led to some aquifers being overused, resulting in highly significant drawdown. Figure 21 illustrates the overuse of a small karstic aquifer located north of Ouled Mimoun through boreholes (SBA1 and SAB2). The piezometric level falls synchronously across the different boreholes

---

[1]This yield is what hydrogeologists often call "critical operating yield", i.e. the flow rate for which quadratic head losses in the screens become predominant and cause a sharp increase in drawdown and therefore energy costs.

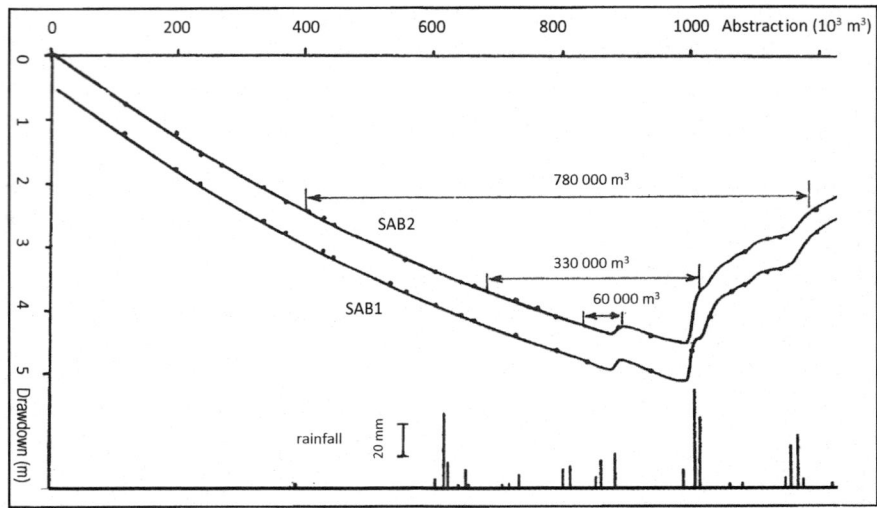

Fig. 21 Mining groundwater (Water abstraction and drawdown) [2, modified]

following the abstraction of 1.4 million m³ of water. This drawdown is not offset by the annual recharge, which has been estimated at 0.78 million m³ using a method developed in the Mounts of Tlemcen [2, 31].

5.4 2011 To Date: Desalination Plants to Limit Groundwater Abstraction

Two desalination plants have been built and commissioned over the past 10 years. Both of these use the reverse osmosis principle and are used for urban water supply.

The Souk Tlata desalination plant began operating in 2011 (with a production capacity of 200,000 m³/day) and is currently supplying 30,000 m³/day (June 2017) to 19 communes and 340,000 inhabitants.

The Honaine plant, which has the same production capacity, is to supply 29 communes (500,000 inhabitants) with 180,000 m³ of water a day (D.R.E. Report, 2017).

By the end of 2017, 45 of the 53 municipalities that make up the wilaya (district) were connected to the desalination plants' supply systems, and three others were in the process of being connected.

The remaining five communes will continue to be supplied with groundwater abstracted from the new well fields installed near Chott el Gharbi, south of the Mounts of Tlemcen.

As a result of the increase in desalinated water production, the use of many of the boreholes located in the Mounts of Tlemcen has been put on hold, helping to progressively restore the original water table level in karstic aquifers.

5.5   Future Proposals for the Integrated Management of Groundwater Resources

5.5.1   Various Options for Water Production

The Mounts of Tlemcen region now has three main water resources:

- Surface water, which collects in rivers on the fringes of the Mounts of Tlemcen and is stored in large dams (Beni Bahdel, Meffrouch, Sidi Abdelli)
- Groundwater abstracted from 200 deep wells mainly drilled into Tlemcen dolostones and Terni dolostones, located in the foothills of the Mounts of Tlemcen
- Two desalination plants (Souk el Tleta and Honaine), which have been installed in 2011 near the coast, a 100 km further north (with a production capacity of 70 million m^3/year each)

Treated wastewater also constitutes a water resource in the region. However, this resource is currently somewhat limited (10 Mm^3/year produced by the Ain el Hout plant, which treats wastewater from the city of Tlemcen). Moreover, this resource cannot be used to replace the three main resources since it can only be used for irrigation.

5.5.2   Costs with These Various Options

The sea water resource, available for desalination, is almost limitless for meeting the current water demand. It is therefore theoretically possible to continue investing in large desalination plants to keep pace with the growing demand for water.

However, gradually replacing surface water and groundwater with desalinated seawater raises economic and environmental difficulties. Investment in such water production facilities is extremely costly (the investment cost for a reverse osmosis plant equates to 250% of the cost of investing in deep tube wells with a similar production capacity). Operating costs are also very high (operating costs for reverse osmosis are 1,000% higher than those for deep tube wells and 3,000% higher than those for surface water sources).

The extremely high operating costs of desalination plants are due to their high power consumption [32]. This constitutes one of the factors of the structural financial imbalance of the drinking water sector in Algeria, where the average water production exceeds the selling price to most users [33].

Moreover, desalination plants consume a lot of energy and therefore have a large carbon footprint (Table 7).

Table 7 Production costs and environmental impact of various water sources

	Reverse osmosis sea water treatment	Groundwater abstraction	Surface water storage (dam) and treatment
Average annual production capacity (Mm^3/year)	400,000 m^3/day for the Souk el Tleta and Honaine plants, i.e. 146 Mm^3/year[a]	200 bore-holes × 2,000 m^3/day, i.e. 146 Mm^3/year	Volume that can be regulated: 209 Mm^3/year[b]
Investment cost	3,885$/1,000 m^3/year[c]	1,644$/1,000 m^3/year[d]	Depends on dam site
Unit costs for water production	0.6$/m^3 [e] 0.5–0.8$/m^3 [g]	0.15$/m^3 [f]	0.2$/m^3 [g]
Energy for production	5.5–8 kWh/m^3 [h]	0.3–0.6 kWh/m^3 [d]	0.05–0.15 kWh/m^3 [h]
Water head and energy to pump water up to Tlemcen	800 m 2.5 kWh/m^3 [d]	150 m 0.375 kWh/m^3 [d]	50 m 0.125 kWh/m^3
Total energy footprint	7–10 kWh/m^3	0.7–1 kWh/m^3	0.2–0.3 kWh/m^3
Environmental impact	Energy wastage (carbon footprint)	Groundwater level depletion	Natural river flow interruption
	Brine release		Land flooding

[a]Source: ADE website
[b]Source: [34]
[c]Source: M. Hocine Necib, Minister of Water Resources, cited in *Maghreb émergent* – 25 December 2017
[d]Source: author's calculation
[e]Source: Hyflux cited in *Le matin d'Algérie* – 7 July 2008
[f]Source: [35]
[g]Source: [36]
[h]Source: [32]

5.5.3 Managed Aquifer Recharge

Groundwater resources in the Tlemcen region have been increasingly abstracted over the last century as the demand for water for irrigation and drinking has grown.

The various stages of this production increase are summarised in Table 8.

We will not dwell on the first three levels which are quite conventional:

- Level 1 – gravity-based spring water use
- Level 2 – construction of reservoir dams downstream main karstic springs
- Level 3 – drilling boreholes to abstract water during low water periods

On the other hand, the following three levels are more innovative and reflect strategies for optimising the exploitation of water resources in the specific context of karst:

Table 8 Increasing levels of groundwater resources use

	Management model	Examples in the Mounts of Tlemcen	Pros	Cons
Level 1	Water collection from natural spring	Gravity flow irrigation systems along Oued Chouly	Very little environmental impact	No flow regulation (all flood water is untapped and lost to the sea)
Level 2	Water storage downstream of main springs	Beni Bahdel, Meffrouch, Sidi Abdelli, Sikkak dams	This option makes it possible to store (and make use of) flood flow	No use of the regulating capacity of karstic aquifers
Level 3	Well drilling and groundwater abstraction; sustainable abstraction yield, i.e. < dynamic reserves	Most boreholes completed before 1980 met this criterion	This option makes it possible to valorise karst storage capacity at low cost (as an additional reservoir)	The drop in the piezometric level leads to the gradual drying up of springs
Level 4	Deep drilling and mining of aquifers, i.e. abstraction yield > dynamic reserves	This management method has been used with boreholes where excessive drawdown was observed (e.g. SAB1, SAB2 and BH2 wells)	This option makes it possible to cope with severe water deficits over a number of years	Large groundwater level depletion leads to the progressive drying up of the shallowest boreholes and increases running costs (power)
Level 5	Management of karst aquifers as strategic reserves	This option has not been implemented in the region so far; it would involve building additional high-yield wells that would not be put into operation but that would be reserved for periods of water shortages	This option allows water to be stored in reservoirs that are less vulnerable to pollution than dam lakes	This option implies making investments today that will not be developed until a few years from now
Level 6	Managing aquifer recharge of karst aquifers with surface water or treated wastewater	This option has not yet been implemented in the region; it would consist of using karst aquifers as backup reservoirs to store dam overflow or treated wastewater outside of irrigation periods	This option makes it possible to develop resources that would otherwise be lost at sea. The implementation of MAR is simpler in karstic aquifers than in porous aquifers	The rate of return on such investments is uncertain, and further economic analysis is recommended. The environmental risks of MAR are not negligible in karst aquifers

- Level 4 – Deep drilling and mining of aquifers, i.e. abstraction yield > renewable reserves. This option is often used for large productive aquifers located in arid or semiarid regions (e.g. Nubian Sandstone Aquifer System in Libya/Egypt). In the Tlemcen region, this management method has been used since 1985 to small karst aquifers where renewable reserves were considered as limited, but where high yield boreholes were available (e.g. SAB1, SAB2 and BH2 wells). As a consequence of mining groundwater, significant drawdown has been observed. It was considered for a while as an acceptable option able to cope with regional water shortages. South of the Mounts of Tlemcen, the new Chott el Gharbi water scheme is based on the same principle because the planned high abstraction rate (110,000 m^3/day) is most probably higher than the renewable reserves.
- Level 5 – Management of karst aquifers as strategic reserves. One of the specific features of karst aquifers is their network of natural drains (cavities and fissures) that allow global exploitation of the resource with a limited number of boreholes. It is thus possible to equip a karst aquifer at a lower cost, without putting it into operation but preserving it as a strategic water resource, which will only be mobilised in the event of a water crisis. This option would involve building additional high-yield wells that would not be put into operation but whose use would be reserved for periods of water shortages.
- Level 6 – Managing aquifer recharge. This option consists in recharging karst aquifers with surface water or treated wastewater, i.e. using karst aquifers as backup reservoirs. This solution is especially suitable for storing water that would otherwise be lost, such as dam overflow or the flow at wastewater treatment plant outlets outside the irrigation period (a major constraint with the reuse of treated wastewater for agriculture is that the demand for irrigation water varies throughout the year, while the production of treated wastewater remains steady). This option is easier to implement in a karstic aquifer than in a porous aquifer because the risk of clogging the injection wells is lower. However, the environmental risks involved are greater, because transfer speeds in karst can be very high [37].

5.6 Vulnerability Mapping and Protection Perimeters for Vulnerable Water Resources

In general, groundwater quality is good in karstic aquifers in the Mounts of Tlemcen. However, water can be contaminated by human activity. The resulting deterioration in quality can be assessed by monitoring water physicochemical and bacteriological features.

5.6.1 Potential Pollution Hotbeds in the Mounts of Tlemcen

Human activities can have a potentially polluting impact on groundwater quality, including activities such as:

- Agriculture that involves the intensive use of fertilisers and pesticides. When improperly used (intensively or at the wrong time), pesticides can infiltrate the soil and seep into the aquifer.
- Rearing cattle, sheep and poultry that produce large quantities of nitrogen-rich droppings that require careful management.
- Sanitation, using a sewer system that is in poor condition.
- Discharging water into the natural environment without pretreatment. This is the most dangerous source of groundwater pollution. We have noted that there is a lack of wastewater treatment plants within the study area. Urbanisation and its resulting activities can have a major adverse effect on water quality when there is no thorough environmental knowledge of the karstic areas.
- Hydrocarbon poorly managed storage.
- Finally, many quarries exploit the limestones and dolomites of the Mounts of Tlemcen. Once abandoned, they are transformed into wild dumping sites.

5.6.2 The Protection of Groundwater Resources in Algerian Legislation

Law No. 83-17 of 16 July 1983 [38], including the Water Code (in chapter II of heading VI), was the first to introduce water resources protection. This law stipulates that "any supply worked on water for domestic purposes and intended for human consumption must be protected against any accidental or deliberate cause likely to degrade the quality of the water". According to Article 110, this protection consists of setting up a perimeter around a geographical area. Within this perimeter, any activity likely to undermine the quality of water resources is prohibited or must be regulated. These activities are listed in Article 111 of the same law. However, Article 112 identifies facilities where protection must be in places, such as dams, water reservoirs, wells, boreholes and vulnerable groundwater areas. In 2005, the Water Act [39] devoted a whole chapter to protecting and preserving water resources. This sets out two types of the perimeter for both qualitative and quantitative protection. This law also lists all prohibited or regulated activities within these perimeters. Subsequently, Executive Decree No. 07-399 of 23 December 2007 [40] on qualitative water resources protection perimeters sets out the conditions and methods for creating these perimeters. Article 10 of this decree specifies what needs to be included in technical studies to determine these perimeters. For example:

- Determining the geological and hydrogeological characteristics of the aquifer system
- Assessing the vulnerability of a water resource to pollution.

Article 15 of the same decree stipulates that "based on the results of the approved technical study, creating and defining qualitative protection perimeters must be pronounced by a decree made by the competent regional wali when these perimeters affect a single wilaya".

Finally, Executive Decree No. 10-73 of 6 February 2010 [41] covers the quantitative protection of groundwater aquifers with the aim of defining methods for setting out quantitative protection areas for aquifers that have been or are currently being overused.

Drilling for groundwater in the Mounts of Tlemcen, which started at the beginning of the 1980s and continues today (170 drilling campaigns), does not always involve protective perimeters. Protection is only required for the drilling campaigns listed in Wilaya Decree No. 276 of 12 February 2014. This is not particularly effective as the perimeters defined are not based purely on scientific criteria. It should be noted that this decree concerns about 36 boreholes and 18 springs located throughout the Mounts of Tlemcen.

5.6.3 Mapping the Vulnerability and Pollution Risks of Karst Aquifers in the Mounts of Tlemcen

Numerous methods to determine the vulnerability of groundwater have been developed around the world. These range from the most complex, using numerical modelling of groundwater flow, taking into account the physical, chemical and biological processes in the flooded area, to methods that balance different criteria affecting groundwater vulnerability [42].

To date, mapping has only been carried out for the groundwater's intrinsic vulnerability to contamination. This takes the geological, hydrogeological and hydrological characteristics of the region into account and does not include the nature of contaminants. Mapping the intrinsic vulnerability of water resources is based on the concept that the soil's surface is the contamination source and the piezometric surface is the contamination target. The water's path corresponds to the different intersecting soil layers between the ground and groundwater surfaces.

Several intrinsic vulnerability mapping methods have been applied to the Mounts of Tlemcen. These include the simplified OCPK method [43] in the Tlemcen region [44]; the COP method for the Meffrouch basin [20]; and the RISK method for the Ghar Boumaza karst system [45] and for the region of A. Fezza.

The parameters these models use, such as geological formations, flow, infiltration conditions and precipitation, are relatively easy to identify and map. However, the lack of a soil map for the regions studied is a significant shortfall. Consequently, the soil parameter (S) is frequently determined from other criteria, including the type and extent of covering vegetation and outcropping geological formations. In addition to thickness measurements made in situ, soil samples are taken and analysed in the laboratory to calculate grain size.

1. The mapping of pollution vulnerability in the Tlemcen region was carried out over an area of 200 km². Four degrees of vulnerability were identified: extreme, high, moderate and low. The resulting map revealed that 15 of the 28 boreholes abstracting water from the karst, and which are mostly used to supply drinking water, are located in highly vulnerable areas. However, there is no protection perimeter for these.
2. The COP method was applied to map the Meffrouch Basin's vulnerability to groundwater pollution. This is an area that extends over 90 km² and is mostly made up of karstic carbonate formations. The resulting vulnerability map shows that the Meffrouch Basin has five degrees of vulnerability (very high, high, moderate, low and very low). Through this method 27% of the total catchment area is deemed to be in a high to very high vulnerability zone. This percentage reveals that anthropogenic activities can have a considerable impact on ground-water in the region. Creating a vulnerability map can help prevent pollutants from reaching this vital resource by delineating areas that need protection.
3. The application of the RISK method was used for the Ghar Boumaza karst system. This method is easy to implement and does not require much data. It was well-suited to the study site. However, its weighting system can be modified to highlight the most significant factor(s). The resulting map showed that over 66% of the system's surface is in a high vulnerable zone, nearly 5% is in an area of very high vulnerability, more than 29% is in an area of moderate vulnerability and only 0.02% of the total area has a low vulnerability (Fig. 22).

The vulnerability mapping by the multicriterion methods detailed above made it possible to obtain very useful documents as to the highlighting of the zones sensitive to the pollution and thus to protect. The limited data available does not allow the use of more sophisticated methods.

Hazard assessment, as a pollution potential resulting from the human action, is based on land cover data. The hazards identified through the Mounts of Tlemcen have already been detailed at the beginning of paragraph g. Their mapping at a scale of 1:50000 makes it possible to obtain an unclassified hazard map (H). According to the European COST 620 approach [46], these hazards are classified and weighted according to their dangerousness (Hi). This ranking is possible by the following relation:

$$Hi = H \cdot Qn \cdot Rf$$

Hi: hazard index (or degree of harm) that ranges from 0 to 120 and is classified into five classes that are materialised by a colour code (Table 9).

H: the value of the weight of the hazard which varies from 10 to 100.

Qn: a ranking factor that can vary between 0.8 and 1.2. It allows for the comparison between the hazards of the same category. Therefore, the diversities in harmfulness within each hazard category will be mainly due to variability in the amount of harmful substances [46].

Fig. 22 Vulnerability map
of the karstic system of Ghar
Boumaza by the RISK
method [45]

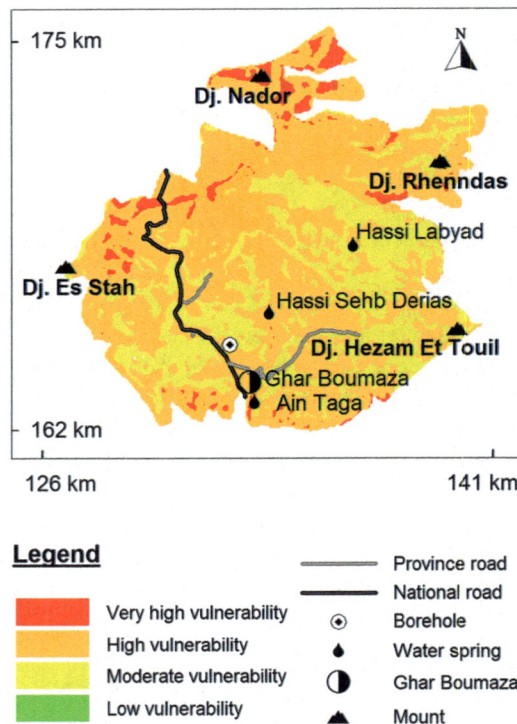

Table 9 Hazard index classified [47]

Hazard index	Hazard index class	Hazard level	Colour
0–24	1	None or very low	Blue
>24–48	2	Low	Green
>48–72	3	Medium	Yellow
>72–96	4	High	Orange
>96–120	5	Very high	Red

Rf: reduction factor which can take a value between 0 and 1. It will take the value 0 in the case where the hazard is secure, and it will take the value of 1 otherwise [47].

The combination of the vulnerability map and the classified hazard map provides a pollution risk map. The identification of high-risk areas allows a better dimensioning of protection perimeters of catchment works (source capture and boreholes).

6   Conclusions

The Tlemcen Mountains conceal nearly 2,000 km^2 of limestone and dolomite outcrops from the Jurassic and Cretaceous periods which dominate the surrounding areas and thus constitute a regional water tower, better watered than the Tafna plain or the High Plains of El Aricha.

Extensive drilling works demonstrated that these rocks are well karstified at depth, whereas the surface karstic forms are relatively modest. The deep aquifers of the Tlemcen Mountains contain significant permanent reserves (estimated at over 6 billion m^3) and benefit from significant recharge (200 million m^3 per year). They thus constitute the main groundwater resources to the west of the Oran meridian.

Deep drilling in well-karstified aquifers presents specific difficulties (total loss of drilling mud, absence of cuttings), but these difficulties were overcome, and 270 boreholes were drilled, leading, during the 1980s and 1990s, to very intensive water abstraction, aiming to supply water to the wilaya of Tlemcen and Ain Temouchent and even part of the wilaya of Oran.

Such intensive abstraction has led to an excessive piezometric level drawdown for some aquifers. The recent commissioning of two large seawater desalination plants has made it possible to reduce groundwater withdrawals and stabilise and even restore piezometric levels.

The water quality of these aquifers is good, but it should not be concluded that they are not very sensitive to pollution. On the contrary, the maps we have produced (using the COP and the RISK methods) clearly show the general vulnerability of these aquifers. If no large-scale pollution has yet been recorded, it is because the karstic areas are relatively sparsely inhabited. This situation could change with the demographic and economic growth of the Tlemcen wilaya.

7   Recommendations

Taking into account the aquifer drawdown observed during the 1990s, the long-term exploitation of the karst aquifers in the Mounts of Tlemcen must be carefully managed, taking into account both the growing demand for water and all available water resources, including seawater desalination and the reuse of treated wastewater for irrigation.

From an integrated water resources management (IWRM) perspective, karst aquifers can play several different roles: (a) water supply to villages located far from the major regional distribution networks, (b) strengthening supply to the localities of the High Plains and (c) a role of strategic reserve, thanks to their large permanent reserves (6 billion m^3) and high borehole productivity.

In order for these aquifers to play this strategic reserve role, it is important: (a) to maintain in good working order the wells whose operation has been suspended since the start-up of desalination plants and (b) to define protection perimeters for these wells, taking due account of aquifer vulnerability to surface pollution.

Finally, a managed aquifer recharge (MAR) strategy could be put in place, by organising the reinfiltration into the karst of the overflow from the dams.

References

1. Darnault CJG (2008) Karst aquifers: hydrogeology and exploitation. Overexploitation and contamination of shared groundwater resources. Springer, Berlin, pp 203–226
2. Collignon B (1986) Hydrogéologie appliquée des aquifères karstiques des Monts de Tlemcen. thèse de doctorat, Université d'Avignon
3. Seltzer P (1946) le climat de l'Algérie. Université d'Alger, Institut de Météorologie et Physique du Globe, Carbonnel, Barcelona, p 219, 2 cartes
4. Chaumont M, Paquin C (1971) Carte pluviométrique de l'Algérie du Nord, échelle 1/500 000″ (4 feuilles et notice). Société de l'Histoire Naturelle de Afrique du Nord, Alger
5. Laborde JP (1993) Carte pluviométrique de l'Algérie du Nord au 1/500,000 – notice explicative – Projet PNUD/ALG/88/021. Agence nationale des ressources hydrauliques, Alger
6. Ghenim AN, Megnounif A, Djelloul Smir SM (2014) Evaluation des changements dans la pluviométrie du bassin versant de la Tafna (Nord Ouest de l'Algérie – Secheresse) 24:107–114. https://doi.org/10.1684/sec.2013.0380TO
7. Auclair D, Biehler J (1967) Etude géologique des hautes plaines oranaises entre Tlemcen et Saida. Publication du Service Géologique de l'Algérie, Issue 34
8. Benest M (1985) Evolution de la plate-forme de l'Ouest Algerien et du Nord-Est Marocain au cours du Jurassique supérieur et au début du Crétacé: Stratigraphie milieux de dépôts et dynamique sédimentaire. Thèse université de Lyon
9. Bensaoula F (2006) Karstification, hydrogéologie et vulnérabilité des eaux karstiques - Mise au point d'outils pour leur protection (Application aux Monts de Tlemcen – Ouest Oranais). Thèse d'état, Université de Tlemcen, 216 p
10. Benest M, Bensalah M, Bouabdellah H, Ouardas T (1999) La couverture mésozoïque et cénozoïque du domaine Tlemcénien (Avant pays Tellien d'Algérie occidentale) – Stratigraphie, paléoenvironnement, dynamique sédimentaire et tecto-genèse alpine. Bulletin du Service Géologique de l'Algérie 10(2):127–157
11. Lucas G (1942) Description géologique et pétrographique des monts de Ghar Roubane et du Sidi El Abed. Bulletin du service de la carte géologique de l'Algérie
12. Lucas G (1952) Bordure nord des hautes plaines dans l'Algérie occidentale Primaire Jurassique. Analyse structurale Monographies régionales 1ere serie n° 21
13. Benest M Dubel C, Elmi S (1978) Modalités de l'apparition de la sédimentation carbonatée de plateforme interne sur la frange meridionale du domaine Tlemcénien pendant l'Aalénien et le Bajocien: Les dolomies du Tenouchfi (Algérie Nord-Occidentale) – Livre jubilaire Jacques Flandrin – Docum. Lab. Géol. Fac. Sci. Lyon - H-S.4, pp. 29–69, 10 fig., 6 p l
14. Benest M (1972) Les formations carbonatées et les grands rythmes du Jurassique supérieur des Monts de Tlemcen – CRAS-D-275, pp 1469–1472
15. Elmi S (1970) Rôle des accidents décrochants de direction SSW-NNE dans la structure des monts de Tlemcen (Ouest Algérien). Bulletin de la societé d'histoire naturelle de l'Afrique du Nord 61:3–7
16. Gevin P (1987) Essai de réserve souterraine en vraie grandeur le barrage sur l'oued Meffrouch (Algérie) Bulletin d'hydrogéologie n°7, 217–228. Ed. P. Lang
17. Gautier M (1952) La géologie et les problèmes de l'eau en Algérie – Tome 1 – Eléments de technologie des barrages algériens et de quelques ouvrages annexes – Le barrage des Beni-Bahdel et la conduite d'Oran

18. Bensaoula F (2008) Exportation des carbonates et dissolution spécifique dans le système karstique de Boumaza (Monts de Tlemcen – NO algérien) Karstologia n°52, 31–38.5-2
19. Notenboom J (1980) Biospeleogische expeditie in Algérie, ed. Biosp. werkgroep
20. Bensaoula F, Adjim N, Adjim M, Collignon B, Zeghid K (2016) First application of the COP method to vulnerability mapping in the Meffrouch catchment (Northwest of Algeria) – Larhyss J – ISSN 1112–3680, n°26 – Juin 2016 – 45–59
21. Birebent J (1948) Explorations souterraines en Algérie campagne 1946–1947 – Annales de spéléologie-Tome III-1948-Fascicule 2–3
22. Collignon B (1983) Les principaux phénomènes karstiques de l'Algérie, Spéléologie algérienne 1982–1983, pp 57–66
23. Collignon B (1991) Ghar Boumaza: the underground river Tafna – International Caver – Issue 1, pp 24–30
24. Bensaoula F (2007) Etude de la karstification à partir des données de forages: le cas des monts de Tlemcen (Algérie). Karstologia 49:15–24
25. Bensaoula F, Bensalah M, Achachi (2005) A Etude des circulations d'eaux profondes dans les dolomies du Dogger de Zouia (Bordure occidentale des Monts de Tlemcen, nord-ouest algérien)-Bulletin d'hydrogéologie n° 21(2005), université de Neuchatel-éditions Peter Lang
26. Doumergue F (1926) Carte géologique – feuille de Tlemcen à l'échelle 1/50000
27. Verdeil P, Issadi A (1985) Le thermalisme de l'ouest Algérien dans le cadre de la tectonique globale – 5ème séminaire des sciences de la terre – Alger 2,3,4 Déc 1985
28. Collignon B (1985) Le forage hydraulique dans les roches très karstifiées: étude des difficultés renontrées et des solutions adoptées lorsdes travaux dans les Monts de Tlemcen – Alger – COTEFHYD – Journées techniques du forage hydraulique
29. Bensaoula F, Adjim M, Derni I (2012) Trente années de prospection et de mobilisation des ressources en eau souterraines par forages dans la wilaya de Tlemcen. LARHYSS J 10:91–99
30. Nouaceur Z, Benoît L, Imen T (2013) Changements climatiques au Maghreb: vers des conditions plus humides et plus chaudes sur le littoral algérien? Physio-Géo 7(1):307–323
31. Collignon B (1988) Évaluation des réserves permanentes et renouvelables des aquifères karstiques de l'Ouest de l'Algérie à partir du suivi piézométrique des forages en exploitation. Besançon, 4ème Colloque d'hydrologie en pays calcaires et milieu fissuré -Actes, pp 99–105
32. Vince F, Batz S (2007) Pourquoi et comment réduire les consommations d'énergie du dessalement par osmose inverse? TSM Numero 9:75–86
33. Kertous M (2013) Analyse des déterminants de la demande d'eau potable en Algérie: Une approche par panels dynamiques. Revue des Sciences de l'Eau:193–207
34. Bensaoula F, Adjim M (2008) La mobilisation des ressources en eau: contexte climatique et contraintes socio-économiques. LARHYSS J, juin, 79–92
35. Naggar OM (2003) Analysis of groundwater production cost – Cairo – seventh international water technology conference. UNESCO, Paris
36. Abazza H (2012) Considérations économiques concernant l'approvisionnement en eau par dessalement dans les pays au sud de la Méditerannée. Union Européenne – Mécanisme de soutien à la Gestion Intégrée Durable de l'Eau – Mécanisme de Soutien, Bruxelles
37. Xanke J (2017) Managed aquifer recharge into a karst groundwater system at the Wala reservoir, Jordan. Phd Thesis (Karlsruhe)
38. Journal officiel de la république algérienne démocratique et populaire (1983) Loi portant Code des eaux
39. Journal officiel de la république algérienne démocratique et populaire (2005) Loi relative à l'eau
40. Journal officiel de la république algérienne démocratique et populaire (2007) Décret exécutif n°07-399 de dec. 2007 relatif aux périmètres de protection qualitative des ressources en eau
41. Journal officiel de la république algérienne démocratique et populaire (2010) Décret exécutif n°10-73 du 6 février 2010 relatif aux périmètres de protection quantitative des ressources en eau
42. Bézèlgues S, Des Garets E, avec la collaboration de Mardhel V, Döerfliger N (2002) Cartographie de la vulnérabilité des nappes de grande Terre et de Marie- Galante (Guadeloupe) – Phase 1 Définition de la méthode de détermination de la vulnérabilité – rapport BRGM – 51783-FR, 41p.,7 tab

43. Nguyet VTM, Goldscheider N (2006) A simplified methodology for mapping groundwater vulnerability, contamination hazards and risk: first application in a tropical karst area, Vietnam. Hydrogeol J 1:1–10
44. Bensaoula F, Bensalah M (2007) Cartographie de la vulnérabilité des eaux karstiques de la région de Tlemcen (Algérie). Adaptation et application de l'approche européenne. Bulletin d'hydrogeologie N°22: 59–76. université de Neuchatel-éditions Peter Lang
45. Bensaoula F, Adjim M, Benazzouz B, Khatir O, Bouchama A, Harkat M, Fellah HS (2018) Vulnérabilité et risque de pollution du système karstique de Ghar Boumaza (Monts de Tlemcen, Nord-Ouest algérien) – Application de la méthode R.I.S.K. Karstologia 70:53–62
46. Zwahlen F (ed) (2004) Vulnerability and risk mapping for the protection on carbonate (karst) aquifers-COST 620. European commission, directorate-general XII Science Research and Development, Brussels, p 297
47. Ketelaere DD, Hötzl H, Neukum C, Civity M, Sappa G (2003) Hazard mapping, COST action 620 vulnerability and risk mapping for the protection of carbonate (Karst) aquifers – final report. Université Neuchâtel, Neuchâtel

Part V
Towards a Sustainable Development

Participatory Approaches to Sustainable Development and Management of Soil Resources in Arid Zones of Algeria

A. S. Belouchrani, A. Bouderbala, and M. Hocine

Contents

A. S. Belouchrani (✉) and M. Hocine
Department of Plant Productions, Higher National School of Agronomy, Algiers, Algeria
e-mail: amelbelouchrani@yahoo.fr

A. Bouderbala
Department of Earth Sciences, University of Khemis Miliana, Khemis Miliana, Algeria

Abdelazim M. Negm, Abdelkader Bouderbala, Haroun Chenchouni, and
Damià Barceló (eds.), *Water Resources in Algeria - Part I: Assessment
of Surface and Groundwater Resources*, Hdb Env Chem (2020) 97: 269–292,
DOI 10.1007/698_2019_401, © Springer Nature Switzerland AG 2019,
Published online: 5 December 2019

Abstract For sustainable development in arid zones, the management of soil resources is essential; this resource management includes several parameters, among others the evaluation of soil fertility and the production of chemical fertility maps of the studied soils. To carry out this work, we chose an irrigable perimeter of Tadjmout (Laghouat). The chemical parameters studied are pH, cation exchange capacity (CEC Cmol$^+$/kg), total calcareous (CaCO$_3$%), salinity (EC ds m^{-1}), total nitrogen (Nt%), nitrogen mineral (Nm mg/kg), exchangeable phosphorus (P) mg/kg, and exchangeable potassium (K) mg/kg. The results showed that the soils of this arid zone have an alkaline pH, CEC relatively medium (it does not exceed 15 Cmol$^+$/kg), and calcareous soil; EC is from low to medium and sometimes high. The soils are poor in total and mineral nitrogen, and the exchangeable phosphorus contents are high, which could be due to phosphate fertilizer, and the exchangeable potassium has low to medium values. These soils require rectification fertilization for nitrogen and potassium and maintenance fertilization for phosphorus. With the geographic information system (GIS), we have developed different thematic maps based on the analytical results obtained. The thematic maps obtained allow us to manage a fertilization plan to improve the chemical fertility of these soils and increase agricultural production in the study area.

Keywords Agriculture production, Chemical fertility, Geographic information system, NPK, Sustainable development

1 Introduction

The objective of this study is to realize thematic fertilization maps and to evaluate the chemical fertility of the soils of the irrigable perimeter of Tadjmout area (Wilaya of Laghouat), for the management of soil resources by the participative approach to the sustainable development in arid zones of Algeria. On the environmental front, this initiative aims to preserve soil resources and protect water resources. The study area is part of the group of pastoral areas of Algeria, where the agriculture is considered as one of the main sectors in this region with a total agricultural area of 20,087.06 ha [1].

In a difficult climatic and economic context, the cartography of the chemical fertility is more than necessary for a durable development. Note that in Algeria, there is very little study on the cartography of the soil fertility; the few studies that exist are about soil classification of irrigated perimeters [2].

The goal of producing chemical fertility maps is to increase agricultural production both in quantity and quality and to manage fertilization plans. The chemical parameters of the soil are numerous. In this chapter, we will treat the chemical elements that come within the framework of the evaluation of the fertility and the realization of thematic maps. The parameters studied are pH, cation exchange capacity (CEC), electrical conductivity (EC), total limestone (CaCO$_3$), exchangeable phosphorus (P$_2$O$_5$), and exchangeable potassium (K$_2$O).

2 Chemical Parameters of the Soil

2.1 The pH of the Soil

The pH is a mode of expression of the H^+ ion concentration of a liquid; it is expressed on a scale of 0–14; low values indicate acidity; values greater than seven correspond to an alkaline (basic) character [3]. The acidity, neutrality, or alkalinity of an aqueous solution can be expressed by the concentration of H_3O, so as to facilitate this expression; the logarithm of the inverse of the H^+ ion concentration is used and is pH [4]. Under a temperate climate, the pH of the same horizon tends to fall in summer and increase in winter. These influences are explained in winter by the dilution of H^+ ions in the soil solution under the effect of the rains, in summer by the production of organic acid due to the biological activity, maximum at this time. The seasonal variation generally reaches about one-tenth of a pH unit but can reach 0.5 pH units in Baize limestone soils (2000). The agronomic interest of the pH is that it plays a role vis-à-vis the assimilability of the main nutrients and trace elements. For noncalcareous soils, the agronomic optimum can be set at pH between 6.5 and 7.5 [3]. The pH is an indicator of the fertility status of land. It gives information on the presence of certain toxic salts, its microbial activity, and on its level of assimilability of the elements by the plant [5]. From a pedological and agronomic point of view, the limitation of acidification and alkalinization is a necessary condition for the environment in the equilibrium of the major soil functions, physical, chemical, and biological. Monitoring acidity and alkalinity of agricultural land, therefore, remains an important aspect of preserving their fertility [6].

2.2 Cation Exchange Capacity (CEC)

Cation exchange capacity (CEC) is the maximum capacity of exchangeable cations that a soil can retain at a given pH, which is the sum of the occupied by cations (Ca^{++}, Mg^{++}, K^+, Na^+, H^+, and Al^{+++}). It also expresses the buffer capacity of the soil, its resistance to change in pH. It is strongly related to clay levels and organic matter. CEC is a generally stable value. It may, however, vary slightly in the long term, if the rate of organic matter decreases. It provides the soil fertility index. It can also be used as a criterion in a soil vulnerability assessment model for nutrient losses to groundwater [7]. The knowledge of the CEC is essential. As, for all soils, this value is also a precise indication of soil fertility; it allows us to have a global idea about the nature of clay minerals.

2.3 Electrical Conductivity (EC)

The electrical conductivity of a soil solution is an index of the levels of soluble salts in the soil; it roughly expresses the concentration of ionizable solutes present in the sample that is to say its salinity degree [4, 8]. This measurement makes it possible to quickly obtain an estimate of the overall content of dissolved salts, using a conductivity meter in an earth/water suspension = 1/5 [9]. In saline, sensitive plants will be affected by this salinity because the high electrical conductivity could harm these crops (delayed plant germination, reduced yield) against the resistant plants that can maintain an acceptable level of yield [10].

2.4 Limestone Total (CaCO₃)

Limestone plays a role of the reserve of calcium in the soil and can be a limiting element for certain cultures, as well by its presence as by its absence. It also plays an important role in the behavior of the soil. It constitutes a reserve of calcium cations, which has the effect of flocculating the clay. Despite its inaccuracy, the determination of the total limestone with Bernard's calcimeter is sufficient for the pedologist if the apparatus is tared before, during and after the manipulations, and if the samples are not exposed to the sun. This method is suitable for serial dosing. The principle of this method is very simple. It is the measurement of CO_2 released, following an excess of hydrochloric acid on a known weight of a soil sample [11]. According to [3, 12], limestone in soil performs several functions, and its efficiency is more remarkable when it is at a more advanced degree of fineness. In the state of gravel and coarse sands, it acts as a reserve of calcium in general quite easily mobilizable, but it is especially when it is in the state of finer particles in the form of fine sand and especially of silts that it intervenes in the physical-chemical and biological properties of soils.

2.5 Phosphorus in the Soil

Phosphorus is a very widespread element in the world, and it represents 0.1–0.2% of the lithosphere. It commonly reaches 1.5–4% in the case of volcanic rocks and 1–3% in crystalline rocks. In sedimentary rocks, grades are generally between 0.3 and 0.8% [13]. Phosphorus is essential for life, but natural levels are low in soils. The amount of phosphorus present in the soil is a direct consequence of the source rock. The total phosphorus content is generally less than 1% [14]. Phosphorus is present in the soil in the following forms: soluble phosphorus; phosphorus fixed on the clay-humic complex; insoluble phosphorus; and organic phosphorus. The plant can assimilate only mineral phosphorus (exchangeable and soluble) [15, 16]. The exchangeable phosphorus consists of all the ions adsorbed on the adsorbent complex of the soil. It participates in constant soil solution exchanges and constitutes the bulk of the food pool [17]. Phosphorus is absorbed by roots mainly in the form of H_2PO_4 orthophosphate and very little in the form of HPO_4^- [18].

2.6 Potassium in the Soil

The lithosphere contains an average content of 1.9% potassium. In the soil, potassium has an average content of 1.2%. It is the fertilizing element whose geochemical behavior may appear, a priori, the most related to the mineralogical composition of soils [19]. Its soil content is closely related to the nature and abundance of minerals in the parent rock and the loss of potassium during pedogenesis [20]. Soil potassium is constituted mainly potassium structural (90–98%), exchangeable potassium which represents 2–8% of total potassium, and potassium in solution 0.1–0.2% [21]. According to [22] potassium in the soil, solution is generally regarded as the source of the plant's potassium feed and is found in the form of salts that are more or less soluble in the soil. The exchangeable potassium corresponds to the potassium lining retained by electrostatic attraction around the electronegative colloids. This fraction represents 1–10% of the total potassium [23].

3 Cartography

The cartography allows us to establish maps for different domains, among other agronomic domains. The International Cartographic Association (ICA) defines cartography as "all studies and scientific, artistic, and technical operations involved in the results of direct observations or the exploitation of documentation, in order to map, and other forms of expression and their use [15, 23, 24]". Thematic mapping is a tool for analysis, decision support, and communication widely used to represent some variables; thematic cartography allows the realization of particular graphic images that reflect the spatial relationships of one or more phenomena, one or more themes [25]. The thematic map is a graphic document based on communication by the signs. It is a visual language. This visual language is specific because it is the opposite of written or spoken language: the eye first perceives a set; it generalizes and then looks for detail [26]. A thematic map is a communication tool that has the same importance as the oral or written language in a society; it is the best way to record and communicate information about the location and spatial characteristics of objects. In order to develop a detailed scale fertility map, we have used the geographic information system (GIS).

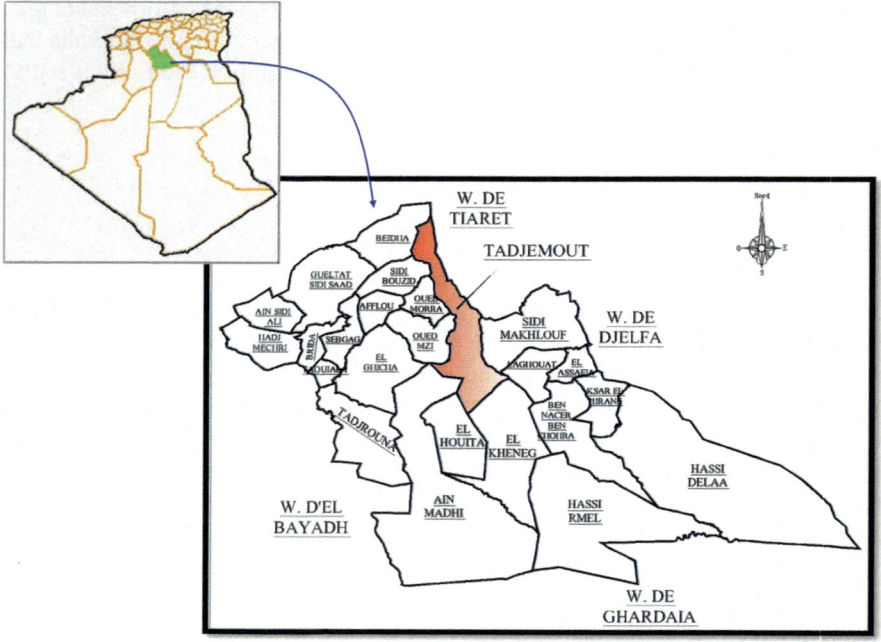

Fig. 1 Location of the study area

4 Location of the Study Area

Soil samples were selected from the perimeter of Tadjmout (Laghouat) by
[27, 28]. A detailed description of the profiles was made on the ground using
material from ENSA. These samples were taken from eight profiles. The study
area is located 48 km northwest from the city of Laghouat and exactly in the
commune of Tadjmout (Fig. 1). This commune is delimited in north by the commune
of El-Beidha; in the south by the communes of Laghouat, Kheneg, and Oued M'zi;
in the west by the communes of Aflou, Sidi-Bouzid, Oued Morra, El-Ghicha, and
Ain Madhi; and in the east by the communes of Sidi Makhlouf, Ain Chouhada,
El Guedid, and El Idrissia. The irrigable perimeter of Tadjmout occupies an area of
approximately 1,400 ha; it is bordered on the south by Oued M'zi, on the east by
Oued Jekidjika, and on the north and west by the old Tadjmout-Laghouat road.

4.1 Climate

The climate of the central part of the steppes is of course like that of all Algerian
steppes of the Mediterranean type contrasted with a long dry and hot summer season

and a cold and rainy winter season. The amount of rainfall is very low and shows great intermittent and interannual variability [29].

4.2 Geomorphology of the Region

It is represented by alluvial terraces developed on recent alluvium formed after the deposit of alluvium M'zi Wadi. The municipality is characterized by flat and relatively uneven terrain. It is a relief type (Dayas) valley(Faids), collecting the waters of wadis descending from the surrounding mountains and by (Chaaba) the runoff bed descending mounds (Argoub). Sometimes, one can meet rocky plateaus and in the northwestern part the presence of exclusively flat microreliefs with sand dunes.

4.3 Hydrography

We found that streams in Tadjmout have irregular flows. Also, the M'zi wadi bed is dry in summer, but it is covered by violent floods during the winter period [28].

4.4 Vegetation

Vegetation plays a role in the differentiation of soils, while soil plays a role in the selection of plants it carries. Some plant species are indicator plants. They reflect the edaphic conditions in which they live because they are well adapted to this or that environment: limestone, acid, rich in nitrate, water, salt, etc. In the studied site, one has the important occupation of the *Atriplex halimus*, which gives an idea on the aridity of our soil. Moreover, the farm is surrounded by *Pinus halepensis* (Aleppo pine) and the *Casuarina* to protect the soil against vulnerability to wind [28]. According to [27, 28] farmers in the town of Tadjmout practice in their orchards several types of crops that are adapted to the climatic and edaphic conditions of the region, among which we have market gardening: especially apple tomato, lettuce, carrot, turnip, etc.; the fruit trees (apricot, pomegranate, apple, vine, etc.); date palm; and cereals (barley, wheat). All these crops are irrigated except the cereal crop, which is conducted in dry and practiced for the self-consumption of the populations as well for the food supplementation of the animals. The surplus of the production is sent to the market. According to climatic variability of the study area, we find that irregular rainfalls are interannual and intra-annual, in most cases a longer dry period and violent winds (sirocco and sandstorms).

4.5 Geology

The area of Laghouat has a bedrock of sedimentary rocks dating from secondary, tertiary, and quaternary. Substrate knowledge is fundamental for determining the original soil material and understanding the resulting geomorphological phenomena [2]. There are two types of bedrock: marly bedrock and bedrock with calcium carbonate.

5 Methodology

The soil samples are representative of about 100 samples from the soil survey of Tadjmout carried out by [28]. Soil samples are air-dried, crushed, and sieved to 2 mm. The methods of the chemical characteristics carried out are pH, with a sol/solution ratio of 2/5 (electrometric method); the cation exchange capacity (CEC) METSON method; electrical conductivity (EC), extraction is done with saturated paste; and total limestone, volumetric method (BERNARD calcimeter).

Analytical data on the chemical characteristics of soils in the Tadjmout area will enable us to produce chemical fertility maps. To establish these fertility maps, we used the computer hardware which is subdivided into two, the hardware that includes the machines used and the software that defines the software and the documentation: topographic map of Tadjmout 1/50,000 and pedological map of Tadjmout 1/25,000.

6 Characterization of the Soils

Soils in dry lands are generally poor in organic matter, fragile, and shallow. Pedogenesis processes occur under semiarid climate conditions, soil formation is influenced not only by precipitation but also by high evaporation, and our study area belongs to this area. According to [27, 28] and according to the French classification (CPCS), the soils of Tadjmout belong to three classes: raw mineral soils, little evolved, and calcimagnetic. The study area is divided into eight sites (see Fig. 2):

Site 1: characterized by one of the poorly evolved, non-climatic soils of wind power, modal comprises profile 1.

Site 2: characterized by little evolved soil, non-climatic, of alluvial contribution, modal comprises the profile 2.

Site 3: characterized by a little evolved soil, non-climatic, of alluvial contribution, deep includes the profile 3.

Site 4: characterized by calcimagnetic soil, carbonate, calcareous brown, accumulation of limestone in the form of friable clusters includes profile 4.

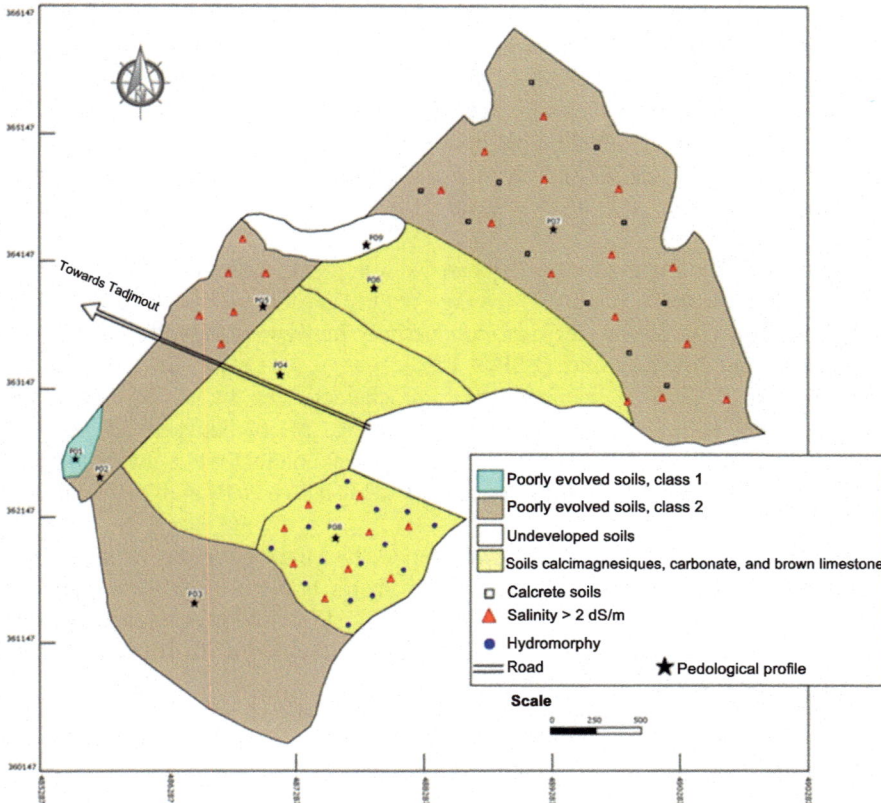

Fig. 2 Location map of profiles (Tadjmout, Laghouat) [28]

Site 5: characterized by a little evolved ground, non-climatic, of alluvial contribution, halomorphy and the hydromorphy include the profile 5.

Site 6: characterized by calcimagnetic, carbonate, calcareous brown, modal, and calcareous encrustation, soil has profile 6.

Site 7: characterized by a little evolved ground, non-climatic, of alluvial contribution, on encroutement of limestone, halomorph includes the profile 7.

Site 8: characterized by a calcimagnetic soil, brown carbonate, limestone, and hydromorphy, salt has the profile 8.

7 Morpho-Analytical Synthesis of the Description of the Profiles of the Area

The morphological descriptions of the profiles were made by [27, 28].

Profile No 1: Slightly evolved soil; characterized by a flat topography (slope = 1%); a sandy texture for the horizon no 1 and sandy-silty for the horizon no 2; a neutral pH between 7.24 and 7.78; low natural fertility (CEC < 15 cmol/kg); low calcareous soil (2–10%); electrical conductivity less than 1 ds m^{-1} so the soil is not salty; and low alkalinity (ESP < 15%).

Profile No 2: poorly evolved soil; characterized by a flat topography (slope < 2%); a silty-sandy texture; an alkaline pH between 7.5 and 8.7; moderately low natural fertility (CEC < 15 cmol/kg); calcareous soil with a grade (10–25%); electrical conductivity between 0.5 and 1 ds m^{-1} (nonsaline soil); and low alkalinity (ESP < 15%).

Profile No 3: poorly evolved soil; located at a slope nil; a sandy loamy texture; an alkaline pH between 7.5 and 8.7; average natural fertility (CEC > 15 cmol/kg); calcareous soil (10–25%); electrical conductivity between 1.57 and 2.34 ds m^{-1} (saline soil); and low alkalinity (ESP < 15%).

Profile No 4: It is a calcimagnetic soil; characterized by a flat topography (slope < 2%); with a clay-silty texture; an alkaline pH of between 7.5 and 8.7; good natural fertility (CEC > 15 cmol/kg); electrical conductivity between 0.5 and 1 ds m^{-1} (nonsaline soil); rich in limestone $CaCo_3$ between 10 and 25%; and low alkalinity (ESP < 15%).

Profile No 5: poorly evolved soil; characterized by a flat topography (slope < 2%); with a silty texture for horizon #1; sandy loam for horizon 2; an alkaline pH of between 7.5 and 8.7; good natural fertility (CEC > 15 cmol/kg); electrical conductivity 2 and 4 ds m^{-1} (very saline soil); rich in limestone $CaCo_3$ between 10 and 25%; and low alkalinity (ESP < 15%).

Profile No 6: It is a calcimagnetic soil; characterized by a flat topography (slope < 2%); with a silty texture for horizon #1; silty-sandy for horizon 2; an alkaline pH of between 7.5 and 8.7; good natural fertility (CEC > 15 cmol/kg); electrical conductivity 1 and 2 ds m^{-1} (saline soil); rich in limestone, $CaCo_3$ between 10 and 25%; and low alkalinity (ESP < 15%).

Profile No 7: It is a calcimagnetic soil; flat ground; with a silty texture; an alkaline pH between 7.5 and 8.7; moderate natural fertility (CEC < 15 cmol/kg); electrical conductivity 2 and 4 ds m^{-1} (very saline soil); rich in limestone $CaCo_3$ between 10 and 25%; and low alkalinity (ESP < 15%).

Profile No 8: It is a calcimagnetic soil; characterized by a zero slope topography; with a silty texture; an alkaline pH between 7.5 and 8.7; good natural fertility (CEC > 15 cmol/kg); electrical conductivity 1 and 2 ds m^{-1} (saline soil); rich in limestone $CaCo_3$ between 10 and 25%; and low alkalinity (ESP < 15%). The soils of Tadjmout are characterized by soils belonging to the arid zone, flat topology, limestone, alkaline pH, and low fertility. Based on the analytical results, the sets of profiles studied have an alkaline pH (greater than 7.5); the studied soils of the profiles (4, 5, 6, 7, and 8) are calcareous soils; we find a limestone level between 15.26% at level (P7H2) and 19.34% at level (P5H1). The cation exchange capacity is relatively average; it does not exceed (15 cmol/kg) for the profiles (1, 2, 3, and 7) and good for the profiles (4, 5, 6, and 8), and it exceeds this value to reach 21.16 cmol/kg for profile 4; this natural fertility can be explained by the high levels of clay. The charge sites are dominated in the ensemble by Ca^{++} cation. According to the OLSEN interpretation standards cited by Hanotiaux (1985), all the profiles contain exchangeable phosphorus contents between low (P5H1, P5H2), medium (P3H1, P7H1, P3H2, P2H2), and high (P1H1, P2H1, P4H1, P6H1, P8H1, P1H2, P4H2, P7H2, P8H2). The potassium content exchangeable according to the standards cited by [30] represents average contents in all profiles except P1H2 and P3H2 which are low in exchangeable potassium.

8   The Correlations Between the Different Chemical Parameters of the Soil

Concerning the correlations between the different chemical parameters of the soil, we notice that there is a highly positive and significant relation between the CEC and the clay rate, a high relationship between Na^+ exchangeable and CEC, and nonsignificant correlation between exchangeable phosphorus and total limestone, while there is a just inverse relationship between CEC and $CaCO_3$ in range between 0 and 3%, because it depends on the solubility of $CaCO_3$ in Ca^{++}. The pH of soils changes as the calcium carbonate content is in range from 0 to 2.5%.

9   Development of Fertility Maps

According to the analytical results obtained from the soils of the study area, we made the fertility maps which are presented in Figs. 3, 4, 5, 6, 7, 8, 9, 10, 11, 12, 13, and 14.

Figure 3 shows that the pH of the first horizon of the studied profiles varies from 7.5 to 8.7, which indicate that the soils are alkaline.

Figure 4 shows that the pH of the majority of the second horizon of the studied profiles varies from 7.5 to 8.7, which indicate that these soils are alkaline; in the other part, the less important area has a pH that varies between 6.5 and 7.5; it is a neutral pH.

Figure 5 shows a CEC of the first horizon of the soils in the studied profiles, with an average between 10 and 25 cmol/kg.

Figure 6 shows a CEC of the second horizon of soils in the studied profiles, with an average from 10 to 25 cmol/kg in the major part of the area. The other part of less important area has a poor CEC, with an average from 5 to 10 cmol/kg.

Figure 7 shows $CaCO_3$ values of the first horizon of the studied profiles, of the majority of soils from 10 to 25%; it indicates the character of limestone soils. The other part of less important area is noncalcareous, with values from 2 to 10%.

Figure 8 shows values of $CaCO_3$ of the second horizon of the studied profiles for the majority of soils from 10 to 25%; they have calcareous character, while the other part of less important area is noncalcareous, where the rates vary from 2 to 10%.

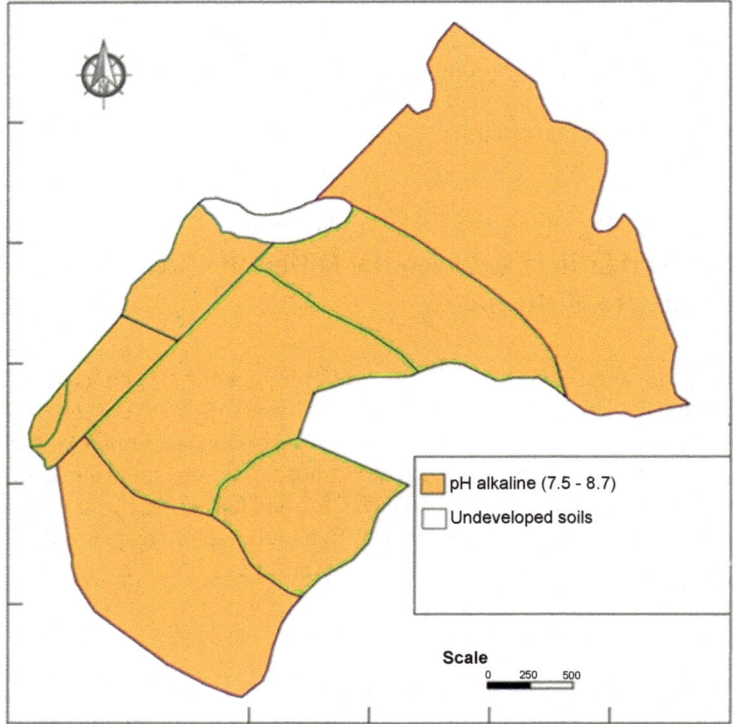

Fig. 3 pH map of the first horizon

Figure 9 shows the electrical conductivity of the first horizon of the studied profiles; we observe a high heterogeneity concerning the salinity, while the majority of the soils are from saline to very saline, and the EC varies from 1 to 4 ds m^{-1}.

Figure 10 shows the electrical conductivity of the second horizon of the studied profiles; we observe a heterogeneity of the distribution of the salinity; the majority of the soils are saline and very saline, and the EC varies from 1 to 4 ds m^{-1}.

Figure 11 shows the exchangeable potassium content of the first horizon of the studied profiles; the soils are moderately provided of exchangeable potassium.

Figure 12 shows the exchangeable potassium content of the second horizon of the studied profiles; the soils are low to medium exchangeable potassium.

Figure 13 shows the exchangeable phosphorus content of the first horizon of the studied profiles. Soils are moderately rich in exchangeable phosphorus, and only a small area is poor in exhaustible phosphorus.

Figure 14 shows the exchangeable phosphorus content of the second horizon of the studied profiles; the soils are moderately rich in exchangeable phosphorus, as well as for the first horizon; only a small area has a very low content of exchangeable phosphorus.

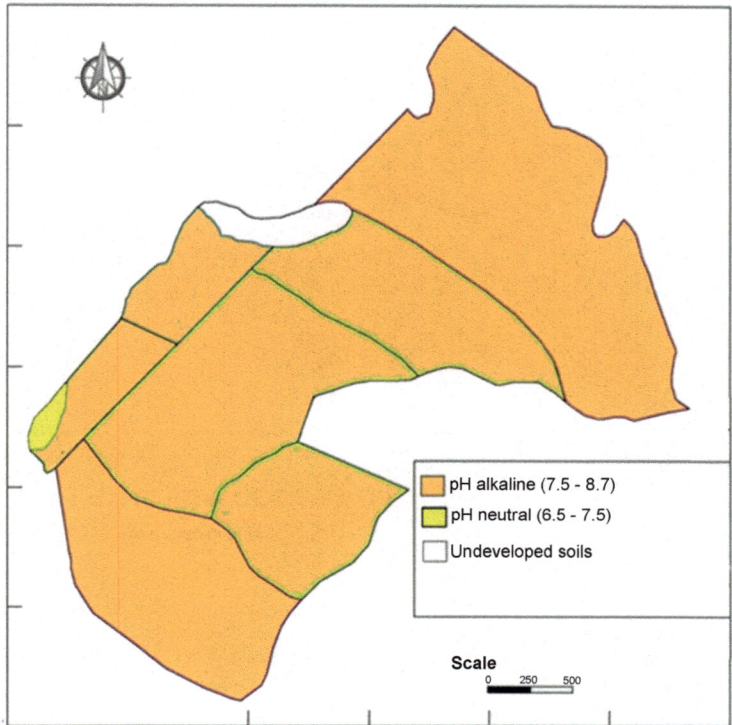

Fig. 4 pH map of the second horizon

According to the maps of the irrigable perimeter of the study area, we find that: the pH map: For the first horizon, the alkaline soils occupy all the areas 1,330.55 ha; for the second horizon, the alkaline soils occupy 1,322.55 ha and 17 ha; and neutral soils occupy 8.38 ha. The CEC map, for horizon 1, the soils have the average CEC, occupying all the parcels 1,330.55 ha, and the horizon 2, the soils have the average CEC occupy an area of 1,322.17 ha, and soils that have weak CEC occupy an area of 8.38 ha. The map of $CaCo_3$, the horizon no 1, the calcareous soils occupy all the parcels 1,330.55 ha, and the horizon no 2, the calcareous grounds occupy an area of 1,322.17 ha; for the noncalcareous soils, it has an area of 8.38 ha. The EC map, for horizon 1, nonsaline soils cover an area of 64 ha, slightly saline soils (268.97 ha), saline soils (297.28 ha), and very saline soils (711.75 ha). Horizon 2, nonsaline soils (268.96 ha), slightly saline soils (52,553 ha), saline soils (395.79 ha), and very saline soils (549.25 ha). The exchangeable potassium map K_2O, concerning the horizon no 1, soils with average K_2O content occupy all areas (1,330.55 ha). For horizon 2, soils have an average K_2O content (1,100.03 ha), and soils with a low content have an area of 230.52 ha. The map of exchangeable phosphorus P_2O_5, horizon 1, soils with a high content of P_2O_5 (618.80 ha), soils moderately rich in P_2O_5 (647.75 ha), and soils low in P_2O_5 (64 ha), horizon 2, the soils have a high content of P_2O_5

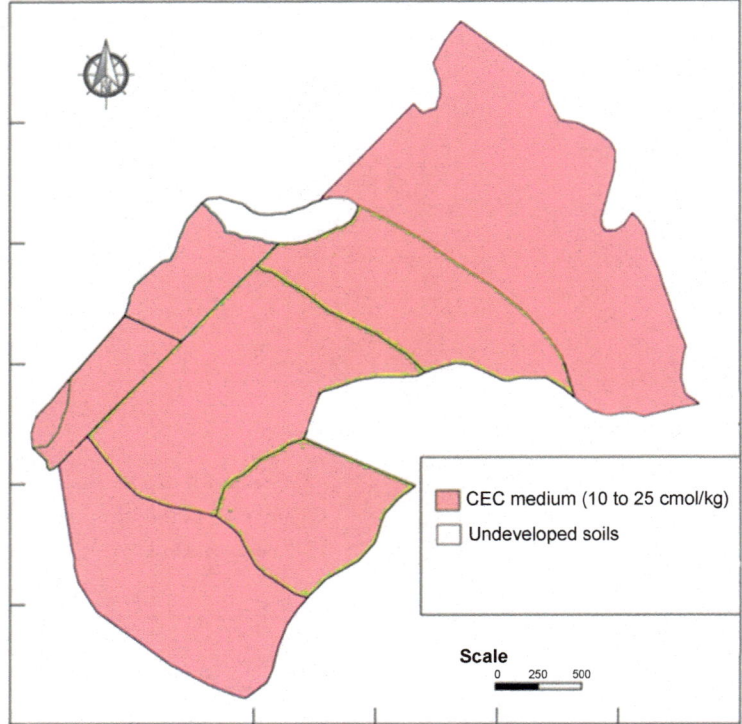

Fig. 5 CEC map of the first horizon

(818.21 ha), soils moderately rich in P_2O_5 (448.34 ha), and soils with a low P_2O_5 content occupy an area of 64 ha.

The previous maps show that the soils studied are alkaline pH, limestone, relatively medium CEC, low to medium salinity (EC), and sometimes very high salinity; they are moderately rich in exchangeable phosphorus and moderately rich in low exchangeable potassium. These soils require fertilization to improve their chemical fertility.

10   Conclusion

The soils of the Tadjmout perimeter belong to the arid zone. The analytical results show that these soils are at alkaline pH; cation exchange capacity (CEC) is relatively medium, which does not exceed 15 cmol/kg, low to medium, and sometimes high salinity (EC); exchangeable phosphorus with values between 26.62 and 34.87 ppm, this high content of exchangeable phosphorus could be due to phosphate fertilizers, and the exchangeable potassium has medium to low values. These soils require fertilization to improve their chemical fertility. The thematic maps allow us to

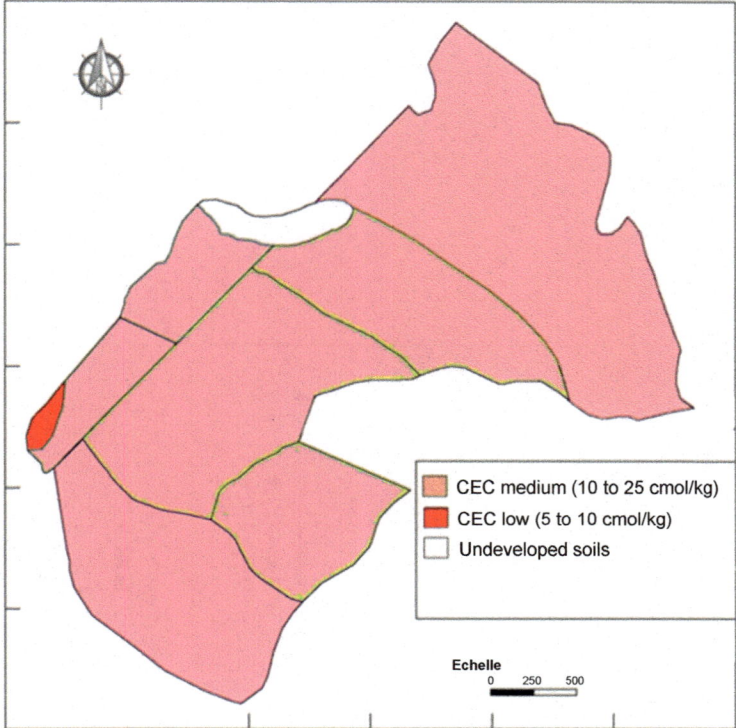

Fig. 6 CEC map of the second horizon

manage a fertilization plan to improve the chemical fertility of these soils, in order to increase agricultural production. It is imperative for the agricultural exploitation of arid soils in the context of sustainable development that further research is based on experimentation and monitoring over a long period.

11   Recommendations

The mapped soils require phospho-potassium and nitrogen fertilizer maintenance to improve their chemical fertility. Before the planting of each crop, the analysis of nitrogen and phosphorus in the soil is necessary in order to avoid pollution of groundwater.

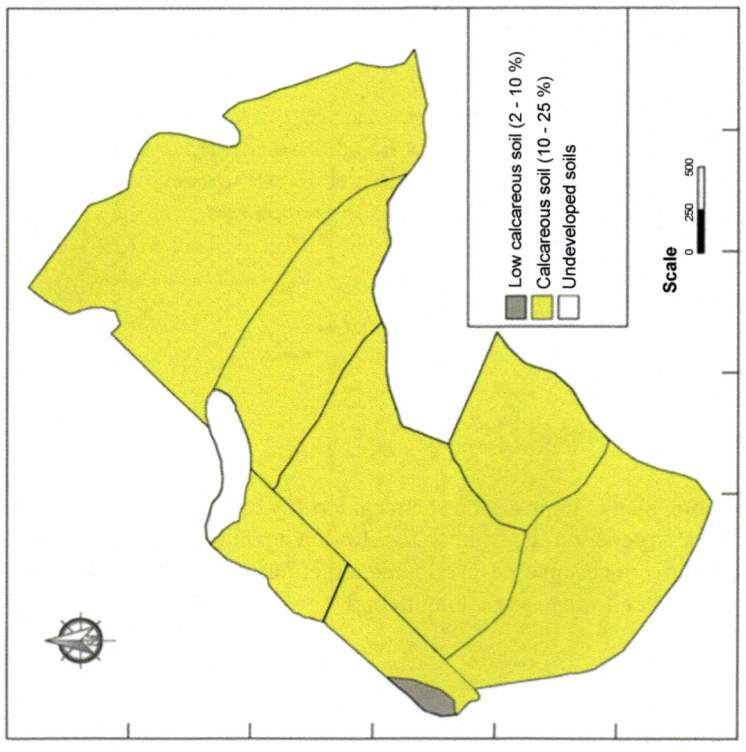

Fig. 7 CaCO₃ map of the first horizon

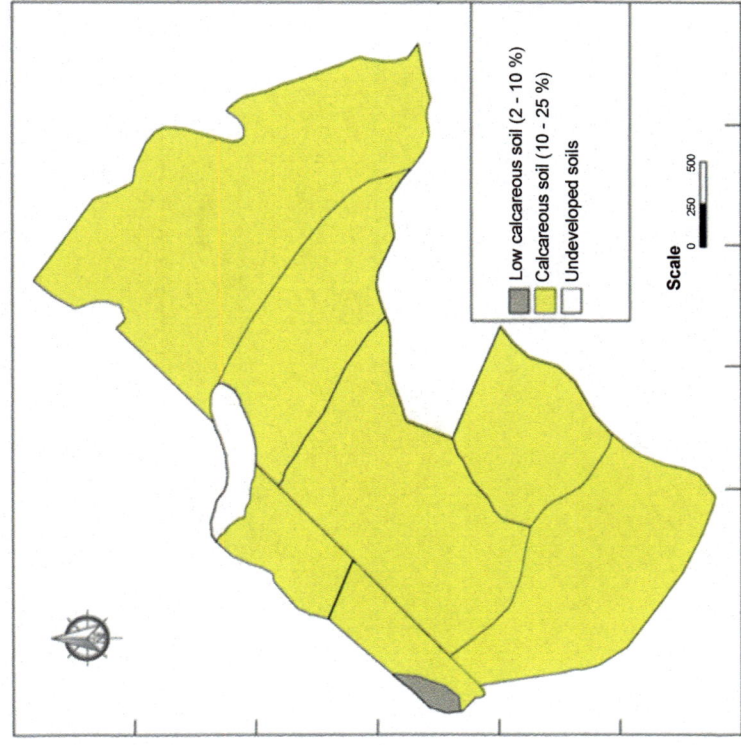

Fig. 8 CaCO₃ map of the second horizon

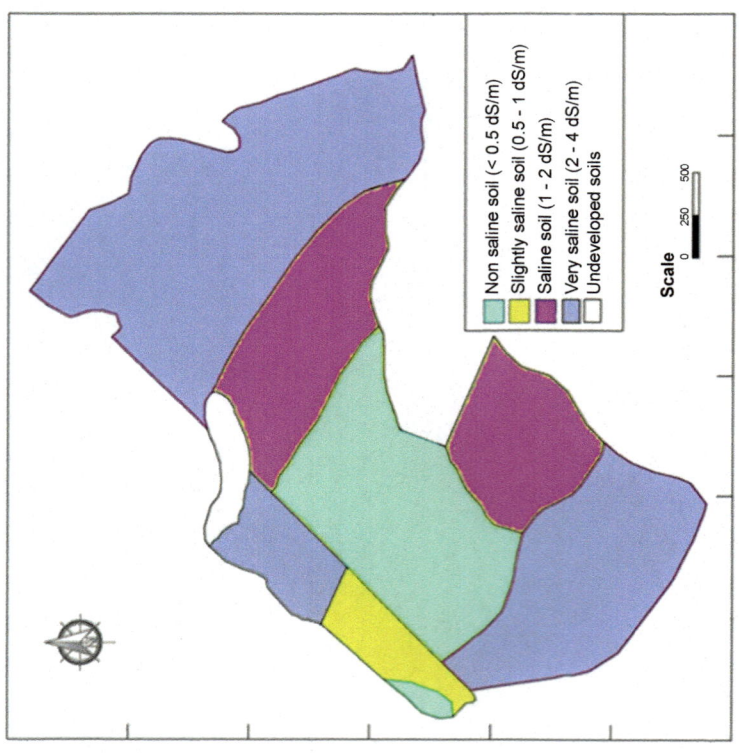

Fig. 9 Electrical conductivity of the first horizon

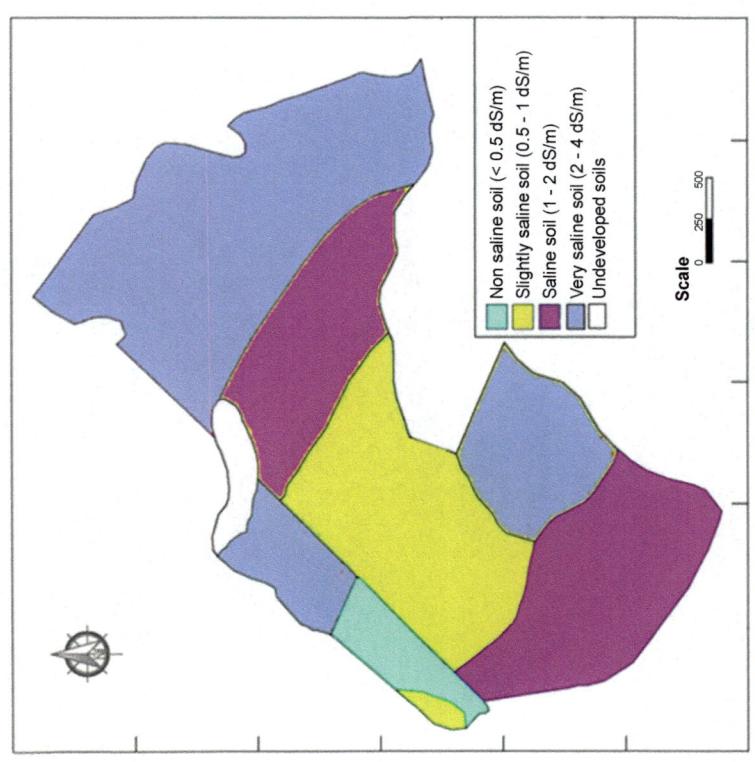

Fig. 10 Electrical conductivity of the second horizon

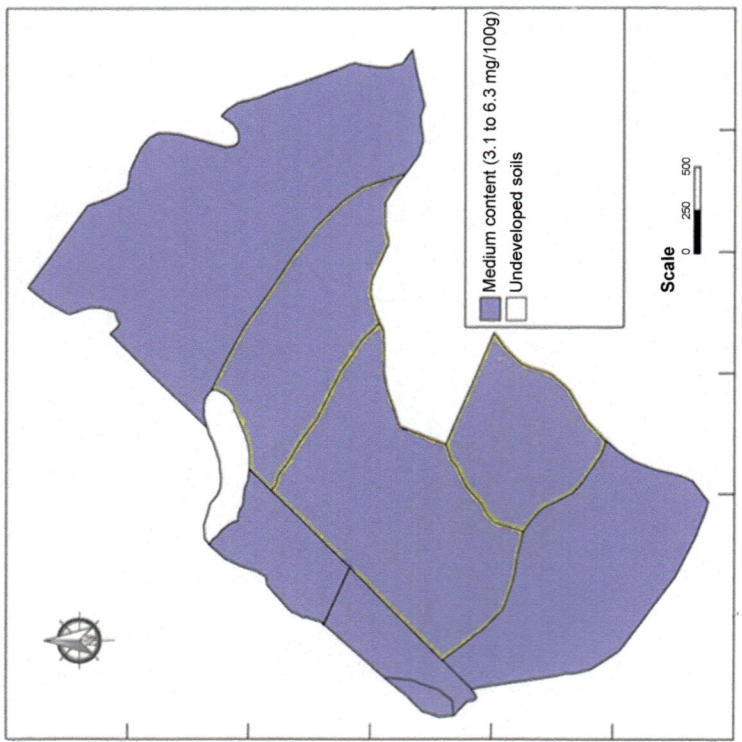

Fig. 11 Exchangeable potassium map of the first horizon

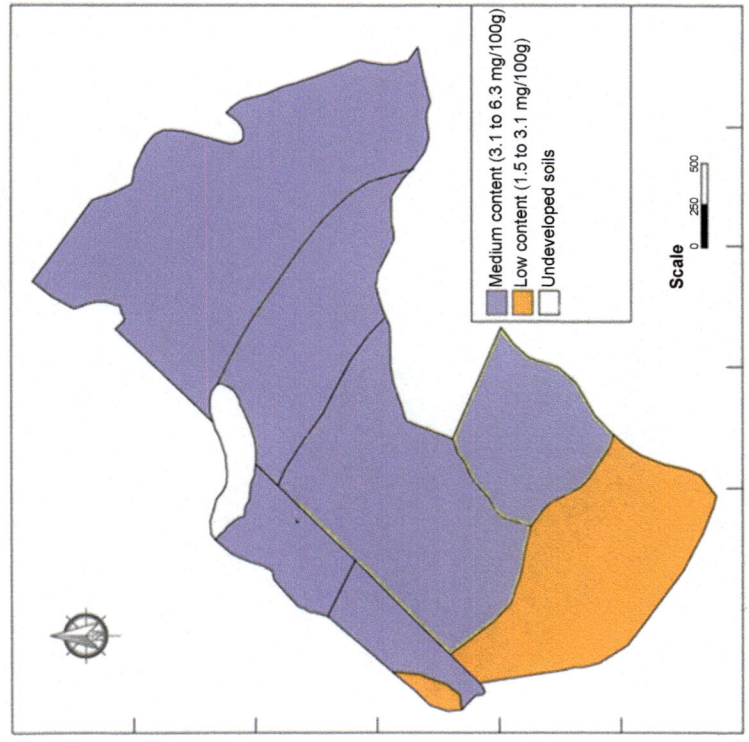

Fig. 12 Exchangeable potassium map of the second horizon

Fig. 13 Exchangeable phosphorus map of the first horizon

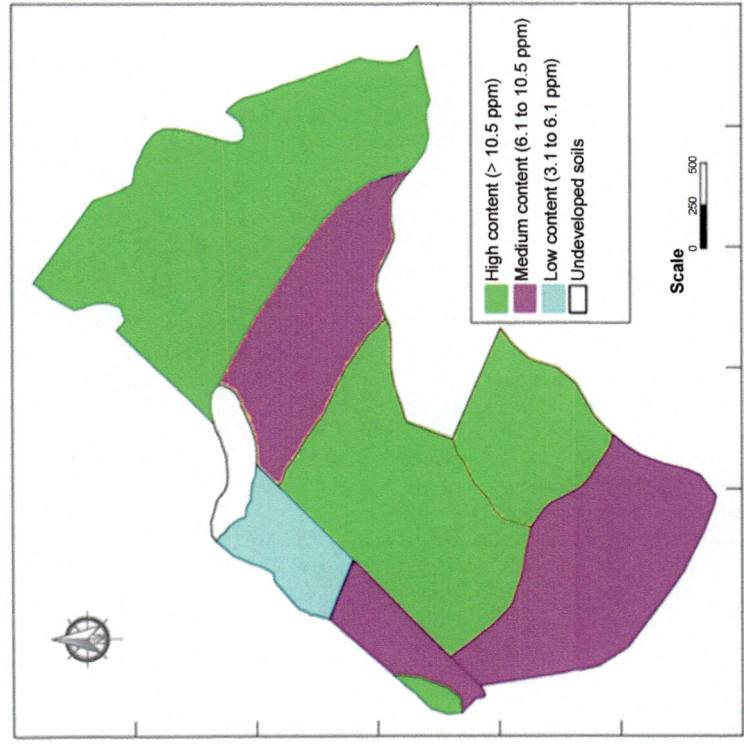

High content (> 10.5 ppm)

Medium content (6.1 to 10.5 ppm)

Low content (3.1 to 6.1 ppm)

Undeveloped soils

Scale

0   250   500

Fig. 14 Exchangeable phosphorus map of the second horizon

References

1. ANDI (2015) National Agency for Investment and Development, Algeria, p 20. (in French)
2. BNEDER (1994) Study of the environment inventory and the possibilities of integrated development in the wilaya of Laghouat, report about the analysis of current situation analysis, p 50. (in French)
3. Baize D (2000) Guide of soil analyzes, 2nd edn. INRA, Paris, pp 87–124. (in French)
4. Lement M (2003) Chemical analyzes of soils (selected methods). Tech. and Doc. edition, p 387. (in French)
5. Pansu M, Gautheryou J (2003) Soil analysis: mineralogical, organic, and mineral, p 993. (in French)
6. Ciesielski H, Sterckeman T, Baliteau JY, Caria G, Goutiers V, Willery JP (2008) Evolution of pH and CEC of the soils in the north of France depending on the amount of liming (CaCO3): influence of organic carbon, p 161. (in French)
7. Martin A, Nolinn M (1991) Soil classification and interpretation. The Pedological Team, Quebec, p 160. (in French)
8. Rhoades JD (1996) Salinity: electrical conductivity and total dissolved salts. Methods of soil analysis, Part 3, Chemical methods. Soil Science Society of America book series, vol 5. American Society of Agronomy, Madison, pp 417–435
9. Guy A (1978) Soil analysis methods. CRDP, Marseille, p 191. (in French)
10. Maas EV (1986) Salt tolerance of plants. Appl Agric Res 1:12–26
11. Baize D (1988) Guide for current analyzes in pedology. INRA, Paris, pp 57–71
12. Gaucher G (1981) Agricultural pedology treatment: the factors of pedogenesis. Ed. Lelotte, Dison, p 730. (in French)
13. Gervy R (1970) Phosphates and agriculture. Ed. DUNOD, p 675. (in French)
14. Bossche V (1999) Becoming phosphorus brought on soils and risk of contamination of surface water: the case of sewage mud. Doctoral thesis, University of RENNE, p 321. (in French)
15. Prevost P (2006) The basics of modern agriculture. Ed. Lavoisier, Paris, p 262. (in French)
16. Morel C (2002) Characterization of the phytoavailability of soil phosphorus by modeling the transfer of phosphate ions between soil and solution. HDR thesis, p 80. (in French)
17. Morel C, Fardeau JC (1990) Fixation of soil vis-à-vis phosphorus and the consequences for phosphate fertilization. Prospectives agricoles, p 147
18. Marschner H (2002) Mineral nutrition of higher plants, 2nd edn. Academic Press, London, p 889
19. Robert M, Trocme S (1975) Potassium in soil. Masson. Ed, p 459. (in French)
20. Quemener J, Bosc M (1988) Note on the determination of exchangeable potassium in "phosphorus and potassium in the soil-plant relationship": consequence on fertilization. Ed I.N.R.A., Paris, pp 109–132. (in French)
21. Villemin (1993) The dynamics of potassium in the soil. Agric Perspect 181:23–27. (in French)
22. Schneidr A (1997) Release and fixation of potassium by loam soil as affected by initial water content and potassium status of soil samples. Eur J Soil Sci 48:263–271
23. Gachon L (1988) Phosphorus and potassium in soil-plant relationships: consequences on fertilization. Ed. Lavoisier, p 566. (in French)
24. Guerphi M (2011) Cartography course. ENSSMAL, Dely Brahim, pp 1–3. (in French)
25. Tremelo ML, Zanin C (2003) Knowing how to make a map – assistance in the design and the realization of a thematic map. Belin-Sup Geography, Paris, p 199. (in French)
26. Monmonier M (2003) How to make the cards. Ed. Flammarion, Paris, p 21. (in French)
27. Charef A (2010) Methodological approach for the evaluation of soil phospho-potassium and nitrogen fertility. Mem. Ing. ENSA, El Harrach, p 61. (in French)
28. Bouzidi A (2010) Tadjmout soils (Laghouat) features, mapping and evaluation. Mem. Ing. ENSA, El Harrach, p 94. (in French)
29. Pouget M (1980) Soil – vegetation relationships in the southern Algerian steppes. Ed. O.R.S.T. O.M., Paris, p 555. (in French)
30. Hanotiaux G (1985) Soil analysis. Department of Soil Science, Faculty of Agricultural Sciences, Gembloux, p 48. (in French)

Scale Inhibition in Hard Water System

Amina Karar and Abdellah Henni

Contents

Abstract The precipitation of an insulating layer of scaling on the walls of the water distribution pipes has serious technical and economic consequences. Various methods were used to prevent the scale formation in water such as the chemical methods in which the germination of the $CaCO_3$ crystals is blocked using the inhibitor. In recent years, a few studies have been focused on the aspects of the surface scaling so that the different mechanisms were proposed to explain the differences between the scaling precipitation in bulk solution and scale deposition at the surface. The water distribution of some Algerian town resulting from the drilling water is supersaturated with respect to calcium carbonate. This causes reducing heat transfer in heat exchanger systems, limiting the efficiency of these

A. Karar (✉) and A. Henni
Lab. Dynamic Interactions and Reactivity of Systems, Kasdi Merbah University, Ouargla, Algeria
e-mail: amina.karar@hotmail.fr

Abdelazim M. Negm, Abdelkader Bouderbala, Haroun Chenchouni, and Damià Barceló (eds.), *Water Resources in Algeria - Part I: Assessment of Surface and Groundwater Resources*, Hdb Env Chem (2020) 97: 293–318, DOI 10.1007/698_2020_530, © Springer Nature Switzerland AG 2020, Published online: 18 July 2020

devices (valves and taps) by decreasing the flow rate in the pipes; this phenomenon is more prevalent at high heating temperature. In this book chapter, we aim to give an overview of the different antiscaling properties in hard water. We also provide the inhibitors used and the researches done on Algerian water.

Keywords Algeria, Calcium carbonate, Hard water, Inhibitor, Scale

1 Introduction

Water plays a central and general role in human activities, development of social economy, and the ecological environment balance. Water contains mineral salts and dissolved or suspended substances. Indeed, during an increase in temperature, removal of dissolved CO_2, an increase in the concentration of certain dissolved salts, or, more generally, a change in chemical equilibrium, dissolved solids can crystallize. A compact and insulating adhesive layer is then formed on the surface of the water pipe. This deposit causes the decrease of heat exchange, the reduction of the partial or total diameter of water pipes, the dysfunction of domestic and industrial installations and equipment, and the inhibition of detergency and ultimately significant financial losses. The primary agents responsible for crystallization fouling are the carbonates and sulfates of calcium or magnesium, barium salts, silicate, and phosphate.

Scaling is essentially linked to the formation of calcium carbonate ($CaCO_3$). Scaling may contain other residues such as algae, calcium sulfate, clays, and the brucite Mg $(OH)_2$. But it is always calcium carbonate that precipitates first, usually in the colloidal form, because its solubility is lower than that of others.

The formation of scale is a very complex phenomenon which can be subdivided into two phases: a crystal of calcium carbonate is formed, and the precipitate is formed around the initiator crystal and then its growth; it is the stack of adsorbed germs that forms $CaCO_3$ growth sites (Fig. 1).

Scaling in natural hard water is a major concern in different facets of industrial processes and domestic installations (Fig. 2). Undesirable scale deposits often cause numerous technical and economic problems such as total or partial obstruction of pipes leading to a decrease in flow rate; reduced heat transfer as scale precipitate is 15 to 30 times less conductive than steel; seizure of valves and clogging of filters; etc. Therefore, it is essential to establish appropriate methods to study this phenomenon and find effective ways to combat it. Among the methods used to inhibit this phenomenon, the use of products to block the formation of tartar by preventing crystal growth (called tartrifuges products): organic molecules such as tannins, humic acids, citric acid, or glutamic acid, there is also chemicals: polyphosphates, phosphonates [1].

The ideal inhibitor would be a compound in a solid form whose solubility would be very low but largely sufficient to ensure a total scaling inhibition. It could thus be

Fig. 1 Adsorption of calcium carbonate germs on a metal wall

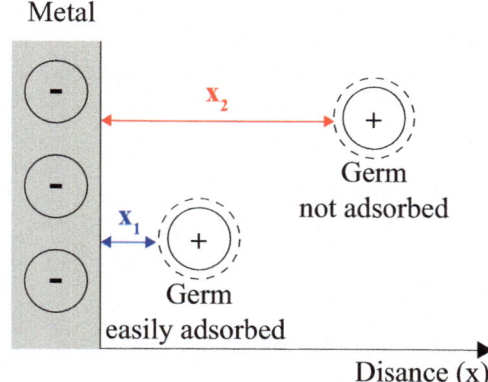

Fig. 2 Pipeline obstructed by scale of Ouargla city [2]

brought into direct contact with the water to be treated without having to worry about its concentration, which would be automatically regulated by its solubility. Environmental requirements impose many challenges in the field of water treatment. Thus, the concept of "green chemistry" has been proposed, and the use of "green" chemicals has become a necessity. It is, therefore, of paramount importance to develop "green" scaling inhibitors in order to combat scaling phenomena that have disastrous or even catastrophic consequences [3].

For decades, various attempts have been made to estimate the scaling power of natural waters and characterize the scaling formation mechanisms. These methods can be roughly divided into two categories: electrochemical methods and chemical methods. For the first one, we can use the chronoamperometry, chronoelectrogravimetry, and electrochemical impedance technique. All these techniques are based on the reduction of the oxygen dissolved in the water. Among the chemical methods, we can enumerate the critical pH method an evaporation method, the LCGE method, the rapid controlled precipitation method, a polymer scaling test, and a continuous test on tubes and others.

2 Scale

In this section, we introduce the definition of scale, the different type of scale, and the influence on industrial processes.

2.1 Definition

Water contains a number of dissolved species that can react under specific conditions to form a precipitate. Some of the precipitated salts have a very low solubility which is called poorly soluble salts. The precipitation reaction can be triggered by changes in conditions such as supersaturation, temperature, pH, or pressure. Precipitation results in deposition and scale formation on all surface materials, including pipe walls, heat transfer surfaces, pumps, etc. This process is also called fouling by crystallization (Fig. 3).

Calcium and magnesium carbonates and sulfates are the main agents responsible for the fouling of crystallization, although barium salts, silicates, and phosphate deposition play an important role in some industries. Deposits formed by water, usually observed, include calcium carbonate, calcium and barium sulfate, silica deposits, iron deposits, magnesium, and calcium phosphates. If we extend our area of interest to wastewater, struvite is the main cause of the mineral scale [4]. Scaling is a phenomenon that occurs when a surface is in contact with incrusting water, likely

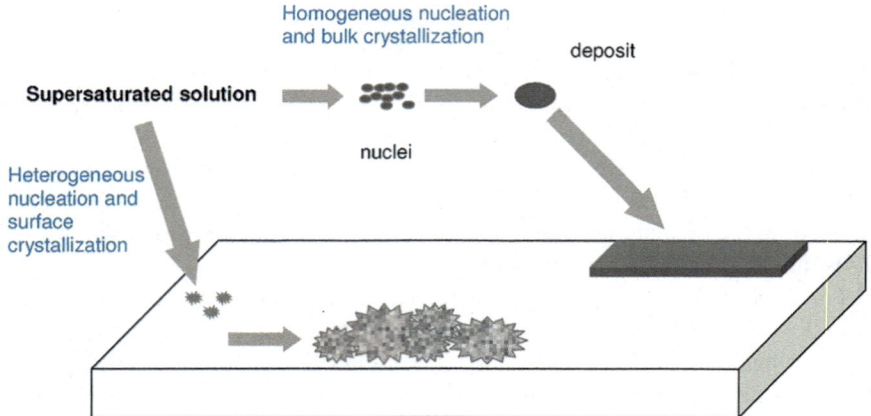

Fig. 3 Schematic illustration of scale formation schemes [6]

to cause the formation of a product of very low solubility in the form of an adherent deposit. In the case of natural waters, the compound likely to precipitate first is calcium carbonate [5].

2.2 Different Type of Scale

2.2.1 Calcium Carbonate $CaCO_3$

Calcium carbonate ($CaCO_3$) is the most common compound (in terms of geographical distribution and abundance) in mineral precipitates of biological origin (marine and geological organisms) [7]. It is an essential element of sedimentary rocks [8] and a major constituent of scale that is found in drinking water pipes and in various structures in contact with natural or distribution water. It may be accompanied by various poorly soluble salts according to:

– The origin of the water
– The treatment he has suffered
– Temperature
– The nature and the state of corrosion of the pipes

It is mainly $CaSO_4$, $Mg(OH)_2$, $Fe(OH)_3$, and salts of various metals [9]. But it is always the calcium carbonate that precipitates first, usually in the colloidal form, because its solubility is lower than that of the others [10].

There are three crystallographic varieties of calcium carbonate:

The Calcite Thermodynamically, calcite (Fig. 4a, b) is the most stable polymorph of calcium carbonate, and it comes in many forms in nature. According to the data of the literature, calcite can present 700 different crystal forms always in the same system and also a certain number of twins (intimate association of two or more crystals according to precise plans).

The Vaterite Unlike calcite and aragonite, there is a very little occurrence of vaterite in natural minerals [11]. In fact, exposed to water, vaterite is very unstable and generally recrystallizes in the form of calcite. The most common facies is in lens (Fig. 4c); but we also encounter facies in the form of "roses of the sands."

The Aragonite Aragonite is orthorhombic pseudohexagonal and is usually in the form of a needle (Fig. 4d). It is metastable at ordinary temperature and transforms into calcite at high temperature.

Calcite is the most stable crystalline form at 25°C. Calcite is the least soluble (pK = 8.35). Aragonite is more soluble than calcite (pK = 8.22); it precipitates hot ($T > 60°C$). The vaterite form is the most unstable of the three crystallographic forms of $CaCO_3$ and makes its identification difficult. The solubility of calcium carbonate increases with temperature (Table 1).

Fig. 4 The SEM images of CaCO$_3$ forms: (**a**) amorphous calcite, (**b**) layered and rhombohedral calcite, (**c**) spherical vaterite, and (**d**) needle aragonite [12]

Table 1 The temperature effect on the solubility of calcium carbonate [12]

	Solubility (mg L^{-1})	
Temperature	à 25°C	à 50°C
Calcite	14.43	14.43
Aragonite	15.28	16.17

Natural waters contain in solution many chemical species at very different concentrations according to their origins. We can classify them into two groups according to:

- The fundamental elements: $CO_{2(free)}$, H_2CO_3, HCO_3^-, CO_3^{2-}, H^+, OH^-, and Ca^{2+}, which participate in the carbonic and calcocarbonic equilibrium.
- The characteristic elements: Mg^{2+}, Na^+, and K^+ for the cations and SO_4^{2-}, NO_3^-, and Cl^- for the anions. These ions do not intervene in the preceding equilibria except by their action on the ionic strength of the solution.

(a) The carbonic equilibrium

When we dissolve CO_2 in water, we have the following equilibrium:

$$CO_2 + H_2O \leftrightarrow H_2CO_3 \text{ with the constant } K_0 = \frac{[H_2CO_3]}{[CO_2]} = 10^{-1.5}$$

At ordinary temperature, so that there is only 3% of carbon dioxide in the form of H_2CO_3

The latter has in solution, the behavior of a diacid according to:

$$H_2CO_3 + H_2O \leftrightarrow HCO_3^- + H_3O^+ \text{ with a constant } K_1 = \frac{[H_3O^+][HCO_3^-]}{[H_2CO_3]}$$

$$HCO_3^- + H_2O \leftrightarrow CO_3^{2-} + H_3O^+ \text{ with a constant } K_2 = \frac{[CO_3^{2-}][HCO_3^-]}{[HCO_3^-]}$$

If we note: $[CO_{2\,(free)}] = [H_2CO_3] + [CO_2]$
We write the balance:

$$2H_2O + CO_{2\,(free)} \leftrightarrow HCO_3^- + H_3O^+ \text{ and we can put } K_3 = \frac{[HCO_3^-][H_3O^+]}{[CO_{2\,(free)}]}$$

$$= \frac{K_1}{\left[1 + \left(\frac{1}{K_0}\right)\right]}$$

In addition, we have the autoprotolysis of water:

$$2H_2O \leftrightarrow OH^- + H_3O^+ \text{ with a constant } K_e = [H_3O^+][OH^-].$$

The previous constants depend on the temperature of the water, the ionic strength of the solution, i.e., the concentration, and the charge of all the ions present (fundamental and characteristic).

(b) The calcocarbonic equilibrium

For water in equilibrium with solid calcium carbonate, the equilibrium:

$$Ca^{2+} + CO_3^{2-} \leftrightarrow CaCO_{3(S)}$$

This equilibrium, called calcocarbonic, is governed by the law of mass action, which establishes between the concentrations of the ions Ca^{2+} and CO_3^{2-} the following relation:

$$K_s = \left[Ca^{2+}\right]\left[CO_3{}^{2-}\right]$$

K_s depends on the temperature and the ionic strength of the solution. It also depends on the crystallographic variety. Under the usual conditions and at ordinary temperature, the $K_s = 10^{-8.3}$ is often adopted.

If the product $[CO_3{}^{2-}][Ca^{2+}]$ is lower than K_s, it will not be possible to precipitate.

Conversely, for having precipitation of calcium carbonate, it is necessary to have:

$$\left[CO_3{}^{2-}\right]\left[Ca^{2+}\right] > K_s.$$

Generally, in a circuit of water, one can write at any point of this circuit:

$$\left[CO_3{}^{2-}\right]\left[Ca^{2+}\right] = \delta \cdot K_s$$

where δ is the local supersaturation coefficient of water. Indeed, three cases can occur at one point:

$\delta < 1$: the water is locally aggressive.

$\delta = 1$: the water is locally at equilibrium.

$\delta > 1$: the water is thermodynamically capable of locally precipitating $CaCO_3$ according to the increasing germination process mentioned above.

The germs of $CaCO_3$ can have varied evolutions:

- The germ will grow and regress and then dissolve.
- The seed will grow and evolve into a crystal. We then go to a phenomenon of sludge generating precipitation.
- The seed will be produced at the level of a metal wall. We will be in a scaling process if the connection with the wall is strong enough.
- The germ will remain suspended in the water and be carried away by the flow.

The solubility of $CaCO_3$ depends on the pH. It is shown in Fig. 5. The solubility of $CaCO_3$ can be increased by the addition of carbon dioxide according to the equilibrium:

$$CaCO_3 + CO_2 + H_2O \leftrightarrow Ca^{2+} + 2HCO_3{}^-$$

2.2.2 Calcium Sulfate

Calcium sulfate is one of the most common scalants in processes involving seawater, such as desalination. It is also often referred to as nonalkaline scale. Pure calcium sulfate is white in color with a similar resemblance to calcium carbonate. It exists as $CaSO_4 \cdot nH_2O$. All these forms are more soluble than calcium carbonate and magnesium hydroxide [13]. The most common is gypsum with a monoclinic prismatic

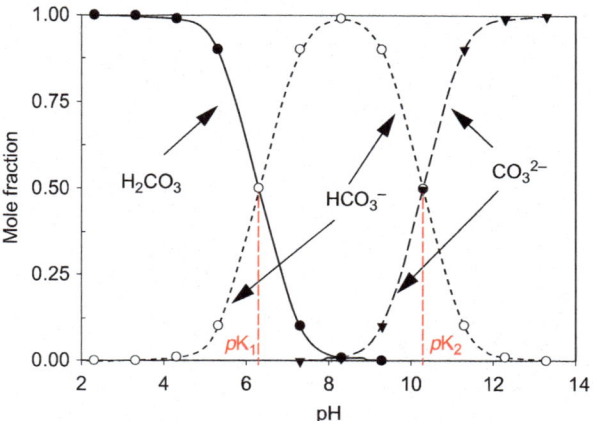

Fig. 5 The distribution of carbonate species as a fraction of total dissolved carbonate in relation to solution pH [12]

crystal structure, with four or eight molecules in the unit cell. It crystallizes as white crystals according to the following reaction [10]:

$$CaCl_2 \cdot 2H_2O + NaSO_4 \rightarrow CaSO_4 \cdot 2H_2O + 2NaCl$$

Layers of gypsum can settle and form a shell with layers of anhydrite. These relatively thin layers (0.6 mm) of calcium sulfate scale on the tubes of low-pressure boilers can cause a massive temperature drop of 180°C [14]. This had previously been shown by Bansal and Müller-Steinhagen [15] for calcium sulfate fouling on plate heat exchangers.

Crystallization of calcium sulfate is a complex phenomenon influenced by a number of parameters, such as temperature, pressure, electrolytes, and dissolved organic matter, and the presence of other minerals [16]. Research by Bansal and Müller-Steinhagen [15] showed that the measured calcium sulfate resistance of a plate heat exchanger was 50% lower than 85°C compared with 90°C. Temperature also influences the formation of polymorphs. Gypsum is the most formed deposit in all reverse osmosis and cooling systems that use moderate temperatures (up to 50°C), while calcium sulfate hemihydrate and anhydrite are the most formed in applications at high temperature [17, 18]. Calcium sulfate is found in many cooling systems, but since most of these systems operate at pH = 7–9, calcium carbonate is the most predominant scale, and calcium sulfate only precipitates when all carbonate is eliminated [19]. Precipitation of calcium sulfate is insensitive to pH, unlike other calcareous deposits [20]. Calcium sulfate may be in the form of hard rock or soft granules [10].

2.2.3 Magnesium Scales

A number of magnesium scales, such as magnesium hydroxide (Mg(OH)$_2$), magnesium carbonate (MgCO$_3$), and hydromagnesite (Mg$_5$(CO$_3$)$_4$(OH)$_2$·4H$_2$O), can be formed in applications using water. Magnesium hydroxide is commonly known as brucite but may also precipitate as amorphous magnesium hydroxide. Magnesium carbonate is known as magnesite (magnesium carbonate anhydrite) [15].

Magnesium hydroxide is the most common of magnesium scales. The magnesium hydroxide is formed in a series of the following reactions [13]:

$$CO_3{}^{2-} + H_2O_5 \rightarrow CO_2 + 2OH^-$$

$$Mg^{2+} + 2OH^- \rightarrow Mg(OH)_2$$

Other magnesium containing deposit is magnesium silicate ($MgSiO_3$) and its hydrated form known as talc. Possibly less known magnesium containing minerals also include iowaite [Mg_6Fe_2 $(OH)_{16}$ $Cl_2 \cdot 4H_2O$], new beryite ($MgHPO_4 \cdot 3H_2O$), serpentine ($3MgO \cdot 2SiO_2 \cdot 2H_2O$), merille [$Ca_{18}Na_2Mg_2$ $(PO_4)_{14}$], and other phosphates [$Mg_3(PO_4) \cdot Mg(OH)_2$, $Mg_3(PO_4)_2 \cdot 8H_2O$, and $Mg_3(PO_4)_2 \cdot 22H_2O$].

The formation of magnesium scales is affected by a number of parameters. The rate of magnesium hydroxide crystallization in seawater depends on temperature, pH, concentration of bicarbonate ions, the development of carbon dioxide, magnesium ion concentration, and total dissolved solids. It has been noted previously that magnesium hydroxide supersaturation is pH dependent and that brucite is formed at a higher pH [21–23]. The effect of temperature is important in the case of nesquehonite, which occurs mainly at temperatures below 42°C, and hydromagnesite, which formed at temperatures between 60 and 90°C [23]. Magnesium hydroxide is a salt of inverse solubility, generally formed at temperatures above 95°C, mainly because of the increased formation of hydroxyl ions at such high temperatures [13].

The two alkaline scales (magnesium hydroxide and calcium carbonate) are closely related, and the presence of magnesium affects the formation of calcium carbonate. It has already been reported that the presence of dissolved magnesium favors the formation of aragonite compared to other polymorphs, with magnesium concentrations in seawater explaining the presence of aragonite rather than calcite when seawater is used [23].

2.2.4 Silica Scales

Large amounts of silicon dioxide and many silicate minerals are present in the Earth's crust, resulting in high concentrations in natural environments. They can be in a soluble form (silicic acid, soluble silicates) or in an amorphous state (colloidal silica) [6]. The solubility of crystalline silica (quartz) is quite low with 5–6 mg L^{-1} at 25°C and pH < 9 with the solubility of amorphous silica ranging from 120 to 150 mg L^{-1} at 25°C and pH < 8–8.5. Amorphous silica is usually classified as dissolved (reactive), colloidal (nonreactive), and particulate (suspended) silica [24]. Ning [25] states that natural waters may contain dissolved and suspended silica concentrations that subsequently form on the surface of boilers, RO membranes, and cooling towers in the form of deposits and scales that may include layers of glass, gels, powders, or nanoscopic particles virtually "invisible." Fouling of silica in water treatment equipment has been dealt with since the earliest developments in industrial water chemistry. As with the previous types of scales, it was reported that a SiO_2 scale layer 0.5 mm thick results in a 90% decrease in heat transfer [16]. Colloidal silica that enters equipment with the feed water can then settle on the boiler tubes or turbines in the form of silicates including SiO_2, $Na_2SiO_3 \cdot 9H_2O$, $NaFeSiO_6$, or $Mg_6[(OH)_8Si_4O_{10}]$ that are not soluble in water and extremely difficult to remove and therefore resulting in losses in turbine efficiency and capacity [26].

Ning [25] has described that "reactive" and "non-reactive" silica in water has a broad spectrum of molecular sizes ranging from silicic acid monomer ($Si(OH)_4$) to dissolved oligomeric forms, through the colloidal polymer suspensions and optionally in the form of silica ($SiO_2)_n$ of silicate particles, n representing very large numbers. If metal hydroxides such as aluminum, iron, magnesium, and calcium are incorporated during dehydration polymerization reactions, metal silicates are formed. It is widely accepted that the polymerization of silica monomers is the formation mechanism of amorphous silica deposits [6, 27].

According to Ning, monomeric silicic acid (the "reactive silica") is the predominant dissolved silica species found in natural waters with concentrations varying from 1–3 mg L^{-1} in mountain lakes to 50–300 mg L^{-1} in well waters in oil production fields. The silicic acid concentration in seawater is reported to range between 1 and 10 mg L^{-1} and is unregulated in municipal drinking water [25].

2.3 Influence of Scale on Industrial Processes

Scale formation affects most industries, as it can occur in a number of industrial processes, such as heating or cooling, desalination, and oil production. Boilers, cooling towers, pipes, tubes, and other equipment used in water-intensive processes pose serious problems, often hampering the overall process and increasing production costs due to associated maintenance costs [28]. The severity of the problem will vary depending on the composition of the water and the operating conditions. The formation of a deposit or layer can take weeks, even months or shorter, leading to major operational problems. This formation of deposits or scale can lead to reduced flow and heat transfer; this leads to additional maintenance or even equipment failure and increased operating costs, not to mention that the deposits formed on the heat transfer surfaces will significantly degrade the performance of the heat exchangers and can lead to a complete failure of the equipment, the mineral deposits affect the nuclear center; chemical, food and beverage industry; the oil and gas industries.

The major sources of concern are cooling water and associated processes due to widespread use in many industries. During cooling applications, flaking is caused by solids concentration, temperature changes, and pressure drop [10]. The other type of recirculating cooling system is an open recirculating system in which water is continuously reused by cooling probably of the greatest interest here. Such systems are frequently used in large central utility stations; in chemical, petrochemical, and petroleum refining plants; in steel mills and paper mills; and in all types of processing plants [21]. Unfortunately, these systems also have the greatest potential for all types of problems, including fouling due to the excessive amount of mineral ions present in the water flowing through the system at high temperatures and a compliment constant fouling potential when replacing evaporated steam. The main types of scales that form in these types of systems include calcium carbonate, calcium sulfate, calcium phosphate, and magnesium silicate [21]. The formation of scale is a major problem in areas such as energy production, including geothermal energy. This process offers many advantages because it is renewable, clean, safe, and flexible, but to make production competitive with other energy resources such as natural gas, it is necessary to minimize operating costs. In this case, the mineral deposits not only reduce the capacities of the wells but also the heat exchange zones, thus limiting the degree of use of the heat transported by the fluids [29].

In many industries, such as mineral processing, increasing scarcity of freshwater and stricter environmental regulations are leading to a mandatory increase in the use of recycled water, leading to more problems with mineral scale formation [30]

3 Scale Inhibition by Chemical Additives

In the face of scale formation problems, several chemical methods based on the use of scale inhibitors have been largely effective [31–33]. Admitting the appearance of crystals as fatal, this treatment directly attacks the time and type of germination. For this purpose, products are used which delay the appearance of the seeds (germination time longer than the residence time of the water in the circuit) and promote the formation of little adherent crystals (homogeneous germination) and/or which decrease the growth rate of crystals [1].

The term "tartrifuges" refers to substances that have one or more of these properties. The tartrifuge effect has been known for a long time. Also, the search for new tartrifuges and their development still remains an approach marked by certain empiricism. For this reason, it is important to specify test conditions in which the effectiveness of a tartrifuge can be appreciated.

3.1 Classification of Different Tartrifuge

The big family of tartrifuges is growing every day, which does not allow us to give an exhaustive list of all the products on offer. However, most of these products can be grouped into families with a common grouping or structure that is the basis of the properties of the tartrifuge. By limiting ourselves to the most classical and the most used, this classification is represented according to Fig. 6.

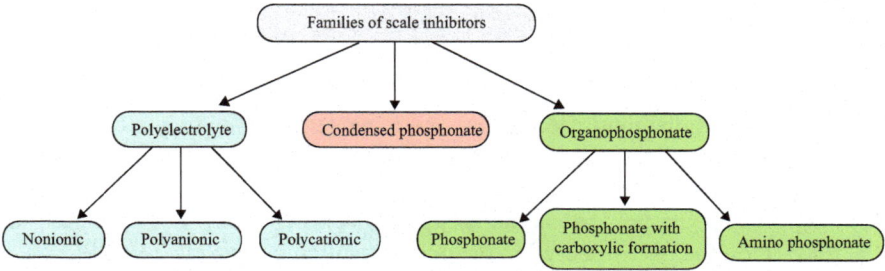

Fig. 6 Classification of different families of scale inhibitors

Fig. 7 The structural formula of sodium tripolyphosphate

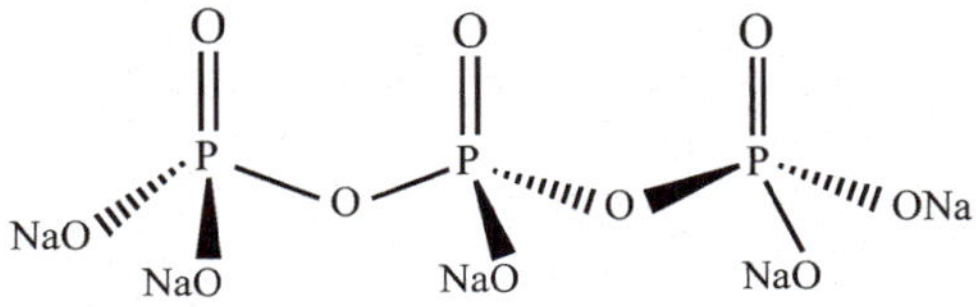

3.1.1 Condensed Phosphates

These products whose properties are known for a long time and which are widely used in Europe, particularly in the formulation of washing powders, hydrolyze easily above 70°C and give orthophosphate (PO_4^{3-}), which once released into the natural environment is considered to be the main cause of the eutrophication phenomena in aquatic environments. In the late 1960s, these products were replaced by organophosphonates.

Sodium tripolyphosphate is the most used. Its structural formula is given in Fig. 7.

3.1.2 Organophosphonates

They are characterized by one or more groups associated with an organic radical. The most used are:

(a) *Phosphonates*: These compounds have an advantage over polyphosphates since the CP bond of 1.87 Å length is more hydrolysis resistant than the PO bond of 1.5 Å in length, and they are easily biodegradable. On the other hand, they have better stability in the presence of chlorine. The best product is 1-Hydroxyethylidene-1,1-diphosphonic acid (HEDP)

(b) *Phosphonates with carboxylic function*: As the name implies, they are composed of two functional groups: -PO_3H_2 and -COOH. These acids have remarkable resistance to hot hydrolysis in the presence or absence of bactericides/oxidants [34].

(c) *Amino phosphonates*: These compounds have both a -PO_3H_2 group and a nitrogen group [34].

3.1.3 Polyelectrolytes

A number of macromolecules have tartrifuge properties. We can distinguish:

(a) *Polycationics*: the most used correspond to quaternary ammonium compounds, for example, polyethylenimines. They are relatively little used because their effectiveness is low [35].
(b) *Polyanionics* are polycarboxylic or sulfonic acids [34, 35]. Let's mention the polyarilics, the polymetacrylics, the polyvinysulfonic ones, etc. [36–38].
(c) *Non-ionic*: they generally have less efficiency than anionics but much higher than cationic, for example, polyacrylamides [35].

3.2 Additional Work on Scale Inhibitors

Several types of scaling inhibitors, mineral or organic, have proven effective and are already marketed. The inhibiting effect of metal ion scaling (Cu^{2+}, Zn^{2+}) results in the formation of mixed carbonates of calcium and copper $Cu_xCa_{1-x}CO_3$ and zinc and calcium carbonates $Zn_xCa_{1-x}CO_3$. The mechanism of inhibition is based on the dehydration energy of Cu^{2+} and Zn^{2+} ions, which is greater than that of Ca^{2+} ions. This makes it possible to block the growth of the seeds formed [39–41]. Metallic iron ions play an important role in inhibiting scaling [42–45].

The most effective is 1,1-hydroxyethylenediphosphonic acid (HEDP) and is active at a concentration of 0.62 μmol L^{-1} (0.2 mg L^{-1}) [46]. A concentration of 9.1×10^{-8} M HEDP is sufficient to completely prevent crystal growth of calcium carbonate for more than 100 h [47].

Rosset and Douville [48] have shown that very low concentrations of organic phosphonate or polycarboxylate inhibitors inhibit the precipitation of calcium carbonate in EL Hamma's borehole water. Effective concentrations of these inhibitors are in the range of 1.1–1.5 mg L^{-1}.

Reddy and Nancollas [49] noted with the Crystal Growth Criterion that HEDP is the best among a range of phosphonates studied, for pure product concentrations of 0.5 ppm. The same result was obtained by Shiliang and Kan [50] when comparing a range of phosphonates as scale inhibitors.

The addition of 0.5–0.6 ppm amino trimethylene phosphonic acid "ATMP" inhibits calcium carbonate encrustation under chronoamperometric conditions [51].

A certain number of organic substances of natural origin have recognized tartrifuge power: this is the case for the metabolites of planktonic algae [52] and humic substances [53].

Abd-El-Khalek et al. [54] studied the antiscaling properties of sodium hexametaphosphate (SHMP) using electrochemical methods as well as a comparison between SHMP and polyacrylic (PPA). This study showed that SHMP is more effective against scale formation than PPA. They also studied the antiscaling properties of the palm leaf extract by chronoamperometry, impedance meter, and optical microscope. The results showed that this extract could be considered as a precipitation inhibitor of $CaCO_3$ with a concentration of 75 ppm on salt water [55].

Gao et al. [56] synthesized the polyaspartic acid derivative (PASP-SEA-ASP) which showed excellent antiscaling properties in marine waters, with a 100% inhibition rate for the assay of 14 mg L^{-1}. This study demonstrates the potential of PASP-SEA-ASP for scale and corrosion inhibition in domestic and industrial facilities using seawater [56].

Touir et al. [57] based their study on the mechanism of action of sodium gluconate on ordinary steel. They evaluated the effect of temperature on the inhibition rate of sodium gluconate, which becomes more important with increasing temperature, which has been explained by the formation of a stable GS-Ca^{2+} complex.

Henghui Huang et al. [58] investigated the scale inhibition behavior of PESA with linear and hyper-branched structure against $CaCO_3$ and $CaSO_4$ scales that were evaluated using static scale inhibition method, and their ability to retard deposition of $CaCO_3$ was also examined. The experimental results showed that, for $CaCO_3$ and $CaSO_4$, the PESA with hyper-branched structure provides a scale inhibiting efficiency as high as 95.9% and 94.3%, respectively, at an inhibitor concentration of 15 mg L^{-1}.

4   Natural Inhibitors of Calcium Carbonate Scaling

Nowadays, one of the major axes of research is to find economic and environmentally friendly inhibitors. Research is underway for new environmentally compatible inhibitory formulations called "green inhibitors." A chemical is defined as "green" based on three criteria: toxicity, bioaccumulation, and biodegradation [35, 59]. The advance on green scale inhibitors is surprising, and numerous researches and discoveries have emerged. These green inhibitors may be petrochemical derivatives, or using "natural" organic molecules or plant extracts, such inhibitors could be used in various fields such as energy, water, and the food industry.

4.1   Petrochemical Derivatives

In this section, we took the initiative to make a selection of interesting green organic molecules from petrochemicals. In a recent article, Hasson et al. [60] reported that the most promising green scale inhibitors relied primarily on polyaspartic acid (PASP).

Polyepoxysuccinic acid (PESA) is a green scale inhibitor developed in the 1990s in the United States. Static experiments performed by Sun et al. at 30°C [61] showed that at a concentration of 10 mg L^{-1}, the scale inhibition ratio of PESA to $CaCO_3$ at three different concentrations of calcium (40, 100 and 200 mg L^{-1}) was greater than

90%. Liu et al. [62] showed that the anticalcination performance of PESA was higher than that of PASP. Therefore, PESA can be considered as an interesting alternative to PASP in water treatment technology. However, the performances of these molecules in field tests are limited [63, 64], and these molecules deserve to be studied further. However, it should be noted that the inhibition performance of PASP was studied in a pilot cooling water plant by Laborelec (Belgium) [64]. The effectiveness of the tested PASP would be similar to that of low molecular weight polyacrylate, a polymer used in some industrial processes. Laborelec has implemented a PASP scale control in a power plant [60]. PASP was developed in the early 1990s and has many applications, such as scale and corrosion inhibition, water softening, and green chemical formulations for detergent formulations [60]. PASP has non-nitrogenous, non-phosphorous, and biodegradable characteristics, which makes it a good green inhibitor. Indeed, the degradation properties of PASP have been examined by Thombre et al. [65] in the laboratory, Gao et al. indicated by the biodegradability of PASP prepared from poly(succinimide) by 70% thermal condensation in 1 month [66] Martinod et al. [67]. This study describes the effect of PASP on the desquamation of breasts reproducing the composition of the North Sea, where the concentration of Ca^{2+} was 14,225 mg L^{-1}. In addition, PASP significantly reduced the growth rate of $CaCO_3$ crystals by blocking the active sites of growth. Recently, Liu et al. used in diluted mineral water (concentration of Ca^{2+} 253 mg L^{-1}) at 80°C [62].

Carboxymethyl inulin (MIC) is a biodegradable and nontoxic polysaccharide-based polycarboxylate, obtained from inulin by chemical synthesis. It has significant inhibitory effects on the crystallization of calcium carbonate by the presence of carboxylic acid groups in its structure. Verraest et al. [68] investigated for the first time the effect of small amounts of MIC between 0.1 and 200 mg L^{-1} on the crystallization of calcium carbonate and concluded that MIC is a good inhibitor of $CaCO_3$ by changing the morphology of the crystals formed. They also showed that the MIC could influence the growth rate of calcium carbonate seed crystals by using constant composition experiments to study the growth kinetics of constant supersaturation seed crystals. The authors suggested that MIC molecules with a high degree of substitution and a high degree of polymerization were the most effective. Figure 8 presents the structures of some green antiscalant from petrochemistry.

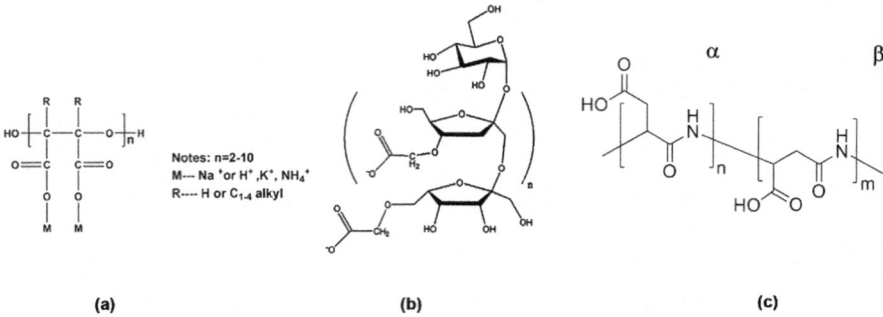

Fig. 8 Structures of some green antiscalant from petrochemistry: (**a**) polyepoxysuccinic acid, (**b**) carboxymethyl inulin, and (**c**) polyaspartic acid

4.2 Natural Organic Molecules

Environmental requirements impose many challenges in the field of water treatment. Thus, the concept of "green chemistry" has been proposed. In their study of the inhibition of the growth of calcite crystals by a humic substance [69], Hoch et al. [70] performed constant composition experiments in a sealed reactor. Solutions with equal molar concentrations of calcium and carbonate have been implicated. They quantified the kinetic effects of humic acid and fulvic acid from different sources on the crystalline growth of calcite by adding seed crystals to a supersaturated solution.

Kumar et al. [71] test the performance of pteroyl-L-glutamic acid (PGLU) scale inhibition by conducting static and dynamic experiments, a specific form of commercially available vitamin M. PGLU is found in almost all leafy vegetables and is essential for many bodily functions, including nucleotide synthesis. Scale inhibition at 100% [72, 73] was obtained by pot experiments at 70°C with 120 mg L^{-1} PGLU. The synthetic waters tested were representative of those found in an oil field at sea. The waters contain high amounts of calcium and bicarbonate. At 90 and 110°C, higher concentrations of PGLU, between 160 and 200 mg L^{-1}, were required to inhibit the formation of the latter result which was confirmed by dynamic tube blocking tests performed at 110°C, where the minimum concentration of PGLU for scale inhibition was 160 mg L^{-1}. The authors suggested using FTIR, XRD, and SEM that scale inhibition occurred by formation of a soluble complex with calcium ions and deformation of $CaCO_3$ crystalline morphology. They concluded that PGLU could be an excellent green chemical for scale inhibition in oil wells.

The authors have shown that certain hydrophobic aquatic organic acids derived from higher plants in the Florida Everglades were able to reduce the rate of calcite growth from a concentration of 0.2 mg L^{-1}. The growth of calcite was almost completely inhibited with a concentration of 5 mg L^{-1} in dissolved organic matter. An SEM study suggested that crystal growth sites were blocked by adsorbent ligands. In a recent work, Gauthier et al. [74] investigated the calcium carbonate scale ability of humic acid in synthetic water (100 mg L^{-1}) at 35°C, purchased as sodium salt. Humic acid also had a significant effect on scaling at a concentration as low as 0.2 mg L^{-1}. The associated efficiency was 78%, as determined by the resistivity response of the FCP. This work may explain the ability of raw rivers to scale, depending on water quality and season, flow [74]. These results could have an impact on many industrial facilities supplied with raw river water, knowing that humic substances are not used as inhibitors in technical applications.

Reddy et al. [75] used the constant composition technique to determine the growth rate of calcite in the presence of citric acid at 25°C and pH 8.55. A solution with constant calcite supersaturation was seeded with calcite crystals synthetically prepared in the presence of citric acid. Then they made a titration of Ca^{2+} ions remaining in solution order to follow the crystallization of calcium carbonate. The authors showed that citric acid exhibited only a moderate reduction in the growth rate of calcite crystals at concentrations as high as 10 mg L^{-1} and did not lead to a reduction in growth rate in the range of 0.01–0.1 mg L^{-1} (total calcium

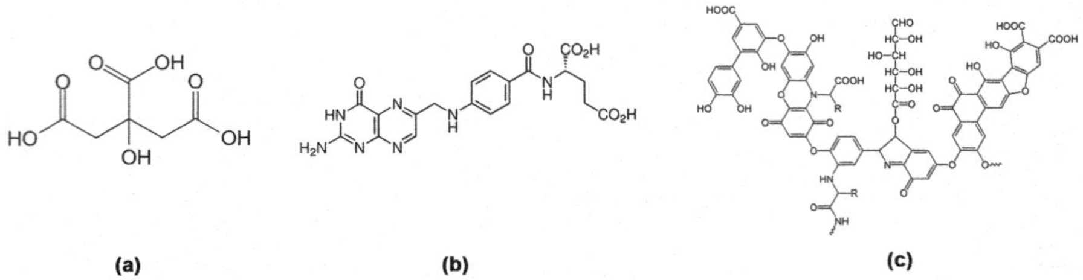

Fig. 9 Structures of some natural organic molecules: (**a**) citric acid [77], (**b**) pteroyl-L-glutamic acid [73] and (**c**) humic acid [76]

concentration 76 mg L^{-1}). This could be explained by the fact that citric acid is a linear polycarboxylic acid. The authors suggested that cyclic and rigid polycarboxylic acids, such as tetrahydrofuran-2-carboxylic acid or cyclopentane-tetracarboxylic acid, were much more effective in descaling [75].

The structures of some green antiscalant from petrochemistry are presented in Fig. 9.

4.3   Plant Extracts as Scale Inhibitors

The most promising alternatives to "natural" organic molecules are the use of plant extracts as scale inhibitors. Recently, the scale inhibition properties of plant extracts have been investigated.

Certain authors suggested that fig leaf extract may complex the cations present in the brine solution or disperse the suspended solids through adsorption. They carried out the same study concerning the olive leaf extract [76]. Indeed, olive leaves contain many phenolic molecules including oleuropein, the most abundant biophenols in olive leaves [77] and caffeic acid [78, 79] The concentration of the inhibitor was 10 mg L^{-1}. The authors assumed that these extracts were more effective than polyaspartic acid in preventing calcium carbonate formation. Indeed, the percentage of inhibition was 16.7% for both soy-based polymer and polysaccharides from sea weeds, whereas it was only 6.6% for polyaspartic acid.

A second strategy was to consider plants containing well-known compounds that can complex calcium cations. In this respect, polyphenols or polysaccharides, which have hydroxyl and/or carboxyl functional groups that interact with divalent ions such as Ca^{2+} or Mg^{2+}, are very good candidates. Abdel-Gaber et al. [80] studied the antiscaling properties of *Punica granatum* hull and leaf extract in alkaline brine at 25°C using conductivity measurements, electrochemical impedance spectroscopy, and chronoamperometry in conjunction with SEM, EDX, and optical microscopic examinations. According to chronoamperometry measurements, the current density increased from 87% when the concentration of the *Punica granatum* extracted from hull increased from 10 to 100 mg L^{-1}. This indicated that the extract is an efficient antiscalant. It must be noticed that the aqueous extract at 50 mg L^{-1} concentration was still effective as a scaling inhibitor after 28 days of storage at 5°C.

Castillo et al. reported inhibition results of calcium carbonate scale performed with aloe vera in Venezuelan oilfields [81, 82]. The scale inhibitor was obtained by dissolving aloe vera gel in the water at a concentration in the range 5–50% wt/wt. This solution contains polysaccharides [83] that can complex with Ca^{2+} ions. Some field tests were carried out on Venezuelan oil wells with water containing high bicarbonate ions (total calcium concentration of 535.4 mg L^{-1}). Information on the inhibitory performance of aloe vera is unfortunately limited. Weekly inspections of coupons performed during field tests with inhibitors (20 or 30 days of duration) led the authors to define the recommended concentration for the inhibitor. The aloe vera solution was reported to provide very effective scale inhibition with a concentration of 15.2 mg L^{-1}. Pressure and temperature were also recorded through the entire field test and remained almost constant during the tests. This indicated the absence of precipitated solids in the system in the presence of aloe vera.

5 Research on Scale Inhibitors in Algeria

Unfortunately, the work done so far on the scale and scale inhibitors in Algeria are not many, but we cannot fail to mention some researchers who worked on ambitious projects. Figure 10 shows the location of hard waters already studied in Algeria (Bordj Bou Arreridj, Setif, Constantine, Tebbessa, Aïn M'lila, and Ouargla).

Ghizellaoui et al. [84] studied the effect the temperature and the concentration of NaOH and K_3PO_4 inhibitors on hard water in Constantine (Hamma), as well as the effect of CaOH, Na_2CO_3, and KH_2PO_4 concentration as inhibitors on Fourchi drilling water [85].

A new green inhibitor, based on the aqueous extract of *Paronychia argentea* (PA), for the reduction of $CaCO_3$ formation on metal surfaces, have developed by Belarbi et al. [86]. They tested the PA extract at different temperatures and with the addition of several biocides. The results concluded that 70 ppm is necessary to completely inhibit scale at 20 and 45°C. However, its efficiency decreases at 60°C [86]. Another green inhibitor has been tested by Kahoul et al. [87] scaling power of Hammam drinking water. The addition of olive leaf extract to the Hammam water even at low concentration (20 ppm) prevents scale coverage of the surface, indicating that the extract can be used as a good antiscalant.

Karar et al. [88] have published an interesting paper; this paper focuses on the study of the glutamic acid (GA) for reducing $CaCO_3$ scale formation on metallic surfaces in the water of Bir Aissa region. This study showed that at 30 and 40°C, a complete scaling inhibition was obtained at a GA concentration of 18 mg L^{-1} with 90.2% efficiency rate. However, the efficiency of GA decreased at 50 and 60°C. This team also investigated the inhibitive effect of citric acid (CA), sodium citrate (SC), and their mixture (CA–SC) on the $CaCO_3$ scale. The electrochemical study showed that CA provides a slight inhibition of $CaCO_3$ deposit at a concentration of 70 ppm

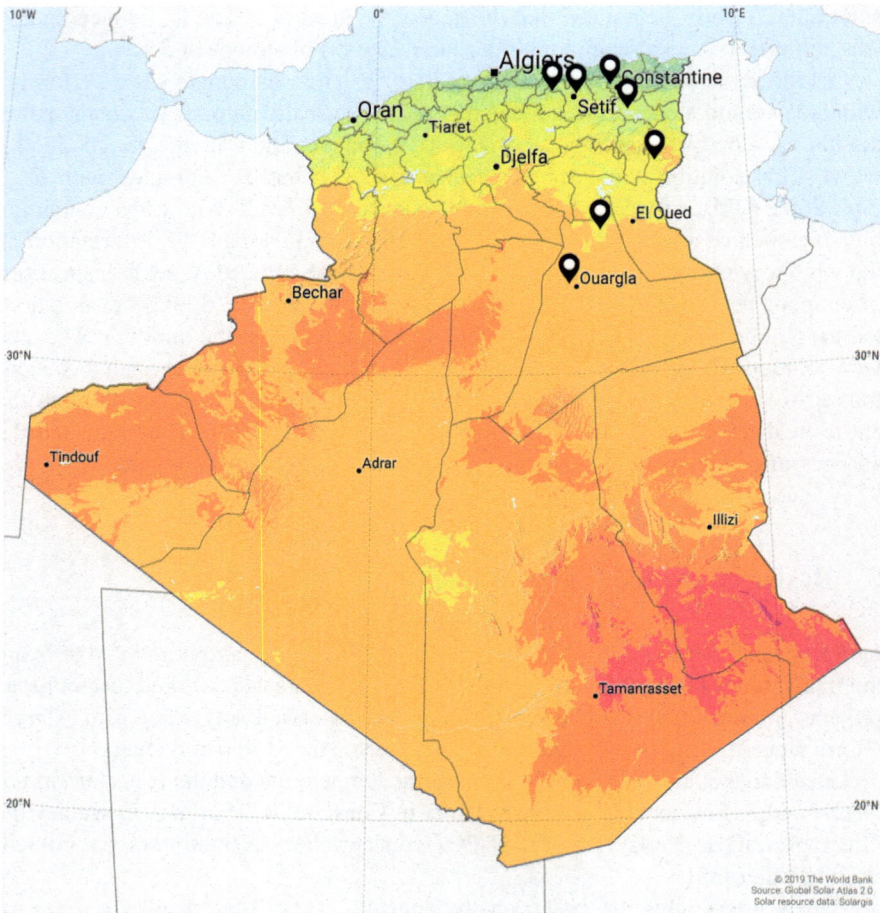

Fig. 10 Location of hard waters already studied in Algeria

on stainless steel surface. The use of SC alone inhibits very little formation of scale. The use of the mixture (50% of CA and 50% of SC) with small concentration led to a significant inhibition of the $CaCO_3$ formation [89, 90]. The works done so far on scale inhibitors in Algeria are given in Fig. 11.

The effect of mineral inhibitor on the precipitation of $CaCO_3$ in two Algerian groundwaters has been examined by Bendaoud et al. [91]. Chronoamperometry tests show that the antiscale treatment with mineral phosphates (KH_2PO_4) is more efficient for the Hamma (Constantine) than Negrine water (Tebbessa) with 2 mg/L. Tested inhibitors affect strongly the nucleation growth kinetics in the case of Hamma water and only the scale compactness in Negrine water. This was attributed to the difference in the chemical composition of the studied waters.

Djallal et al. [92] studied the electrochemical behavior of a carbon steel electrode against scale of barium sulfate. Tests conducted on water in the absence of scale

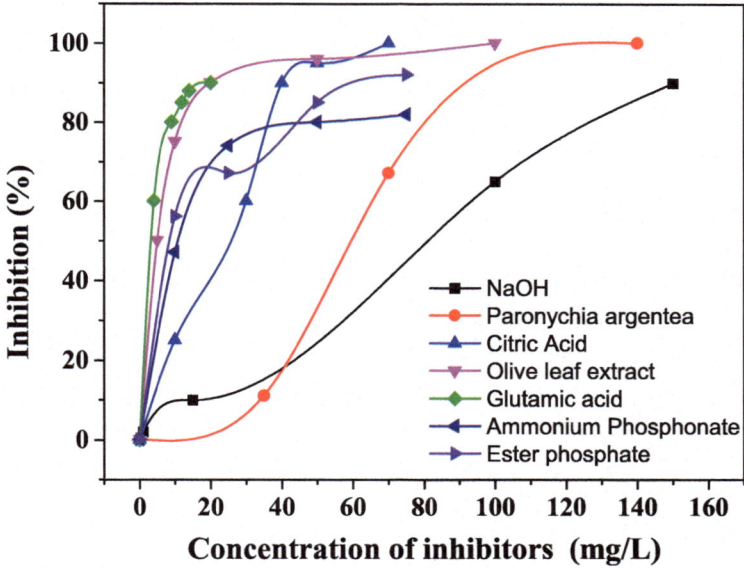

Fig. 11 The works done so far on scale inhibitors in Algeria [84–89]

inhibitors revealed the formation of a thick layer on the surface composed mainly of barium sulfate $BaSO_4$. Scale of barium sulfate naturally forms and evolves with immersion time. In the presence of inhibitors, the electrochemical impedance spectroscopy spectra show a decrease in the resistances and reveal that this inhibiting effect is a tendency to the formation of small quantities of precipitated barite solid on the surface of the electrode and that their effectiveness of inhibition increases with increasing concentration. This study shows that the inhibitor (phosphonate) is much effective at a low dose and at a high concentration. Its efficiency is limited because of the apparition of the micelles. However, the polyacrylate combined with phosphate ester has a satisfying inhibiting effect which increases with increasing of inhibitor concentration. Figure 10 shows the location of hard waters already studied in Algeria.

6   Conclusion

In recent years, many efforts have been made to generate green inhibitors, either from plant extraction or by using natural organic molecules. Green inhibitors obtained from natural products, especially by exploring the international pharmacopoeia, are clearly a growing field. Such green inhibitors could be advantageously used in situations where the use of organic materials would be limited by environmental regulations and/or application difficulties (i.e., toxicity). The Algerian researchers are not sparing themselves to find new inhibitors of nature and health, knowing that the country is full of inexhaustible resources and new discoveries are in prospect.

7 Recommendations

Scale in hard water is a major concern for industrial processes and domestic installations. Undesirable scale deposits cause many technical problems with serious safety and economic consequences, such as:

- Total or partial obstruction of pipes leading to a decrease in the flow velocity
- The reduction of heat transfer due to the precipitate of calcium carbonate whose conductivity is 15–30 times less than steel
- The clogging of the filters

Energy production in nuclear power plants is often limited by scale in cooling towers. In Great Britain, non-productive expenses related to scaling have been estimated at 600 £ million a year. These same expenses are about 1.5 billion euros per year in France. Therefore, it is important to establish appropriate methods to study this phenomenon and find effective ways to combat it.

This study aims to shed light on the phenomenon of scaling and inhibitors that can fight or reduce the formation of scale on industrial and domestic facilities, among the recommendations necessary to minimize this phenomenon:

- Make a complete study on the phenomenon of scaling in the entire territory of Algeria, and educate industry and the government on the risk of scaling and the use of harmful scaling inhibitors.
- Focus research on more economical and environmentally friendly green inhibitors based on toxicity, bioaccumulation, and biodegradation.
- Improve the approach to the study of scaling in Algeria, look for new hard water sites to test inhibitors, and reduce the risk of scaling.

References

1. Leroy P (1995) Mechanism of precipitation of calcium carbonate. Tribune de l'eau 573:31–46. (translated from French)
2. Ketrane R, Leleyter L, Baraud F, Jeannin M, Gil O, Saidani B (2010) Characterization of natural scale deposits formed in southern Algeria groundwater. Effect of its major ions on calcium carbonate precipitation. Desalination 262:21–30
3. Franck HUI (2011) Research on performance evaluation and anti-scaling mechanism of green scale inhibitors by static and dynamic methods, PhD thesis, Paris University, Paris
4. Amjad Z, Demadis ND (2015) Mineral scales and deposits (scientific and technological approaches). Elsevier, Amsterdam

5. Stumm W, Morgan JJ (1996) Aquatic chemistry, chemical equilibria and rates in natural waters. Wiley, New York
6. Antony A, Low JH, Gray S, Childress AE, Le-Clech P, Leslie G (2011) Scale formation and control in high pressure membrane water treatment systems: a review. J Membr Sci 383:1–16
7. Günther C, Becker A, Wolf G, Epple M (2005) In-vitro synthesis and structural characterization of amorphous calcium carbonate. Z Anorg Allg Chem 631:2830–2835
8. Pascal P (1958) New mineral chemistry treaty. Tome IV, Masson, Paris (translated from French)
9. Zidoune M (1996) Contribution to the knowledge of scaling mechanisms by various electro-chemical methods. PhD thesis, Paris VI University, p 160 (translated from French)
10. Dugniolle E (1994) The general scale problem in pipes. Tribune de l'eau 567:6–8. (translated from French)
11. Grasby S (2003) naturally precipitating vaterite (μ-$CaCO_3$) spheres: unusual carbonates formed in an extreme environment. Geochim Cosmochim Acta 67:1659–1666
12. Al Omari MMH, Rashid IS, Qinna NA, Jaber AM, Badwan AA (2016) Calcium carbonate. In: Profiles of drug substances. Excipients and related methodology, vol 41. Elsevier, Amsterdam, pp 31–132
13. Patel S, Finan MA (1999) New antifoulants for deposit control in MSF and MED plants. Desalination 124:63–74
14. Smith C, Coetzee PP, Meyer JP (2003) The effectiveness of a magnetic physical water treatment device on scaling in domestic hot-water storage tanks. Water SA 29:231–236
15. Bansal B, Müller-Steinhagen H (1993) Crystallization fouling in plate heat exchangers. J Heat Transfer 115:584–591
16. Trivedi TJ, Shukla J, Kumar A (2014) Effect of nitrate salts on solubility behaviour of calcium sulfate dehydrate (gypsum) in the aqueous sodium chloride system and physicochemical solution properties at 308.15 K. J Chem Eng Data 59:832–838
17. Amjad Z (2013) Gypsum scale formation on heated metal surfaces: the influence of polymer type and polymer stability on gypsum inhibition. Desalin Water Treat 51:4709–4718
18. Muryanto S, Bayuseno AP, Sedion W, Mangestiyono W (2013) Influence of flow rates and copper (II) ions on the kinetics of gypsum scale formation in pipes. Int J Technol 3:217–223
19. Strauss SD, Puckorius PR (1984) Cooling-water treatment for control of scaling, fouling, corrosion. Power 128:S1–S24
20. Shih WY, Rahardianto A, Lee RW, Cohen Y (2005) Morphometric characterization of calcium sulfate dihydrate (gypsum) scale on reverse osmosis membranes. J Membr Sci 252 (1–2):253–263
21. Glade H, Kromer K, Will S, Loisel K, Nied S, Detering J, Kempter A (2013) Scale formation and mitigation of mixed salts in horizontal tube falling film evaporators for seawater desalination. Proceedings of international conference on heat exchanger fouling and cleaning, Budapest, Hungry, June 9–14, 2013
22. Glade A, Ulrich J (2003) Influence of solution composition on the formation of crystalline scales. Chem Eng Technol 26:277–281
23. Andrews A, Davé B, López-Serrano P, Tsai SP, Frank R, Wilk M, Koutsakos E (2008) Effective scale control for sea water RO operating with high feed water pH and temperature. Desalination 220:295–304
24. Hermosilla D, Ordóñez R, Blanco L, De la Fuente E, Blanco A (2012) pH and particle structure effects on silica removal by coagulation. Chem Eng Technol 35:1632–1640
25. Ning RY (2010) Reactive silica in natural waters e a review. Desalin Water Treat 21:79–86
26. Vidojkovic S, Onjia A, Matovic B, Grahovac N, Maksimovic V, Nastasovic A (2013) Extensive feedwater quality control and monitoring concept for preventing chemistry-related failures of boiler tubes in a subcritical thermal power plant. Appl Therm Eng 59:683–694
27. Nishida A, Shimada Y, Saito T, Okaue Y, Yokoyama T (2009) Effect of aluminium on the deposition of silica scales in cooling water systems. J Colloid Interface Sci 335:18–23

28. Amjad Z, Koutsoukos PG (2014) Evaluation of maleic acid based polymers as scale as scale inhibitors and dispersants for industrial water applications. Desalination 335:55–63

29. García AV, Thomsen K, Stenby EH (2005) Prediction of mineral scale formation in geothermal and oilfield operations using the extended UNIQUAC model, part I. Sulfate scaling minerals. Geothermics 34:61–97

30. Dávilla-Pulido GI, Uribe-Salas A (2014) Effect of calcium, sulfate and gypsum on copper activated and non-activated sphalerite surface properties. Miner Eng 55:147–153

31. Lin W, Colin C, Rosset R (1990) Caractérisation du pouvoir incrustant d'une eau par chronoampérométrie au potentiel optimal d'entartrage. TSM-L'eau 12:613–620

32. Rosset R, Sok P, Poindessous G, Ben Amor M, Welha K (1998) Characterization of the compactness of calcium carbonate deposits in geothermal waters of southern Tunisia by impedancemetry. C R Acad Sci Paris t.1:751–759. (translated from French)

33. Tlili MM, Elfil H, Ben Amor M (2001) Chemical inhibition of scaling. Cahiers Eur Sci Assoc Water Health 6:29–39. (translated from French)

34. Trade Association of Research and Production (1993) Oil and natural gas commits technicians circuits seawater treatment and materials, vol 95. Editions Tech, Paris, pp 99–195. (translated from French)

35. Roques H (1989) Theoretical foundations of chemical treatment of water, vol 1, Edition lavoisier TEC et DOC 217–508. (translated from French)

36. Geffroy C, Persello J, Foissy A, Cabane B, Tournilhac F (1997) The frontier between adsorption and precipitation of polyacrylic acid on calcium carbonate. Rev Institut Fr pétrole 52:183–190

37. Guofeng C, Yao N, Aksay LA, Groves JT (1998) Biomimetic synthesis of macroscopic-scale calcium carbonate thin films, evidence for a multistep assembly process. J Am Chem Soc 120:11977–11985

38. Boggavarapu S, Chang J, Calvert P (2000) A test for mineralization inhibition for calcium salts using agarose hydrogels. Sci Eng C 11:47–49

39. Ghizellaoui S, Lédion J, Chibani A (2004) Study of the inhibition of the scaling power of Hamma waters by rapid controlled precipitation and an accelerated scaling test. Desalination 166:315–327. (translated from French)

40. Lédion J, François B, Vienne J (1998) Characterization of the scaling power of water by fast controlled precipitation. J Eur Hydrol 28:15–35. (translated from French)

41. Urion E, Lejeune G (1950) The role of zinc in inhibiting scaling. Revue de l'Eau 37(2):23–28. (translated from French)

42. Peters RW, Stevens JD (1982) Effect of iron as a trace imurity on the water softening process. AICHE Sympos Ser 78:46–67

43. Katz JL, Parsiegla IK (1995) Calcite growth inhibition by ferrous and ferric ions. In: Amjad Z (ed) Mineral scale formation and inhibition. Plenium Press, New York, pp 11–21

44. Katz JL, Reick MR, Herzog RE, Parsiegla KI (1993) Calcite growth inhibition by iron. Langmiur 9:1423–1430

45. Pernot B, Euvard M, Simo P (1998) Effet of iron and manganese on the scaling potentiality of water. J Water SRT-Aqua 47:21–29

46. Rosset R, Mercier D, Douville S (1997) La mesure du pouvoir entartant des eaux par méthodes électrochimiques et les procédés antitartre. Ann Fasl Exp Chim 90:41–65

47. Nancollas G, Kzmierczak G, Schuttringer E (1981) A controlled composition study of calcium carbonate crystal growth: the influence of scale inhibitors. Corrosion 37:76–81

48. Rosset R, Douville S, Ben Amor M, Welha K (1999) Inhibition of scaling by the geothermal waters of southern Tunisia. Rev Sci Eau 12(4):753–764. (translated from French)

49. Xu G, Yao N, Aksay LA, Groves JT (1998) Biomimetic synthesis of macroscopic-scale calcium carbonate thin films, evidence for a multistep assembly process. J Am Chem Soc 120:11977–11985

50. Shiliang H, Amy T, Thompson MB (1999) Inhibition of calcium carbonate precipitation in NaCl brines from 25 to 90 °C. Appl Geochem 14:17–25

51. Zidoune M, Khalil A, Sakaya P, Colin C, Rosset R (1992) Demonstration of the anti-encrusting effect of aminotri-(methylenephosphonic acid) by chronoamperometry and chronoelectrogrovimetry. C R Aca Sci Paris 315:795–799. (translated from French)

52. Ladel J, Leroy P (1997) Demonstration of the inhibitory effect of planktonic algae metabolites on the precipitation of calcium carbonate in natural waters of superficial origin. J Eur Hydrol 28:69–86. (translated from French)

53. Bailly JR, Couffin N, Lebugle A, Domingueza M, Revel JC, Roque H (1998) Contribution to the study of phenomena of scaling. Tribune de l'eau 592:3–36. (translated from French)

54. Abd-El-Khalek DE, Abd-El-Nabey BA (2013) Evaluation of sodium hexametaphosphate as scale and corrosion inhibitor in cooling water using electrochemical techniques. Desalination 311:227–233

55. Abd-El-Khalek DE, Abd-El-Nabey BA, Mervat A-KMA, Ramadan SR (2016) Investigation of a novel environmentally friendly inhibitor for calcium carbonate scaling in cooling water. Desalin Water Treat 57:2870–2876

56. Gao Y, Fan L, Ward L, Liu Z (2015) Synthesis of polyaspartic acid derivative and evaluation of its corrosion and scale inhibition performance in seawater utilization. Desalination 365:220–226

57. Touir R, Dkhireche N, Touhami ME, Bakri ME, Rochdi AH, Belakhmima RA (2014) Study of the mechanism action of sodium gluconate used for the protection of scale and corrosion in cooling water system. J Saudi Chem Soc 18:873–881

58. Huang H, Yao Q, Jiao Q, Liu B, Chen H (2018) Polyepoxysuccinic acid with hyper-branched Structure as an environmentally friendly scale inhibitor and its scale inhibition mechanism. J Saudi Chem Soc 23:61–74

59. Lui D (2011) Research on performances evaluation and anti-scaling mechanism of green scale inhibitors by static and dynamic methods. PhD thesis, Ecole nationale supérieur d'Arts et Métiers, Paris, pp 19–20

60. Hasson D, Shemer H, Sher A (2011) State of the art of friendly "green" scale control inhib itors: a review article. Ind Eng Chem Res 50:7601–7607

61. Sun Y, Xiang W, Wang Y (2009) Study on polyepoxysuccinic acid reverse osmosis scale inhibitor. J Environ Sci Suppl 21:S73–S75

62. Liu D, Dong W, Li F, Hui F, Lédion J (2012) Comparative performance of polyepoxysuccinic acid and polyaspartic acid on scaling inhibition by static and rapid controlled precipitation methods. Desalination 304:1–10

63. Schweinsberg M, Hater W, Verdes J (2003) New stable biodegradable scale inhibitor formulations for cooling water: development and field tests. 64th International Water Conference, Pittsburgh, PA, Oct 19–23

64. Girasa W, De Wispelaere M (2004) Polyaspartate, an New Alternative for the Conditioning of Cooling Water. 14th International Conference on the Properties of Water and Steam, Kyoto, Japan, Aug 29–Sep 3

65. Thombre SM, Sarwade BD (2005) Synthesis and biodegradability of polyaspartic acid: a critical review. J Macromol Sci 42:1299–1315

66. Gao Y, Liu Z, Zhang L, Wang Y (2010) Synthesis and performance research of biodegradable modified polyaspartic acid. Third international conference on bioinformatics and biomedical engineering, Beijing, China, Jun 11–12

67. Johannsen FR (2003) Toxicology profile of carboxymethyl inulin. Food Chem Toxicol 14:49–59

68. Verraest DL, Peters JA, Bekkum H, Rosmalen GM (1996) Carboxymethyl inulin: a new inhibitor for calcium carbonate precipitation. J Am Oil Chem Soc 73:55–62

69. Stevenson FJ (1994) Humus chemistry: genesis, composition, reactions. Wiley, New York

70. Hoch AR, Reddy MM, Aiken GR (2000) Calcite crystal growth inhibition by humic substances with emphasis on hydrophobic acids from the Florida Everglades. Geochim Cosmochim Acta 64:61–72

71. Kumar T, Vishwanatham S, Kundu SS (2010) A laboratory study on pteroyl-L-glutamic acid as a scale prevention inhibitor of calcium carbonate in aqueous solution of synthetic produced water. J Pet Sci Eng 71:1–7

72. NACE Standard TM 0197-97 (1997) Laboratory Screening Test to Determine the Ability of Scale Inhibitors to prevent the Precipitation of Barium Sulfate and/or Strontium Sulfate from Solution (for Oil and Gas Production Systems), Item No 21228, NACE International

73. Amjad Z (1995) Mineral scale formation and inhibition. Plenum Press, New York

74. Gauthier G, Chao Y, Horner O, Alos-Ramos O, Hui F, Lédion J, Perrot H (2012) Application of the fast controlled precipitation method to assess the scale-forming ability of raw river waters. Desalination 299:89–95

75. Reddy MM, Hoch AR (2001) Calcite crystal growth rate inhibition by polycarboxylic acids. J Colloid Interface Sci 235:365–370

76. Abdel-Gaber AM, Abd-El-Nabey BA, Khamis E, Abd-El-Khaled DE (2011) A natural extract as scale and corrosion inhibitor for steel surface in brine solution. Desalination 278:337–342

77. Bonoli M, Bendini A, Cerretani L, Lercker G, Tosci TG (2004) Qualitative and semiquantitative analysis of phenolic compounds in extra virgin olive oils as a function of the ripening degree of olive fruits by different analytical techniques. J Agric Food Chem 52:7026–7032

78. Maciejewska G, Zierkiewicz W, Adach A, Kopacz M, Zapala I, Bulik I, Cies'lak-Golonka M, Grabowski T, Wietrzyk J (2009) A typical calcium coordination number: physico-chemical study, cytotoxicity, DFT calculations and in silico pharmacokinetic characteristics of calcium caffeates. J Inorg Biochem 103:1189–1195

79. Lee OH, Lee BY, Lee J, Lee HB, Son JY, Park CS, Shetty K, Kim YC (2009) Assessment of phenolics-enriched extract and fractions of olive leaves and their antioxidant activities. Bioresour Technol 100:6107–6113

80. Abdel-Gaber AM, Abd-El-Nabey BA, Khamis E, Abd-El-Rhmann H, Aglan H, Ludwick A (2012) Green anti-scalant for cooling water systems. Int J Electrochem Sci 7:11930–11940

81. Castillo LA, Torin EV, Garcia JA, Carrasquero MA, Navas M, Ailoria A (2009) New product for inhibition of calcium carbonate scale in natural gas and oil facilities based on Aloe Vera: application in Venezuelan oilfields. Latin American and Caribbean Petroleum Engineering Conference, Cartagena de Indias, Colombia, 31 May–3 June

82. Viloria A, Castillo L, Garcia JA, Carrasquero Ordaz MA,Torin EV (2011) Process using Aloe for inhibiting scale. US Patent US 8039421 B2

83. Woodward C, Davidson EA (1968) Structure-function relationships of protein polysaccharide complexes: specific ion-binding properties. Proc Natl Acad Sci U S A 60:201–205

84. Menzri R, Ghizellaoui S (2012) Chronoamperometry study of the Inhibition of groundwater scaling deposits in Fourchi. Energy Procedia 18:1523–1532

85. Labiod K, Ghizelloui S (2012) Contribution to the inhibitors methods study of the scaling: chemical, electrochemical processes in the presence of Ca(OH)$_2$, Na$_2$CO$_3$ and KH$_2$PO$_4$. Energy Procedia 18:1541–1556

86. Belarbi Z, Gamby J, Makhloufi L, Sotta B, Tribollet B (2014) Inhibition of calcium carbonate precipitation by aqueous extract of Paronychia Argentea. J Crys Growth 386:208–214

87. Aidoud R, Kahoul A, Naamoune F (2017) Inhibition of calcium carbonate deposition on stainless steel using olive leaf extract as a green inhibitor. Environ Technol 3:14–22

88. Karar A, Naamoune F, Kahoul A, Belattar N (2016) Inhibitory effect of glutamic acid on the scale formation process using electrochemical methods. Environ Technol 37:1996–2002

89. Karar A, Naamoune F, Kahoul A (2016) Chemical and electrochemical study of the inhibition of calcium carbonate precipitation using citric acid and sodium citrate. Desalin Water Treat 57:16300–16309

90. Karar A, Naamoune F (2018) Inhibition of calcium carbonate precipitation by citric acid. Mater Biomater Sci 1:019–023

91. Bendaoud YB, Ghizellaoui S, Tlili M (2012) Inhibition of CaCO$_3$ scale formation in ground waters using mineral phosphates. Desalin Water Treat 38:271–277

92. Labraoui-Djallal K, Bounoughaz M (2016) Evaluation efficiency of barium sulfate scale inhibitors by electrochemical impedance spectroscopy. Int J Electrochem Sci 11:1777–1788

Part VI
Conclusions

Update, Conclusions, and Recommendations for "Assessment of Surface and Groundwater Resources in Algeria"

Abdelazim Negm, El-Sayed Ewis Omran, and Damia Barcelo

Contents

Abstract The present situation in Algeria's water resources is framed by problems related to water sustainability. Natural resources are at the core of Algeria's sustainable development and are critical to socioeconomic growth. This chapter captures

A. Negm (✉)
Water and Water Structures Engineering Department, Faculty of Engineering, Zagazig University, Zagazig, Egypt
e-mail: amnegm85@yahoo.com; amnegm@zu.edu.eg

E.-S. E. Omran
Soil and Water Department, Faculty of Agriculture, Suez Canal University, Ismailia, Egypt

Institute of African Research and Studies and Nile Basin Countries, Aswan University, Aswan, Egypt
e-mail: ee.omran@gmail.com

D. Barcelo
ICRA, Catalan Institute for Water Research, Girona, Spain
e-mail: dbcqam@cid.csic.es

Abdelazim M. Negm, Abdelkader Bouderbala, Haroun Chenchouni, and   321
Damià Barceló (eds.), *Water Resources in Algeria - Part I: Assessment of Surface and Groundwater Resources*, Hdb Env Chem (2020) 97: 321–336,
DOI 10.1007/698_2020_563, © Springer Nature Switzerland AG 2020,
Published online: 8 July 2020

the assessment of surface and groundwater resources in Algeria (in terms of findings and suggestions) and provides ideas extracted from the volume cases. In addition, some updated findings from a few recently published research works related to the water resources covered themes are presented. This chapter offers a number of suggestions to protect the resources for the current issues experienced by Algeria.

Keywords Algeria, Assessment, Deserts, Environment, Groundwater, Resources, Surface water, Sustainability

1   Introduction

Water is one of Algeria's most significant raw materials. Because of its significance, in the context of sustainable management of this invaluable resource for the future, Algeria has taken quantitative and qualitative adaptive steps. Algeria has substantial surface water resources primarily in the country's north where rainfall is more favorable. An important part is already captured and exploited for national use, industry, and agriculture by a number of medium and big dams. Before the first dams were built in the early twentieth century, groundwater was the only reliable source of water available, either in the form of springs or shallow wells. The rest came as spate irrigation from surface waters diverted from seasonal rivers. Using shallow wells and springs, shallow renewable aquifers have been recognized. Poor management is associated with inadequate resource understanding, proliferation of illicit wells, and bad coordination among the multiple groundwater-responsible authorities. Groundwater resources are considered to be a much more reliable source of water than surface water and are considered to be the primary source of water during the drought. It is easier to access and is invaluable as a source of drinking and irrigation water of excellent quality.

This book holds much promise and potential for water resources in Algeria. This book addresses the question of how science and technology can be mobilized to make that promise come true. Therefore, the book intends to address the following main themes.

– Current Status of Water Resources
– Climate Change Impact and Hydrogeological Investigations
– Evaluation of Evapotranspiration Models
– Aquifer Characterization and Assessment of Groundwater Resources
– Toward a Sustainable Development

The next section provides a summary of the significant results of some of Algeria's latest (updated) water resource research. Then, the main suggestions for scientists and decision-makers and the key findings of the book chapters are summarized. The update, conclusions, and recommendations presented in this chapter come from the data presented in this book.

2   Update

The following are the major update for the book project based on the main book theme:

2.1   Current Status of Water Resources

Three chapters are identified in the book related to the current status of water resources. The first study discusses the origin and quality of groundwaters of the Taoura syncline aquifer (northeastern Algeria). The study concerns the Taoura scheme, defined by the interference in the syncline of two aquifers. A sedimentary formations porous medium aquifer, which also overlays a karstic/fractured aquifer. The covers filling the syncline hide the karstic formation. The groundwaters of the Taoura syncline aquifer are stored in both the above aquifers. Also, the first level is porous, and characterizing the sheet contained in the Mio Plio Quaternary formations against the second medium is of the cracked type and refers to the deep karstic sheet. The karst aquifers are little developed in the Maghreb. In Algeria, the karst has a limited extension. It is present in some regions in the north such as Tlemcen [1]. Groundwater from Taoura karst aquifer is used to supply drinking water to several localities and to irrigate numerous perimeters. In this way, karst groundwater is subject to a double risk, the first is related to its overexploitation, and the second is due to the exposure of the groundwater to pollution by return flows from irrigation.

The second assesses the impact of toxic metals on water quality around an Abandoned Iron Mine, Bekkaria, Algeria. Many findings have shown that deposits of slag heaps and mine waste around the mine are also a significant cause of contamination and can be readily mobilized [2, 3]. The mines are generally abandoned without rehabilitation after the cessation of the operation, which leads to an extension of the pollution process. So, the quality of the water is threatened even after the cessation of exploitation. Once stopped mining no initiative of environmental protection was taken. This had reflected negatively on the environment; indeed during long years, the spoil heaps remained deposited on the soil surface, upstream of the wadi and the aquifer, directly exposing to the effects of pollution. This research evaluates the water quality of the wadis and wells in order to highlight the impacts of these spoil heaps on the water quality of this region [4].

The third is to evaluate the impacts of pesticides on soil and water resources in Algeria. According to the United Nations research, more than half of the world's population growth is expected to happen in Africa by 2050 [5]. The region is poorly endowed with two important natural resources, although blessed with big petroleum and gas reserves: productive land and affordable renewable water resources. Only 6% of the land in the region is arable, and there is restricted to freshwater supply accessible [6]. Pesticides (herbicides, insecticides, fungicides, etc.) are vital instruments for agriculture; they assist combat damaging insects and weed and thus contribute to economic food production in large numbers [7]. By comparison, if these herbicides are misused, their residues can be very harmful to soil, water, and the environment and ultimately to human health. This is because most insecticides are persistent and, therefore, toxic owing to their lipophilic characteristics [8]. Glyphosate and 2.4-D, particularly in the irrigable perimeter of Bou Namoussa from 1968, are the most common herbicides in Algeria [9]. The importance of fertilizing components like nitrogen and assimilable phosphorus greatly affects crop output. For these reasons, studying the effect of herbicides on soil and water is essential.

2.2 Climate Change Impact and Hydrogeological Investigations

Two approaches are identified in this book, which correlated climate change impact and hydrogeological investigations. The first study identified is the analysis of flood characteristics in the context of climate variability in northern Algeria: case of the Cheliff watershed. Floods are one of the basic characteristics of a watercourse system. Unfortunately, to draw global conclusions, a lengthy sequence of flood information is not available. Many studies on the genesis and the danger of the flood have been carried out for a few years in the world [10, 11]. Flooding is a common environmental hazard and a leading cause of natural disaster fatalities and economic damages worldwide. The study of the Algerian watercourse floods remains a quasi-unknown field due to the very limited specific indications about data that given in the Algerian hydrological directories [12].

The second study recognized is assessing the climate change impact on water resources and adaptation strategies in the Algerian Cheliff basin. In specific, Algeria has suffered a decline in annual average rainfall. It is anticipated that rainfall will continue to decline in the next century [13]. This situation is particularly noticeable in regions subject to a semiarid climate regime. This is particularly the case in the Cheliff basin, which is one of the largest basins in northern Algeria. It is affected by water shortage due to the expansion of industrial and agricultural activities with population growth, on the one hand, and the reduction of water resources caused by extreme droughts, on the other hand. Recent studies over the past few years have revealed a declining rainfall trend in most Algerian regions [14] and a significant decrease in flow to dams and significant groundwater level drop and high vulnerability to groundwater pollution [15]. At present, the major concern of the country is to predict, with scientifically accepted margins of uncertainty, the potential impacts of climate change predicted by the IPCC on water resources. In this context, it is necessary to put in place sound adaptation strategies aimed at minimizing the negative impacts that climate change would bring in the future.

2.3 Evaluation of Evapotranspiration Models

Three potential practices for evaluation of evapotranspiration models are identified. The first is the assessment of projected precipitations and temperatures change signals over Algeria based on regional climate model – RCA4 simulations. Algeria is the largest African and Mediterranean country. It is located in the southern seashores of the Mediterranean Sea. The future rainfall evolution may be critical for human activities since increased temperatures may further exacerbate droughts and water shortages. Given the ongoing aridification and/or sometimes very abrupt climate change, the current distribution of climate conditions at the global scale will be reorganized. Some climates will disappear completely, while others will appear in some regions. This region, which includes Algeria, has been considered as the region for which there is the widest consensus between projections and model types used by the Intergovernmental Panel on Climate Change (IPCC) about future decreases in total rainfall. However, IPCC model resolution, which ranges from 100 to 200 km, does not allow a sufficient level of regional detail, and a higher resolution can be achieved by using regional climate models. Two scenarios are primarily used for future projections, namely, RCP4.5 and RCP8.5. Since the precipitations and temperatures are used to define climate zones, the future change of these two climate variables is also evaluated by the shift in surface area of each climate zone as defined in Köppen–Geiger classification [16].

The second is the comparison of evolving connectionist systems (ecos) and neural networks for modeling daily pan evaporation from Algerian dams' reservoirs. Evaporation (*EP*) from dams' reservoirs measured using pans is one of the most important methods adopted for quantifying the *loss of water* through *evaporation*. *Black-box artificial intelligence techniques (AI) have been developed as alternative approaches for quantifying evaporation, and several kinds of models have been proposed worldwide.* Some other important investigations can be found in the literature which highlights the importance of data-driven models in estimating pan evaporation [17]. Recently, Eray et al. [18] introduced two machine learning methods at the first time in the area of pan evaporation modeling: (1) evolving connectionist systems, the dynamic evolving neural-fuzzy inference systems named (DENFIS), and (2) multi-gene genetic programming (MGGP). To the best of our knowledge, only the study conducted by Eray et al. [18], *no* others *investigations* have addressed the application of any evolving connectionist systems (ECoS) for modeling daily EP.

The third is the new formulation for predicting daily reference evapotranspiration (et$_0$) in the Mediterranean region of Algeria country: optimally pruned extreme learning machine (OPELM) versus online sequential extreme learning machine (OSELM). This chapter aims to investigate the capabilities and usefulness of two new data-driven techniques: optimally pruned extreme learning machine (OPELM) and online sequential extreme learning machine (OSELM) newly applied and compared for predicting daily reference evapotranspiration (ET$_0$) in the Mediterranean region of Algeria. Nowadays, reference evapotranspiration (ET$_0$) is one of the

most important components of the hydrological cycle that has received great importance and has paid great attention by researchers worldwide. Recently, Wu and Fan
[19] proposed the extreme gradient boosting (XGBoost) model for modeling daily
ET_0 in China. In another study, Huang et al. [20] introduced the CatBoost model as a
new gradient boosting decision tree for modeling daily ET_0 in china.

2.4 Aquifer Characterization and Assessment of Groundwater Resources

Two approaches related to aquifer characterization and assessment of groundwater
resources are identified. The first approach is related to studying the water resources
in coastal aquifers of Algeria face climate variability: case of alluvial aquifer of
Mitidja in Algeria. Algeria is considered as a vulnerable country in the world
regarding its water resource availability, especially in front of the changing climate
conditions. The water supply is the main task challenge of the public institutions
under the severe natural conditions of climate variability represented by the decrease
of rainfall with the increase of evaporation and also the different anthropogenic
pollutions. There has been a lot of concern about the impact of climate variability on
the socioeconomic activities in Algeria since the 2000s [21]. The potential impacts
of climate variability on water resources and food security are receiving growing
attention from some researchers, especially in arid and semiarid regions that face of
high water demands for agricultural, domestic, and environmental uses [22]. Understanding climate variability is vital to the ecosystems and our environment, particularly with regard to the changes affecting the sustainability and availability of
groundwater resources [21].

The second approach is associated with the assessment of groundwater resources
in the Jurassic Horst (Western Algeria). In Algeria, karst aquifers play a very
important role in supplying the country's largest springs with water. For thousands
of years, they have fed many cities and countless villages during the summer season,
when most wadis dry up. In western Algeria, surface water resources are very
limited, as rainfall is scarce. It is, therefore, tempting to try to mobilize groundwater
resources. This is not possible everywhere, as the geological context is often
unfavorable. The most extensive outcrops consist in Cretaceous and Cenozoic
clay, marl, and marly limestone. Such types of rock are not suitable for groundwater
abstraction. Hydrogeological prospecting focuses on the reliefs where the Jurassic
geological formations, and in particular the limestones and dolostones of the Malm,
are exposed. This article focuses on the karst hydrogeology, water reserves, vulnerability, and constraints for groundwater resource management in the western section
of Oranese Meseta: the Mounts of Tlemcen which extend over 4,000 km^2, 50% of
which are limestone and dolomite outcrops. These aquifers were used to bring about
significant improvements in urban water supply during the 1980s and the 1990s.
However, increasing population growth and urbanization are now jeopardizing the
sustainable management of the aquifers, and there is also a need to ensure water

resource protection. Sustainable water resource management means making the best use of the various water resources available (dams, wells, wastewater reuse, and desalination) and considering the specific value of the karstic aquifers in the regional context (good quality water, low turbidity, huge storage volume, limited investment needs).

2.5 Toward a Sustainable Development

Two methodologies are used toward sustainable development. The first method is the scale inhibition in hard water system. The precipitation of an insulating layer of scaling on the walls of the water distribution pipes has serious technical and economic consequences. Various methods were used to prevent the scale formation in water such as the chemical methods in which the germination of the $CaCO_3$ crystals is blocked using the inhibitor. In recent years, a few studies have been focused on the aspects of the surface scaling so that the different mechanisms were proposed to explain the differences between the scaling precipitation in bulk solution and scale deposition at the surface. The water distribution of some Algerian town resulting from the drilling water is supersaturated with respect to calcium carbonate. This causes reducing heat transfer in heat exchangers systems, limiting the efficiency of these devices (valves and taps) by decreasing the flow rate in the pipes; this phenomenon is more prevalent at high heating temperature. The primary agents responsible for crystallization fouling are the carbonates and sulfates of calcium or magnesium, barium salts, silicate, and phosphate. Scaling is essentially linked to the formation of calcium carbonate ($CaCO_3$). Scaling may contain other residues such as algae, calcium sulfate, clays, and the brucite $Mg(OH)_2$. But it is always calcium carbonate that precipitates first, usually in the colloidal form, because its solubility is lower than that of others. Scaling in natural hard water is a major concern in different facets of industrial processes and domestic installations. Calcium and magnesium carbonates and sulfates are the main agents responsible for the fouling of crystallization, although barium salts, silicates, and phosphate deposition play an important role in some industries. Deposits formed by water, usually observed, include calcium carbonate, calcium and barium sulfate, silica deposits, iron deposits, and magnesium and calcium phosphates. If we extend our area of interest to wastewater, struvite is the main cause of the mineral scale [23].

The second method is the participatory approaches to sustainable development and management of soil resources in the arid zones of Algeria. For sustainable development in arid zones, the management of soil resources is essential; this resource management includes several parameters, among others the evaluation of soil fertility and the production of chemical fertility maps of the studied soils. To carry out this work, we chose an irrigable perimeter of Tadjmout (LAGHOUAT). The objective of the study is to realize thematic fertilization maps and to evaluate the chemical fertility of the soils of the irrigable perimeter of Tadjmout area (Wilaya of Laghouat), for the management of soil resources by the participative approach to the sustainable development in arid zones of Algeria. On the environmental front, this initiative aims to preserve soil resources and protect water resources. The study area is part of the group of pastoral areas of Algeria, where the agriculture is considered as one of the main sectors in this region with a total agricultural area of 20,087.06 Ha [24].

3 Conclusions

Throughout the course of this book project, the editorial teams achieved several findings drawn from this book. The chapter draws significant lessons from the book cases in relation to methodological concepts, specifically the water resources covered topics in Algeria. This chapter provides the present problems faced by the water resources in Algeria. These results are essential to improving water resources in Algeria. The following results could be stated on the basis of the materials mentioned in all parts of this volume:

1. The research concern an area characterized by the presence of two superimposed aquifers, of which the first is porous and the second (deeper) is fractured and karstic. The top of karst formation is, indeed, sometimes close to the ground surface, sometimes very far from it. The presence of faults favors the communication of the deep karst with the surface and the superficial aquifer, whose waters can be drained into the karst aquifer. The piezometric maps carried output in evidence a recharge of the superficial aquifer by the borders. This lateral contribution influences the chemical composition of the groundwaters of both the aquifers. The doubt about the hydraulic connections, which could not be resolved by hydraulics, led to examining the water quality in order to define the origin of groundwaters of the karst aquifer. Results of the PCA elaborations prove an opposition between the calcium bicarbonate waters at the origin of the observed mineralization and the rest of the elements. The interpretation of cross-plots related the main ions shows good correlations between HCO_3^- + SO_4^{-2} and Ca^{2+} + Na^+ and between Na^+ and Cl^-. The presence of sodium in water is linked to the process of ion exchange. This tendency is confirmed by the BEI indices. The present study revealed that the water quality of the study area remains influenced by natural factors, particularly the geological formations present at the zone level. We also noticed a decrease in water levels in the wells. This is explained by the draining and or climatic hazards.

2. The conclusions of the work concern the effects of the spoil heaps deposited upstream of a Wadi, and an aquifer system is presented. The taking away carried out went up that water of the wadis, and the surface aquifer is charged in ETM. The concentrations observed in water in the wadis remain however very high compared to water of the wells. This distribution would be due to the trapping of the ETM which are made at the level of soils separating the two levels from water. To confirm the origin of the ETM, the Sr^{2+}/Ca^{2+} ratio was studied to show the influence of the gypsiferous formations on the water quality. The results obtained by the mathematical model carried out confirm this relation well.

3. Algeria, with its arid and semiarid climate, is highly vulnerable to climate change with desertification as a major concern. Overall, temperature and evaporative demand are expected to increase for Algeria. In Algeria, even if global climate change is not taken into consideration, water shortages are a significant acuity issue in many parts of the nation. Agricultural research in Algeria will face many difficulties mainly determined by the momentum produced by economic reforms, which are themselves conditioned by two main limitations: climate changes and its effects such as growing drought and flooding and water scarcity. From now on, the last will be a significant determinant in defining all the elements of Algeria's food safety policy. The globalization of the economy causes powerful entropy on the markets and excellent instability of the prices of agricultural products due to both political and climatic hazards. Of course, some constraints worsen the current situation in Algeria: first, a mismatch between the criteria and the available resources. A geographical imbalance between requirements and resources. Second, rupture of the risk of sustainable development.

4. The study of the Algerian wadis floods remains a quasi-unknown field as only it is the first one carried out in northern oust of Algeria. It is based on the flow data recorded at the gauging stations of Cheliff watershed, which are available and considered in the present study. The statistical study of the recorded flows of the different stations localized in the watershed of Cheliff shows the highest values of flows which were recorded in the station of Pontebba (1,300 m^3/s) and in the station of Ain Hamara (878 m^3/s). Also, the most remarkable events were observed during the 1970s for the stations of Arib Cheliff, Rahouia, and Djnane Ben Ouadha, the 1980s for the station of Pontebba, and finally during the 1990s the stations of Ain Amara, Sidi AEK Djilali, and El Ababsa. We report here that the autumn floods characterize the 1990s and the winter floods for the 1980s. In general, the different hydrometric stations show that there are two distinct periods in terms of duration and flood power. The first is the most important in terms of duration; it characterizes the spring season, with few floods sometimes. The other period is very short. It corresponds to the autumn season. It is characterized by flash floods. The flow is very abundant in March and October in second place. The analysis of rainfall and the monthly flows showed a relatively temporal concordance, where the rainiest months are usually the most abundant inflow. In this region of upper and middle Cheliff, the average maximum flow reaches 400 m^3/s, if the rainfall is around 500 mm. The fact of precipitation falls almost on the whole Basin, so the flows are directly influenced by certain parameters like the state of the soil, the vegetation cover, and the air temperature.

5. Climate change in the study area has had an adverse effect on the precipitation cycle, including all water resources. The reduction in precipitation produced a downward trend in water inflow as shown by recording annual decreases from 1,025 to 815 Mm^3 between 1968–2001 and 2009. The consequences of water shortages are changes in the environmental balance which will consequently affect various human activities, especially the available water supply for

domestic and industrial consumption as well as for the agricultural economy. Climatic scenarios agree on a decrease in annual precipitation, averaging between 10% and more than 30% at the end of the twenty-first century. These results reflect the availability of surface water resources, which will tend to decrease, with longer and more severe periods of low water. The Cheliff Basin is considered particularly vulnerable to acute water scarcity in the coming years. A large water deficit due to population growth and an increase in water demand by different sectors of the economy is also expected (e.g., agriculture, industry). Faced with these challenges, the country has implemented a new water resource management policy through the construction of new dams and use of unconventional water resources. It is hoped that the current investigation will help policymakers to make better decisions in developing water management strategies for the watersheds of Algeria.

6. The spatial variability and temporal evolution of precipitation and temperature over Algeria were analyzed over 1951–2100 using a set of observational data and nine RCM RCA4 simulations from the CORDEX-Africa program. The analysis is done at annual and monthly time scales that lead to the three main findings: First, over the historical period 1951–2005, the long-term trends of precipitation and temperature are characterized by an increase in annual mean temperature of about +0.02°C/year, in the western part of the country and of +0.04°C/year, in the eastern part, and in the same time a 0.5–1.5 mm/year decrease in annual mean precipitation in the northern part of the country. Second, as far as the future evolution (2005–2100) of the temporal variability of annual and monthly precipitation and temperature in Algeria is concerned, the study shows that all models project an increase in temperature and a decrease in precipitation during the 1945–2100 periods especially under RCP8.5 scenario. Finally, the current decrease in precipitation and increase in temperature and the anticipated shrinking of the surface area of the temperate climate zone will lead to numerous problems related, among other things, to food security and displacement of local populations in Algeria. These considerations must be included in future socioeconomic development plans. The rate at which such changes will occur in the future in Algeria's three climate zones should deserve special attention.

7. The proposition and development of robust models are the main objectives of any modeling strategies, and it is necessary to have a model with sufficient accuracy. Until now, several kinds of models have been proposed for modeling daily and monthly pan evaporations. The existing models, which have proved their effectiveness in practice, must also be followed with other modeling approaches. To keep up with the continuously growing demand for models for the nonlinear and complex process, neurofuzzy approaches have proven their efficiency. Starting from this statement, a new kind of neurofuzzy models called DENFIS is proposed for modeling daily EP from dam reservoir of Algeria, using easily measured daily climatic variables such as T_{max}, T_{min}, RH, and W_S. The proposed DENFIS model is compared to the standard MLPNN and MLR. According to the obtained results, a number *of major* conclusions can be

drawn. Firstly, compared to the standard MLPNN and MLR, we have obtained only a marginal improvement of the results for all stations, and overall, MLPNN was found to be the best model. Secondly, the best performance in validation phase was obtained using the MLPNN model with only three input variables: T_{max}, T_{min}, and RH. However, at Jijel station, the best performance was obtained using MLPNN with four input variables. Thirdly, DENFIS_OF model provided better accuracy for all input combination compared to DENFIQ_ON model. *Fourthly, and finally,* as the models were applied only for two stations, in the future, extending the investigation to other stations can help to draw more robust conclusions.

8. A new modeling approach of predicting daily ET_0 in the Mediterranean region of Algeria is proposed. The proposed model is based on the use of newly data-driven paradigm proposed during the last few years and called the extreme learning machines (ELM). One of the novelties of the present work is the inclusion of the Julian day as an input variable, and the performances of the models were improved a lot. Although the ET_0 can be estimated in different ways, through direct measurement *using a lysimeter or using empirical and semi-empirical formulas, the* use of data-driven models that use climatic variables as inputs is very welcome. Another important conclusion of the present study is that the OP-ELM provides more accurate estimates than the OS-ELM for all four stations.

9. Water resources in Algeria, particularly in the Mitidja plain, are limited, vulnerable, and unequally distributed spatially. This sensitive situation inevitably requires new actions to exploit these resources in a rational way. The purpose of this work is to study the spatial variability of some chemical properties of groundwater in the Mitidja plain by a parametric approach of the quality index in order to evaluate the quality of water for drinking and irrigation purposes, in a relationship with climate variability. Analysis of the chemical data showed the predominant of groundwater facies: sodium chloride, calcium bicarbonate, and mixed facies. In addition, the salinity of the water in this plain varies from average to very high salinity. It can be said that the groundwater in this aquifer is facing a huge risk of nitrate pollution and seawater intrusion and becomes unfit for domestic and agricultural use. It appears that the northeastern part of Mitidja is the most vulnerable to the salinity because the contents of some elements greatly exceed the international standards set by WHO. Thus, this work will be a statement of place, and it can be a tool for decision-making regarding the exploitation of this resource from the groundwater quality and the risk of contamination, whether for human health.

10. The Tlemcen Mountains conceal nearly 2,000 km^2 of limestone and dolomite outcrops from the Jurassic and Cretaceous periods which dominate the surrounding areas and thus constitute a regional water tower, better watered than the Tafna plain or the high plains of El Aricha. The deep aquifers of the Tlemcen Mountains contain significant permanent reserves (estimated at over 6 billion m^3) and benefit from significant recharge (200 million m^3 per year). They thus constitute the main groundwater resources to the west of the Oran meridian. Deep drilling in well-karstified aquifers presents specific difficulties (total loss

of drilling mud, absence of cuttings), but these difficulties were overcome, and 270 boreholes were drilled, leading, during the 1980s and 1990s to very intensive water abstraction, aiming to supply water to the Wilaya of Tlemcen and Ain Temouchent and even part of the Wilaya of Oran. Such intensive abstraction has led to an excessive piezometric level drawdown for some aquifers. The recent commissioning of two large seawater desalination plants has made it possible to reduce groundwater withdrawals and stabilize and even restore piezometric levels. The water quality of these aquifers is good, but it should not be concluded that they are not very sensitive to pollution. On the contrary, the maps we have produced (using the COP and the RISK methods) clearly show the general vulnerability of these aquifers. If no large-scale pollution has yet been recorded, it is because the karstic areas are relatively sparsely inhabited. This situation could change with the demographic and economic growth of the Tlemcen Wilaya.

11. In recent years, many efforts have been made to generate green inhibitors, either from plant extraction or by using natural organic molecules. Green inhibitors obtained from natural products, especially by exploring the international pharmacopeia, are clearly a growing field. Such green inhibitors could be advantageously used in situations where the use of organic materials would be limited by environmental regulations and/or application difficulties (i.e., toxicity). The Algerian researchers are not sparing themselves to find new inhibitors of nature and health, knowing that the country is full of inexhaustible resources and new discoveries are in prospect.

12. The soils of the Tadjmout perimeter belong to the arid zone. The analytical results show that these soils are at alkaline pH cation exchange capacity (CEC) relatively average, which does not exceed 15 Cmol/kg, low to medium and sometimes high salinity (EC) and exchangeable phosphorus: the values are between 26.62 and 34.87 ppm; this high content of exchangeable phosphorus could be due to phosphate fertilizers; exchangeable potassium has average to low values. These soils require fertilization to improve their chemical fertility. The thematic maps allow us to manage a fertilization plan to improve the chemical fertility of these soils, in order to increase agricultural production. It is imperative for the agricultural exploitation of arid soils in the context of sustainable development and further research based on experimentation and monitoring over a long period.

4   Recommendations

The ability to adapt to future issues is a key component of Algeria's water resources. We contend that to accomplish this objective, water resources need integrated flexibility. The editorial teams observed certain aspects that could be explored for further enhancement throughout the course of this book project. Based on the contributors' results and findings, this chapter offers a number of recommendations that provide suggestions for future researchers to go beyond the scope of this book.

1. The abandoned Bekkaria mine in Algeria is considered as a source of pollution for the immediate environment, especially by a former ore deposit on the ground. For that it is strongly recommended to remove the mineral just around the mine and make a good drainage of surface water from the mine. A cleaning of the wadi nearby can avoid the trapping of chemical elements in the soil. Reforestation companions around the mine can increase the soil's purifying power. Undertake an abandoned mine management strategy in the region.

2. Besides, the weakness of our resources is overused by the poor spatial and temporal distribution of these resources; Algeria suffers from the soil erosion and siltation of dams and losses due to outdated distribution and poor management. Among the biggest challenges that will face Algeria until 2023 are the necessity to ensure sustainable management of natural resources and ecosystems, the necessity to ensure the food security of the nation and citizens, the resolution of the throbbing questions of employment through the development of a productive and competitive economy, and the establishment of the foundations of effective governance of both the economy and society.

3. For future studies, a number of convergence points have been recognized by analyzing the strategic directions of studies in Algeria and their comparison with those of the European Union, based on the significant problems for Horizon 2023. These relate to the research planification based on major challenges (food security, climate change, and water economy). Priority research themes include: first, food security, sustainable agriculture, fight against climate change, efficient use of natural resources. Second, inclusive, innovative, and secure societies. Third, integrating SMEs in the process of research and innovation (industrial primacy pillar). Fourth, willingness to mobilize industry stakeholders and engineering sciences, most directly concerned and most likely to integrate scientific knowledge in an innovation perspective. Fifth, necessity to develop innovation in a direction favorable to smart, sustainable, and inclusive growth. Finally, a greater role for social sciences in the development of research to address all societal challenges.

4. The use of flood forecasting models is of great importance to watercourse managers. The existence of foreseeable natural risk, over a region, must lead the decision-makers to prohibit or admit, under certain conditions, certain modes of occupation or land use. Also, provide, if necessary, the implementation of collective safeguarding and protection measures. The preventive measures to be adopted aim to ensure, on the one hand, better control of the flood hazard, and on the other hand, a limitation of the vulnerability of the people, the goods, and the exposed activities.

5. Future work on the impacts of climate change would benefit from the integration of groundwater and surface water resources. Better assessment of the effect of climate change on water resources at a regional scale would require the application of the relevant regional models and the use of weather generator data for other future climate scenarios based on changes of different meteorological parameters. The results of these models may be used to simulate the runoff pattern by using a suitable hydrological model. The findings of such studies are

important in the preparation of regional water management plans. Other benefits could be regionalization of parameters of hydrological models to assess hydrological behavior in ungagged basins. Development of new drought forecasting tools based on new approaches is needed.

6. Results obtained in the present investigation highlighted a number of interesting points that may warrant further study. Firstly, the application of DENFIS models for modeling daily EP must be extended to other sites for in-depth analysis. Secondly, the selection of several other climatic variables as inputs can help to improve the accuracy of the models. Thirdly and finally, the models must be applied for modeling EP at monthly time steps.

7. The applicability of the OPELM and OSELM models has been demonstrated in four sites with relatively similar climatic conditions. However, the applicability of the proposed models in other different regions needs to be investigated for robust conclusions. More in-depth investigation of the proposed approaches with several different inputs variables will be helpful for further analysis.

8. We can present some recommendations for the managers in the goal to protect the alluvial aquifer of Mitidja: implementation of a new monitoring network well distributed in this plane. Respect the perimeter of protection map when installing new drilling wells, in order to reduce the quantitative and qualitative degradation of groundwater resources. Implement future polluting anthropogenic activities in low vulnerability areas to limit their negative impact on water resources in the groundwater. It is necessary to control the industrial discharges and to sensitize the farmers to use fertilizers in a rational way – review the pricing of water, especially industrialists to be economical and protective toward the water. A hydrogeological modeling study of the alluvial aquifer and the transfer of nitrates in this aquifer are very important in order to manage the groundwater in the alluvial aquifer properly.

9. The aquifer drawdown observed during the 1990s, the long-term exploitation of the karst aquifers in the Mounts of Tlemcen must be carefully managed, taking into account both the growing demand for water and all available water resources, including seawater desalination and the reuse of treated wastewater for irrigation. From an integrated water resources management (IWRM) perspective, karst aquifers can play several different roles: (a) water supply to villages located far from the major regional distribution networks, (b) strengthening supply to the localities of the high plains, and (c) a role of strategic reserve, thanks to their large permanent reserves (6 billion m^3) and high borehole productivity. In order for these aquifers to play this strategic reserve role, it is important (a) to maintain in good working order the wells whose operation has been suspended since the start-up of desalination plants and (b) to define protection perimeters for these wells, taking due account of aquifer vulnerability to surface pollution. Finally, a managed aquifer recharge (MAR) strategy could be put in place, by organizing the re-infiltration into the karst of the overflow from the dams.

10. Scale in hard water is a major concern for industrial processes and domestic installations. Undesirable scale deposits cause many technical problems with

serious safety and economic consequences, such as total or partial obstruction of pipes leading to a decrease in the flow velocity; the reduction of heat transfer due to the precipitate of calcium carbonate whose conductivity is 15 to 30 times less than steel; and the clogging of the filters. Energy production in nuclear power plants is often limited by the scale in cooling towers. In Great Britain, nonproductive expenses related to scaling have been estimated at 600£ million a year. These same expenses are about 1.5 billion euros per year in France. Therefore, it is important to establish appropriate methods to study this phenomenon and find effective ways to combat it.

11. Also one of the recommendations is to shed light on the phenomenon of scaling and inhibitors that can fight or reduce the formation of scale on industrial and domestic facilities. Among the recommendations necessary to minimize this phenomenon: First, make a complete study on the phenomenon of scaling in the entire territory of Algeria, educate industry and the government on the risk of scaling and the use of harmful scaling inhibitors. Second, focus research on more economical and environmentally friendly green inhibitors based on toxicity, bioaccumulation, and biodegradation. Third, improve the approach to the study of scaling in Algeria, look for new hard water sites to test inhibitors, and reduce the risk of scaling.

12. The mapped soils require phospho-potassium and nitrogen fertilizer maintenance to improve their chemical fertility. Before the planting of each crop, the analysis of nitrogen and phosphorus in the soil is necessary in order to avoid the pollution of groundwater. The missing information on the impact of fertilizers and pesticides on groundwaters in Algeria needs additional efforts be devoted in the future to carry out comprehensive monitoring of pesticides used in Algerian agriculture to preserve the quality of groundwaters.

References

1. Zerrouki H (2014) Aspects quantitatifs et qualitatifs de la source de Bouakkous: impact le champ captant d'ain chabro (zone semi-aride Tebessa). Doctorat de l'université d'Annaba, 168 p
2. Alhamed M, Wohnlich S (2014) Environmental impact of the abandoned coal mines on the surface water and the groundwater quality in the south of Bochum, Germany. Environ Earth Sci 72(9):3251–3267
3. Jahanshahi R, Zare M (2015) Assessment of heavy metals pollution in groundwater of Golgohar iron ore mine area, Iran. Environ Earth Sci 74(1):505–520
4. Papp DC, Cociuba I, Baciu C, Cozma A (2017). Composition and origin of mine water at Zlatna gold mining area (Apuseni Mountains, Romania). Procedia Earth Planet Sci 17:37–40
5. United Nations, Department of Economic and Social Affairs, Population Division (2017) World population prospects: the 2017 revision, 2017
6. Malabo MP (2018) Water-wise: smart irrigation strategies for Africa. Dakar
7. Anikwe MAN, Okonkwo CI, Mbah CN (2003) Yield of soybean. Tropicultura 5:22–27
8. Kanissery RG, Gerald-Sims K (2011) Biostimulation for the enhanced degradation of herbicides in soil. Appl Environ Soil Sci 10:10

9. Yamamoto Y, Sukchan S (2017) Land suitability analysis concerning water resource and soil property. In: Analysis concerning water resource JIRCAS working report no. 30

10. Hirpa FA, Pappenberger F, Arnal L, Baugh CA, Cloke HL, Dutra E et al (2018) Global flood forecasting for averting disasters worldwide. In: Global flood hazard: applications in modeling, mapping, and forecasting. Wiley, Hoboken, pp 205–228

11. Zischg AP, Hofer P, Mosimann M, Röthlisberger V, Ramirez JA, Keiler M, Weingartner R (2018) Flood risk (d) evolution: disentangling key drivers of flood risk change with a retro-model experiment. Sci Total Environ 639:195–207

12. Maref N, Seddini A (2018) Modeling of flood generation in semi-arid catchment using a spatially distributed model: case of study Wadi Mekerra catchment (Northwest Algeria). Arab J Geosci 11(6):116

13. Navarra A, Tubiana L (2013) Regional assessment of climate change in the Mediterranean volume 1: air, sea and precipitation and water. Springer, Berlin

14. Elmeddahi Y, Mahmoudi H, Issaadi A, Goosen MFA, Ragab R (2016) Evaluating the effects of climate change and variability on water resources: a case study of the Cheliff Basin in Algeria. Eng Appl Sci J. https://doi.org/10.3844/ajeassp

15. Meddi M, Boucefiane A (2013) Climate change impact on groundwater in Cheliff-Zahrez basin (Algeria). APCBEE Proc 5:446–450

16. Kalognomou EA, Lennard C, Shongwe M, Pinto I, Favre A, Kent M, Hewitson B, Dosio A, Nikulin G, Panitz HJ, Büchner M (2013) A diagnostic evaluation of precipitation in CORDEX models over southern Africa. J Clim 26(23):9477–9506

17. Pammar L, Deka PC (2017) Daily pan evaporation modeling in climatically contrasting zones with hybridization of wavelet transforms and support vector machines. Paddy Water Environ 15:711–722. https://doi.org/10.1007/s10333-016-0571-x

18. Eray O, Mert C, Kisi O (2017) Comparison of multi-gene genetic programming and dynamic evolving neural-fuzzy inference system in modeling pan evaporation. Hydrol Res. https://doi.org/10.2166/nh.2017.076

19. Wu L, Fan J (2019) Comparison of neuron-based, kernel-based, tree-based and curve-based machine learning models for predicting daily reference evapotranspiration. PLoS One 14(5): e0217520

20. Huang G, Wu L, Ma X, Zhang W, Fan J, Yu X, Zeng W, Zhou H (2019) Evaluation of CatBoost method for prediction of reference evapotranspiration in humid regions. J Hydrol 574:1029–1041

21. Bouderbala A (2018) Effects of climate variability on groundwater resources in coastal aquifers (case of Mitidja plain in the North Algeria). In: Groundwater and global change in the western Mediterranean area. Springer, Cham, pp 43–51

22. Bhattarai U (2017) Impacts of climate variability on biodiversity and ecosystem services: direction for future research. Hydro Nepal J Water Energy Environ 20:41–48

23. Amjad Z, Demadis ND (2015) Mineral scales and deposits (scientific and technological approaches). Elsevier, Amsterdam

24. ANDI (2015) Agence nationale de Développement de l'Investissement. ANDI, Algérie, p 20

Printed by Printforce, the Netherlands